Global Power Shift

Series Editor

Xuewu Gu, Center for Global Studies, University of Bonn, Bonn, Germany

Managing Editor

Hendrik W. Ohnesorge, Center for Global Studies, University of Bonn, Bonn, Germany

International Advisory Board

Luis Fernandes, Pontificia Universidade, Rio de Janeiro, Brazil
G. John Ikenberry, Princeton University, Princeton, NJ, USA
Canrong Jin, Renmin University of Beijing, Beijing, China
Srikanth Kondapalli, Jawaharlal Nehru University, New Delhi, India
Dingli Shen, Fudan University, Shanghai, China
Kazuhiko Togo, Kyoto Sanyo University, Tokyo, Japan
Roberto Zoboli, Catholic University of Milan, Milano, Italy

Ample empirical evidence points to recent power shifts in multiple areas of international relations taking place between industrialized countries and emerging powers, as well as between states and non-state actors. However, there is a dearth of theoretical interpretation and synthesis of these findings, and a growing need for coherent approaches to understand and measure the transformation. The central issues to be addressed include theoretical questions and empirical puzzles: How can studies of global power shift and the rise of 'emerging powers' benefit from existing theories, and which alternative aspects and theoretical approaches might be suitable? How can the meanings, perceptions, dynamics, and consequences of global power shift be determined and assessed? This edited series will include highly innovative research on these topics. It aims to bring together scholars from all major world regions as well as different disciplines, including political science, economics and human geography. The overall aim is to discuss and possibly blend their different approaches and provide new frameworks for understanding global affairs and the governance of global power shifts.

More information about this series at http://www.springer.com/series/10201

S. Mahmud Ali

China's Belt and Road Vision

Geoeconomics and Geopolitics

S. Mahmud Ali
The Institute of China Studies
University of Malaya
Kuala Lumpur, Malaysia

ISSN 2198-7343 ISSN 2198-7351 (electronic)
Global Power Shift
ISBN 978-3-030-36243-0 ISBN 978-3-030-36244-7 (eBook)
https://doi.org/10.1007/978-3-030-36244-7

© Springer Nature Switzerland AG 2020
This work is subject to copyright. All rights are reserved by the Publisher, whether the whole or part of the material is concerned, specifically the rights of translation, reprinting, reuse of illustrations, recitation, broadcasting, reproduction on microfilms or in any other physical way, and transmission or information storage and retrieval, electronic adaptation, computer software, or by similar or dissimilar methodology now known or hereafter developed.
The use of general descriptive names, registered names, trademarks, service marks, etc. in this publication does not imply, even in the absence of a specific statement, that such names are exempt from the relevant protective laws and regulations and therefore free for general use.
The publisher, the authors and the editors are safe to assume that the advice and information in this book are believed to be true and accurate at the date of publication. Neither the publisher nor the authors or the editors give a warranty, expressed or implied, with respect to the material contained herein or for any errors or omissions that may have been made. The publisher remains neutral with regard to jurisdictional claims in published maps and institutional affiliations.

This Springer imprint is published by the registered company Springer Nature Switzerland AG
The registered company address is: Gewerbestrasse 11, 6330 Cham, Switzerland

Acknowledgements

This book would not see the light of day without the contributions made over several years by myriad benefactors. Many variously offered advice, information, enthusiastic encouragement, statistical data and analyses. Several aided and abetted my labours during field trips to and across China, Pakistan and Malaysia. Others extended guidance towards useful sources. A few would rather not be identified by name, but others must be: Prof. Dato' A. Majid Khan, Prof. Fan Gaoyue, Prof. Datuk Danny Wong Tze Ken, Prof. Robert Sutter, Prof. Rosemary Foot, Prof. Orville Schell, Prof. Zhang Yuanpeng, Prof. Kerry Brown, Prof. Michael O'Hanlon, Prof. Arne Westad, Martin Wolf, Mike Billington, Saliya Wickramasuriya, Peter Ngeow and Peter Chang at the ICS, Marc Fitzpatrick and the IISS, Nilanthi Samaranayake and David Finkelstein at the CNA, Dr. Borhanuddin Ahmad, Commodore Jawad Ahmed, Brigadier Shad Mahmood and academic associates too numerous to be individually named.

I acknowledge the support I received from colleagues at the Institute of China Studies (ICS), University of Malaya, in Kuala Lumpur and Springer's editorial and production teams in Heidelberg and elsewhere. Their unstinting patience made smooth publication possible. I must also recognise the quiet perseverance and fortitude of Selina, my long-suffering wife. This is a far from exhaustive list. All share any credit that this work may garner. However, I alone am responsible for any guilt of commission or omission that discerning readers may discover.

Structure and Organisation

The work elucidates terrestrial and maritime BRI projects, and Western and allied scepticism (Bulloch 2018; Friedman 2017; Yamada and Palma 2018), explores Chinese and adversarial thinking colouring the arguments' dialectics, examines the historical evolution of the BRI framework, studies cases central to its core components, i.e., the 'Silk Road Economic Belt' (SREB) and the 'Twenty-First Century Maritime Silk Road' (MSR) constructs and analyses the geoeconomic and geopolitical aspects and implications of China's 'trillion-dollar project of the century'. The book is organised in seven chapters:

Chapter 1 **Introduction** summarises the historical background of the Silk Road construct and how, since 2013, China's Leader, Xi Jinping, has used this framework to pursue an initiative comprising terrestrial and maritime connectivity, infrastructure-building, trade, regulatory coordination and financial convergence between China on the one hand, and the rest of Eurasia, Europe, Africa, Oceania and Latin America on the other. The Belt and Road Initiative (BRI), specifically its globe-girdling scale, massive funding outlays, alleged lack of transparency and purported geostrategic motives concealed underneath its geoeconomic garb, deepened China-rooted anxiety among members of the US-led tacit coalition being forged to counteract the 'China threat' said to be challenging the 'rules-based' post-Cold War global order. Intensifying Sino-US strategic rivalry painted a contextual backdrop for the polarised BRI discourse, reinforcing the dialectic dynamic powering the *transitional fluidity* afflicting the *system*.

Chapter 2 **Fear Factor: Strategists Versus Bankers** reviews the literature on Beijing's BRI vision—examining Chinese, Western and allied policy-statements, leadership remarks and academic and think tank analyses. It summarises the key arguments posited by BRI's champions and critics to illuminate the nature and roots of anxiety afflicting critical practitioners and their advisers. The chapter also examines contrasting postulates advanced by major Western financial/commercial organisations, juxtaposing two distinct perspectives—'strategists' and 'bankers'—to establish BRI's purported geoeconomic and geopolitical pros and cons and, on that basis, expose the widely varied lenses through which the BRI blueprint is viewed at home and abroad.

Chapter 3 **China's Belt and Road: An Evolving Network** consolidates the accretive imagery presented by the proponents and managers of Beijing's dynamic BRI vision, to enable an appreciation of the enterprise's variegated, eclectic, even diffuse, nature. For an understanding of the physical manifestations of BRI's 'final' impression as espoused by Xi Jinping, the CPC Politburo Standing Committee (PBSC), the State Council and the National Development and Reform Commission (NDRC), the chapter focuses on the Silk Road Economic Belt (SREB) undertaking, reviewing the narrative's variations and modifications refined by successive Chinese leaders since 1994. It finally examines longitudinal SREB-MSR connectors in an empirical preview of BRI's purpose and prospects.

Chapter 4 **East Meets West: BRI's Eclectic Origins** reviews significant contributions to Beijing's BRI vision made by myriad non-Chinese institutions, organisations and corporations, whose actions, alongside China's developmental drivers, socio-economic priorities and planning processes, laid down the bed on which the BRI vision germinated. The chapter posits that rather than a singular strand of covertly revisionist geostrategic thought precipitating BRI, slowly crystallising commercial considerations, multilateral socio-economic objectives, regional integrationist politico-economic imperatives and, occasionally, visionary imagination, converged from diverse origins. Mutually reinforcing accretive trends, coinciding after the Cold War, were accelerated by the *Great Recession*'s severity. Beijing's responses to that crisis, informed by multilateral Western advice, and aided by multinational innovation, shaped Xi Jinping's policy inheritance, the latter being the fountainhead of his BRI proclamations.

Chapter 5 **Case Study 1: The China–Pakistan Economic Corridor** examines the 'flagship' network of energy production, transmission and distribution, transport and communications, industrial and agricultural, and infrastructure and connectivity projects announced by Chinese Premier Li Keqiang and President Xi Jinping during consecutive visits to Pakistan, and revised upwards since then. While proponents eulogise its transformative effect on Pakistan's economic and state-building prospects, critics underscore its allegedly hidden objectives, its challenges and uncertain future. The chapter analyses CPEC's domestic, bilateral, regional and systemic ramifications and illuminates a complex combination of positive and negative outcomes characterising such potentially transmogrifying visions, both for Pakistan and Sino-Pakistani relations, and for the South Asian regional *sub-system* and the wider *system* itself.

Chapter 6 **Case Study 2: The Twenty-First Century Maritime Silk Road** reviews Chinese and external documentation explaining both economic and security motivations driving a revival of China's oceanic interests, especially since China's accession to the World Trade Organisation (WTO), when its primarily maritime commerce exploded, transforming global trade patterns and resource flows. Within a decade, China's output expanded to the world's second highest, altering geoeconomic balances, eroding long-familiar assumptions and threatening to redraw the geopolitical order. MSR proposes a series of ports linking coastal China to Asia, Africa and Europe via the SCS, the Indian Ocean, the Pacific and the Mediterranean. Against the backdrop of a competitive pushback by Beijing's critics, the chapter asks, what does the MSR seek to achieve, how and why? What likely challenges confront it, and what are its prospects?

Chapter 7 **Conclusion**. The final chapter, seeking answers to the prefatory questions, summarises the foregoing examination. It is impossible to predict BRI's final outcome before all its planned projects are operational, but the evidence presented above suggests much anxiety informing Western analyses is rooted in a fear of the unknown precipitated by *systemic transitional fluidity* flowing from China's rapid renascence, rather than a rational evaluation of the BRI blueprint per se. An uncertain transition from the 300-year-old trans-Atlantic-centric political–economic order catalysed fearful suspicions and a profound *displacement anxiety* stemming from accretive disruptions backstopping Beijing's BRI vision. Strategists' angst-ridden, zero-sum prophecies draw greater attention than bankers' reasoned analyses. A lack of objective detachment could promote self-fulfilling assumptions. This work aims to address that risk.

References

Bulloch D (2018) After a brief silence, skeptics of China's BRI are speaking up again. Forbes, 18 Apr 2018
Friedman G (2017) Here's why China's OBOR is doomed to fail. Business Insider, 19 July 2017
Yamada G, Palma S (2018) Is China's belt and road working? A progress report from eight countries. Nikkei, Gwadar, 28 Mar 2018

Contents

1 Introduction 1
  1.1 Beijing's Belt and Road Initiative 2
  1.2 The Initiative Triggers a Polarising Discourse 3
  1.3 A Schizophrenic Response to China's
      'National Rejuvenation' 5
  References 8

2 Fear Factor: Strategists Versus Bankers 11
  2.1 Xi Jinping's 'New Silk Road' 11
  2.2 Belt and Road Initiative's Basic Design 16
  2.3 Conflicting Perspectives 19
      2.3.1 China-Rooted Angst 20
      2.3.2 A Darkening Landscape 25
      2.3.3 Warriors to the Fore 28
      2.3.4 The Butter Versus Guns Argument 32
  2.4 The Bankers' View 35
      2.4.1 The Asian Development Bank Perspective 40
      2.4.2 The United Nations Vision 42
  2.5 Belt and Road Initiative's Adversarial Strategic Backdrop .. 45
      2.5.1 Beltway–*Zhongnanhai* Dialectics 45
      2.5.2 Polarisation Cemented 50
      2.5.3 Belt and Road Initiative's Hostile Landscape 53
  References 56

3 China's Belt and Road: An Evolving Network 69
  3.1 A 'New Silk Road' Emerges 69
  3.2 The Silk Road Economic Belt's Accretive Evolution 72
  3.3 Eurasia's Post-Soviet Coalescence 74
  3.4 The 'Shanghai Spirit' Paradigm 78
  3.5 Beijing's Regional Developmental Drivers 85

	3.6	In the *Great Recession*'s Wake	89
	3.7	Xi Jinping's 'New Era' of Reform and Opening-up	94
	3.8	The Silk Road Economic Belt's Longitudinal Spurs	97
		3.8.1 The Lancang–Mekong Economic Corridor	98
		3.8.2 The Bangladesh–China–India–Myanmar Economic Corridor	103
		3.8.3 The Balkan Silk Road	111
	References	114

4 East Meets West: BRI's Eclectic Origins 123
    4.1 The New Silk Road's United Nations-Driven Germination 123
    4.2 US Visionaries Imagine a New Silk Road 130
    4.3 Brussels Builds a Trans-Eurasian 'Land-bridge' 134
    4.4 Multinationals Revive the Silk Road 137
    4.5 The World Bank Pushes China's Global Integration 144
    4.6 Japan and the US Proclaim New Silk Road Visions 150
        4.6.1 Tokyo Thinks up the New Silk Road 150
        4.6.2 Washington's New Silk Road Proposal 158
        4.6.3 The US' Lower Mekong Initiative 164
    References 167

5 Case Study 1: The China–Pakistan Economic Corridor 175
    5.1 China-Pakistan Economic Corridor, a Belt and Road 'Flagship' 175
    5.2 China–Pakistan Economic Corridor Deepens Threat Perceptions 182
        5.2.1 India's Critique of the Corridor 182
        5.2.2 Western Anxiety 185
    5.3 China–Pakistan Economic Corridor's 'Long-Term Plan' 188
        5.3.1 Connectivity 190
        5.3.2 Energy 191
        5.3.3 Trade and Industry 191
        5.3.4 Agricultural Development and Poverty Alleviation 192
        5.3.5 Tourism 192
        5.3.6 People's Livelihood and Non-governmental Exchanges 193
        5.3.7 Financial Cooperation 193
    5.4 The Corridor's Investment and Financing Mechanisms 194
    5.5 The Corridor's Projects in Pakistan 195
        5.5.1 Energy Projects 195
        5.5.2 Major Infrastructure Projects 196
        5.5.3 Mass-Transit Rail Projects 197

	5.5.4	Special Economic Zones	197
	5.5.5	Social Sector Development Projects	198
	5.5.6	Other Major Projects	198
5.6	The Corridor's Early Harvest Projects		198
	5.6.1	Khyber Pakhtunkhwa Province	199
	5.6.2	Punjab Province	199
	5.6.3	Sindh Province	199
	5.6.4	Balochistan Province	200
	5.6.5	Additional 'New Projects'	200
5.7	The China-Pakistan Economic Corridor's Terminal Challenges		200
	5.7.1	The Corridor's Karakoram Highway-Rooted Beginnings	201
	5.7.2	Gwadar's Evolutionary Emergence	205
	5.7.3	Gwadar's Chinese-Funded Projects	208
5.8	The USA's Corridor Critique		210
5.9	China-Pakistan Economic Corridor's Geopolitical Ecology		213
	5.9.1	The Corridor's Multi-layered Implications	214
	5.9.2	Preliminary Questions	215
	5.9.3	First-Order Analytical Issues	215
	5.9.4	Second-Order Analytical Issues	216
	5.9.5	Third-Order Analytical Issues	217
	5.9.6	Fourth-Order Analytical Issues	217
5.10	The Corridor's Insecurity Challenges		218
References			223

6 Case Study 2: The Twenty-First Century Maritime Silk Road ... 231
6.1	Maritime Silk Road Deepens Anxiety		231
6.2	Naval Dynamics		235
	6.2.1	Djibouti Base Jitters	235
	6.2.2	Commercial Rationales and Strategic Mistrust	238
	6.2.3	Duqm Dramatics	241
6.3	Clashing Perspectives, Purposes and Narratives		243
	6.3.1	The Chinese Framework	244
	6.3.2	Maritime History as a Mirror	247
	6.3.3	Hambantota Histrionics	249
	6.3.4	Maldives Malarkey	252
	6.3.5	Maritime Silk Road's Malaysian Misadventure	257
6.4	The Quad's 'Free and Open' Countermoves		259
	6.4.1	Geoeconomics Reinforce Geopolitics	260
	6.4.2	A Quadrilateral Maritime Focus	264

		6.4.3	Quad-China Competition in Bangladesh	265
		6.4.4	The Indian Ocean Region's Big Guns	270
	6.5	Maritime Silk Road's Insecure Origins		271
	References			276
7	**Conclusion**			291
	7.1	The Belt and Road Initiative's Volatile Strategic Backdrop		291
		7.1.1	Economic Outcomes	292
		7.1.2	The Geopolitics of Geoeconomics	294
		7.1.3	Dialogue de Sourds	296
	7.2	Empiricism's Rational Sobriety		299
		7.2.1	The Continuity and Disruptions Marking Xi's China	300
		7.2.2	Beijing's Insecurity Angst	302
		7.2.3	Belt-and-Road's Dialectic Dynamics	305
	7.3	Concluding Observations		307
		7.3.1	Displacement Anxiety and Terminal Fears	308
		7.3.2	The Belt and Road Initiative in the *Systemic* Context	312
		7.3.3	The Outlook	315
	References			320
Index				325

Abbreviations

AAGC	Asia-Africa Growth Corridor
ABC	Australian Broadcasting Corporation
ACMECS	Ayeyawady-Chao Phraya-Mekong Economic Cooperation Strategy
ADB	Asian Development Bank
AI	Artificial Intelligence
AIIB	Asian Infrastructure Investment Bank
ALTID	Asian Land Transport Infrastructure Development
AMTI	Asia Maritime Transparency Initiative
ANU	Australian National University
AP	Associated Press
APEC	Asia-Pacific Economic Cooperation Forum
APFTA	APEC Free Trade Area
ARF	ASEAN Regional Forum
ASEAN	Association of Southeast Asian Nations
ASEM	Asia–Europe Meeting
AT	Arbitral tribunal
AWC	Army War College
B&R	Belt and Road
BAS	Bulletin of the Atomic Scientists
BBC	British Broadcasting Corporation
BCIM	Bangladesh–China–India–Myanmar
BIT	Bilateral investment treaty
BoC	Bank of China
BRF	Belt and Road Forum
BRI	Belt and Road Initiative
C3I	Command, control, communications and intelligence
CASA	Central Asia-South Asia
CCCC	China Communications Construction Company
CCG	China Coast Guard

CCTV	China Central Television
CD	China Daily
CDB	China Development Bank
CEEC	Central and Eastern European Countries
CEO	Chief executive officer
CFO	Chief financial officer
CFR	Council on Foreign Relations
CHEC	China Harbour Engineering Company
CIA	Central Intelligence Agency
CIFIT	China International Fair for Investment and Trade
CIMSEC	Centre for International Maritime Security
CJCS	Chairman of the Joint Chiefs of Staff
CM	Chief Minister
CMC	Central Military Commission
CMPH	China Merchant Port Holdings
CNA	Centre for Naval Analysis
CNO	Chief of Naval Operations
COPHC	China Overseas Ports Holding Company
CPC	Communist Party of China
CPCCC	CPC Central Committee
CPEC	China–Pakistan Economic Corridor
CPV	Communist Party of Vietnam
CRCC	China Railway Construction Corporation
CRI	China Radio International
CRS	Congressional Research Service
CSG	Carrier strike group
CSIS	Centre for Strategic and International Studies
CTF	Carrier task force
CUES	Code for Unplanned Encounters at Sea
DCI	Director of Central Intelligence
DHS	Department of Homeland Security
DIA	Defence Intelligence Agency
DMP	Doraleh Multipurpose Port
DMZ	Demilitarised Zone
DNI	Director of National Intelligence
DoD	Department of Defence
DoJ	Department of Justice
DoS	Department of State
DPP	Democratic People's Party
DW	Deutsche Welle
EAF	East Asia Forum
EAS	East Asia Summit
EBRD	European Bank for Reconstruction and Development
ECRL	East Coast Rail Link
ECS	East China Sea

Abbreviations xvii

EEC	Eurasian Economic Community
EIU	Economist Intelligence Unit
EMP	Electromagnetic pulse
ESCAP	Economic and Social Commission for Asia and the Pacific
EU	European Union
FATA	Federally administered Tribal Areas
FBI	Federal Bureau of Investigation
FDI	Foreign direct investment
FM	Foreign ministry
FMPRC	Foreign Ministry of the People's Republic of China
FOCAC	Forum on China–Africa Cooperation
FoN	Freedom of navigation
FP	Foreign policy
FPRI	Foreign Policy Research Institute
FT	Financial Times
FTA	Free-trade agreement
FTZ	Free-trade zone
FWO	Frontier Works Organisation
FYP	Five-Year Plan
G7	Group of seven developed economies
G20	Group of 20 developed and emerging economies
GDA	Gwadar Development Authority
GDP	Gross Domestic Product
GMS	Greater Mekong Subregion
GNI	Gross national income
GoI	Government of India
GoP	Government of Pakistan
GPA	Gwadar Port Authority
GPS	Global Positioning System
GT	Global Times
HASC	House Armed Services Committee
HP	Hewlett-Packard
IBRD	International Bank for Reconstruction and Development
ICG	International Crisis Group
ICT	Information and communications technology
IDSA	Institute for Defence Studies and Analysis
IFC	International Finance Corporation
IISS	International Institute for Strategic Studies
IMF	International Monetary Fund
INDOPACOM	Indo-Pacific Command
IOR	Indian Ocean Region
IP	Intellectual property
IPR	Intellectual property rights
ISR	Intelligence, surveillance, reconnaissance
IT	Information technology

JCEBS	Journal of Chinese Economic and Business Studies
JCS	Joint Chiefs of Staff
JICA	Japan International Cooperation Agency
JSG	Joint Study Group
K2K	Kunming to Kolkata
KKH	Karakoram Highway
KWh	Kilowatt-hour
LAT	Los Angeles Times
LDP	Liberal Democratic Party
LMC	Lancang-Mekong Cooperation
LMI	Lower Mekong Initiative
LTP	Long-term plan
MCKIP	Malaysia-China Kuantan Industrial Park
MDG	Millennium Development Goals
MEA	Ministry of External Affairs
MiC2025	Made in China 2025
MND	Ministry of National Defence
MoFA	Ministry of Foreign Affairs
MOFCOM	Ministry of Commerce
MPDR	Ministry of Planning, Development and Reform
MSDPRC	Military and Security Developments Involving the PRC
MSR	Maritime Silk Road
MWe	Megawatts electricity
NASIC	National Air and Space Intelligence Centre
NBR	National Bureau of Asian Research
NDAA	National Defence Authorisation Act
NDB	New Development Bank
NDRC	National Development and Reform Commission
NDTV	New Delhi Television
NDU	National Defence University
NI	National Interest
NLD	National League for Democracy
NPC	National People's Congress
NSA	National Security Advisor/Agency
NSC	National Security Council
NSS	National Security Strategy
NYT	New York Times
OBOR	One Belt One Road
ODA	Official development assistance
ODI	Overseas Direct Investment
ONA	Office of Net Assessment
OPIC	Overseas Private Investment Corporation
ORF	Observer Research Foundation
OSD	Office of the Secretary of Defence
OUP	Oxford University Press

Abbreviations

PBSC	Politburo Standing Committee
PD	People's Daily
PLA	People's Liberation Army
PLAAF	PLA Air Force
PLAN	PLA Navy
PLANAF	PLAN Air Force
PMO	Prime Minister's Office
PS	Project Syndicate
PTI	Pakistan Tehreek-e-Insaf/Press Trust of India
R&D	Research and development
RMA	Revolution in Military Affairs
RMB	Renminbi
SAARC	South Asian Association for Regional Cooperation
SASC	Senate Armed Services Committee
SCMP	South China Morning Post
SCO	Shanghai Cooperation Organisation
SCS	South China Sea
SDG	Sustainable Development Goals
SEZ	Special economic zone
SIPRI	Stockholm International Peace Research Institute
SKRL	Singapore–Kunming Rail Link
SLoC	Sea Lines of Communications
SLPA	Sri Lanka Ports Authority
SoD	Secretary of Defence
SOE	State-owned enterprises
SREB	Silk Road Economic Belt
SRF	Silk Road Fund
SSD	Special Security Division
TAPI	Turkmenistan–Afghanistan–Pakistan–India
TAR	Trans-Asian Railway
TEU	Twenty-foot equivalent unit
ToI	Times of India
TRACECA	Transport Corridor Europe Caucasus Asia
UAE	United Arab Emirates
UN	United Nations
UNCTAD	United Nations Conference on Trade and Development
UNDP	United Nations Development Programme
UNESCO	United Nations Educational, Scientific and Cultural Organisation
UNGA	United Nations General Assembly
UNSC	United Nations Security Council
UPI	United Press International
USAF	United States Air Force
USCC	United States-China Economic and Security Review Commission
USN	United States Navy
USPACAF	United States Pacific Air Force

USTR	United States Trade Representative
VoA	Voice of America
WBG	World Bank Group
WEF	World Economic Forum
WFB	Washington Free Beacon
WotR	War on the Rocks
WP	Washington Post
WTO	World Trade/Tourism Organisation
YDRC	Yunnan Development and Reform Commission

Chapter 1
Introduction

China's Belt and Road Initiative (BRI) has drawn passionate critique, notably from political leaders, senior practitioners, academic analysts, media commentators and think tank strategists, from the USA, Europe, Japan, India and Australia. The focus of much vitriol has been on the assertion that BRI is more geopolitical in intent and content, than geoeconomic, as indicated by policy-makers and officials from China and its BRI partner countries. For the analysis in this work, the following definitions are essential to avoid any ambiguity.

> **Geoeconomics** *Economics in its relationship to such geographical conditions as location and natural resources; a condition of economic rivalry among nations.* Collins English Dictionary
> *The use of economic instruments to produce beneficial geopolitical results.*[1]
> Robert Blackwill

> **Geopolitics** *The study of the effect of geographical factors on politics, especially international politics; the combination of geographical and political factors affecting a country or area; politics as they affect the whole world; global politics.* Collins English Dictionary

---

[1] Blackwill (2018) and Blackwill and Harris (2016).

1.1 Beijing's Belt and Road Initiative

Ancient China, famed for its lightweight and glossy silken fabrics, jade, paper, porcelain, salt, spices, sugar, teas and other luxury goods, traded with Europe, Persia and Eurasian lands in between, via numerous trading posts strung along caravan routes. This diffuse skein of trans- and intercontinental trade routes linked the Han Dynasty capital, Xian, to Antioch, a Graeco-Roman metropolis. Starting from around 130 BC, merchants carried gold, silver, other valued metals, ivory, gemstones, cotton, dates, pistachio nuts, saffron powder, livestock, especially camels and, until the fifth century, glass, to China, while returning with Chinese novelties. The route was made famous by the Venetian explorer, Marco Polo (1254–1324), whose travels along these routes to China and Mongolia lasted 24 years, and who for a time was an advisor to the Yuan Dynasty Emperor, Kublai Khan. However, the route was only named the 'Great Silk Road' by the German historian, Ferdinand von Richthofen, in his book, 'China', in 1877.[2]

More recently, in April 2019, Xi Jinping, General-Secretary of the Communist Party of China (CPC), President of the People's Republic of China (PRC) and Chairman, Central Military Commission (CMC), hosted dozens of foreign leaders, ministers and officials at the Second Belt and Road Forum for International Cooperation (BRF) in Beijing. Guests at this meeting included many more national leaders than was the case at the first BRF in 2017, at which Xi formally inaugurated China's Belt and Road Initiative (BRI). Initially named the One Belt One Road (OBOR) scheme, BRI was designed to link 64 Asian, European and African countries to China[3] via new or refashioned transportation and communications networks, pipelines, ports, digital links and trans-border regulatory resonance. By early 2018, the number of BRI partners had reached 'at least 68 countries with an announced investment as high as $8tn' in transportation, energy and telecommunication infrastructure networks linking Asia, Europe and Africa.[4] Oceania and Latin America, too, beckoned.

Notwithstanding Beijing's insistence on BRI/OBOR's collectively beneficial objectives, its purported goals triggered raging controversy among policy- and strategic analytical communities in the USA, Europe, India, Japan, Australia and elsewhere. BRI's champions and detractors advanced starkly contrasting perspectives.[5] Shortly after taking office, President Donald Trump proclaimed Washington's new

---

[2] Voisin (2017), Liu (2012) and Hopkirk (1984).

[3] Afghanistan, Albania, Armenia, Azerbaijan, Bahrain, Bangladesh, Belarus, Bhutan, Bosnia and Herzegovina, Brunei, Bulgaria, Cambodia, Croatia, Czech Republic, Egypt, Estonia, Georgia, Hungary, India, Indonesia, Iran, Iraq, Israel, Jordan, Kazakhstan, Kuwait, Kyrgyzstan, Laos, Latvia, Lebanon, Lithuania, Macedonia, Malaysia, Maldives, Moldova, Mongolia, Montenegro, Myanmar, Nepal, Oman, Pakistan, Palestine, Philippines, Poland, Qatar, Romania, Russia, Saudi Arabia, Serbia, Singapore, Slovakia, Slovenia, Sri Lanka, Syria, Tajikistan, Thailand, Timor-Leste, Turkey, Turkmenistan, Ukraine, United Arab Emirates, Uzbekistan, Vietnam, Yemen. Government of the PRC (2017); six months later, the list was expanded to include Djibouti, Ethiopia, Kenya and South Korea. Hurley et al. (2018, pp. 6–7).

[4] Hurley et al. (2018, p. 1).

[5] Report (2019).

muscularity vis-à-vis Beijing, painting the policy-landscape on which US responses to China generally, and BRI specifically, was fashioned: 'Around the world, we face rogue regimes, terrorist groups and rivals like China and Russia that challenge our interests, our economy and our values. In confronting these dangers, we know that weakness is the surest path to conflict, and unmatched power is the surest means of our defence'.[6] His Vice President, Mike Pence, in a closely watched speech detailing Washington's China policy, announced, 'We will be giving foreign nations a just and transparent alternative to China's debt-trap diplomacy…Be assured: we will not relent until our relationship with China is grounded in fairness, reciprocity, and respect for our sovereignty'.[7]

1.2 The Initiative Triggers a Polarising Discourse

The US Intelligence Community assessed, 'China's leaders will try to extend the country's global economic, political, and military reach while using China's military capabilities and overseas infrastructure and energy investments in the BRI to diminish US influence'. A key concern: 'Successful implementation of the BRI could facilitate PLA access to dozens of additional ports and airports and significantly expand China's penetration of the economies and political systems of participating countries'.[8] Mike Pompeo, then Director, Central Intelligence Agency (CIA), noted, 'The Chinese are very active…We can watch very focused efforts to steal American information, to infiltrate the US with spies—with people who are going to work on behalf of the Chinese government against America'. Pompeo discerned the greatest threat from covert Chinese competition: 'The Chinese are working diligently to put themselves in a position where they are a superpower…We have to do better pushing back against Chinese efforts to covertly influence the world'.[9]

James Mattis, who served as Secretary of Defence until January 2019, mirrored this anxiety in a valedictory report: 'OBOR, which at first included economic initiatives in Asia, South Asia, Africa, and Europe, now encompasses all regions of the world, including the Arctic and Latin America, demonstrating the scope and reach of Beijing's ambition'. Beijing's 'ambitions', when juxtaposed to BRI's alleged purposes, illuminated the *systemic primate*'s angst: 'While some OBOR projects appear to be motivated by economic considerations, OBOR also serves a greater strategic purpose. China intends to use OBOR to develop strong economic ties with other countries, shape their interests to align with China's, and deter confrontation or criticism of China's approach to or stance on sensitive issues'.[10] Since BRI partners could 'develop economic dependencies from over-reliance on Chinese capital',

---

[6]Trump (2018).
[7]Pence (2018).
[8]Coats (2019, p. 25).
[9]Pompeo (2018).
[10]OSD (2018a).

dire implications were anticipated: 'Some OBOR investments could create potential military advantages for China, should it require access to selected foreign ports to pre-position the necessary logistics support to sustain naval deployments' defending Chinese interests in 'the Indian Ocean, Mediterranean sea and Atlantic Ocean'.[11]

President Xi assured a gathering of high-level foreign guests: 'No matter how much progress China has made in development, China will not threaten anyone else, attempt to overturn the existing international system or seek spheres of influence'.[12] His pledges failed to reassure the White House. Trump's Senior Adviser, Matthew Pottinger, representing the President at a Chinese National Day reception at the PRC embassy in Washington, explained why: 'We at the Trump administration have updated our China policy to bring the concept of competition to the forefront. It's right there at the top of the President's National Security Strategy'.[13] The USA was certainly being transparent.

Given the proclaimed scale and extended time frames of China's globe-girdling BRI proposals, empirical evidence essential to reaching reasoned conclusions was limited and ambiguous. Developments—both positive and negative—recorded between the two BRFs offered a data set which could serve as an interim basis for ascertaining what was planned, assessing how plans were translated into 'facts on the ground', and derive rational inferences therefrom. That is the goal of this work. The complexity of this task was dramatically deepened by the USA's suddenly pronounced shift to an adversarial grand-strategic posture vis-à-vis China, formalised in various Trump Administration policy-documents and leadership-level remarks, reinforcing the tenor of the public discourse on China generally and BRI/OBOR particularly.[14]

Dichotomy marked the inaugural BRF, attended by 29 heads of government and 1200 delegates from over 100 countries and multilateral organisations. It was overshadowed by the absence of US President Donald Trump, Japanese Prime Minister Shinzo Abe, and his Indian counterpart, Narendra Modi. With deepening security and economic tensions colouring relations with China, lower-level delegations represented differing depths of scepticism of the substance, intent and prospects of China's 'trillion-dollar project of the century'.[15] Still, underscoring an alignment of interests, Xi announced that participants had signed 'more than 270 cooperation projects or agreements'.[16] Leaders signed a Joint Communique noting convergent policy-orientation, laying the normative foundations for their shared BRI vision.[17]

---

[11] OSD (2018a).
[12] Xi (2018).
[13] Pottinger (2018).
[14] Sevastopulo (2019), Pence (2018), OSD (2018b), Coats (2018), Mattis (2018) and NSC (2017).
[15] White House (2017a, b, c), Han and Sink (2017), Matsui et al. (2017), Yoshino (2017), PTI (2017) and Akbar (2017).
[16] Wu (2017).
[17] FMPRC (2017).

1.2 The Initiative Triggers a Polarising Discourse

However, some participants, while lauding Xi's BRI template, urged transparency, equality and mutuality of benefit during project implementation. Uncertainty reflected much anxiety, even suspicions, among some leaders, senior officials and other practitioners that BRI was Beijing's stratagem designed to non-militarily challenge, and eventually supplant, the USA as the globally dominant *systemic primate*, without offering it an exploitable *casus belli* triggering military conflict which China was unlikely to win.[18] Critics resented Beijing's reticence on widely suspected linkages between China's pursuit of 'the great rejuvenation of the Chinese nation' (*Zhonghua minzu weida fuxing*), a much-publicised leitmotif reportedly driving China's domestic and external engagements, and BRI. Nevertheless, two years later, at the second BRF, senior-level participation, enumeration of projects underway and successes achieved, and proclaimed future plans, highlighted BRI's steady gains in momentum.

1.3 A Schizophrenic Response to China's 'National Rejuvenation'

Japan offered an example. Prime Minister Abe indicated interest in Japan's participation in BRI shortly after sending a leader of his Liberal Democratic Party (LDP) to represent him at the 2017 BRF. And although Narendra Modi condemned BRI's flagship programme—the China–Pakistan Economic Corridor (CPEC)'s alleged violation of Indian sovereignty, Delhi acquiesced in India's inclusion in Beijing's original list of BRI partner states.[19] The latter's 'economic cooperation area', stretching from the Western Pacific to the Baltic Sea, accounted for 62.3% of humanity, 30% of the global GDP and 24% of household consumption.[20] Nonetheless, Abe and Modi swiftly planned to build a new 'Freedom Corridor' stretching from the Asia-Pacific to Africa, one widely perceived as a counterpoint to BRI.[21] And in 2019, after treating nuclear-armed North Korea as the most acutely urgent threat to Japanese national security for years, Tokyo identified China as the source of its gravest collective insecurity.[22]

Still, in addition to China's original 64 BRI partners, another 48 countries in Asia, Australasia, Europe, Africa and Latin America expressed interest in the initiative.[23] Xi reiterated his message that BRI was 'an open, diversified and win-win' proposition designed to open opportunities and benefit all partner states. Even for countries not participating in building networks of roads, railways, power grids, bridges, tunnels,

---

[18]Cavanna (2018), White (2017), Tellis (2017), Blanchard and Flint (2017), Ploberger (2017), Nye (2017), Clarke (2017) and Phillips (2017).
[19]Report (2015) and Chin and He (2016, pp. 1–2).
[20]Chin and He (2016, pp. 2–4).
[21]Chaudhury (2017) and Report (2017b).
[22]Kono and Iwaya (2019).
[23]Chin and He (2016, pp. 5–6).

harbours, pipelines and fibre-optic cables, BRI would serve as an 'open platform' to which they could link their own commerce, and transport and communications networks, to derive whatever gains they thought appropriate to their needs. Beijing envisaged investing $1.2tn in BRI over 2017–2027.[24] Enthusiasts compared BRI to the USA's Marshall Plan for reconstructing war-ravaged Western Europe.[25]

The scale of the BRI *problematique* was reflected in Beijing pouring billions into BRI projects from the Asian Infrastructure Investment Bank (AIIB), China Development Bank (CDB), Silk Road Fund (SRF), Export-Import Bank and the New Development Bank (NDB). By 2016, outward FDI by Chinese state-owned enterprises (SOE) and private firms hit $183.2bn; with BRI-related investment accelerating, this was expected to rise to $750bn by 2022. By the end of 2016, China's three biggest banks had lent $225.4bn to BRI projects.[26] In May 2017, CDB set aside $36.7bn for future BRI projects; it was already funding over 500 projects worth $350bn in BRI regions.[27] By mid-2017, China's Exim Bank had loaned RMB671.4bn to support 1279 BRI projects while the SRF's project-funding topped $80bn.[28] By mid-2017, the Bank of China (BoC) reported having lent over $80bn to 470 projects along BRI alignments, while the Industrial and Commercial Bank of China and China Construction Bank, too, had 'lent billions'.[29] BRI's incorporation into the CPC Constitution in October 2017 suggested Beijing saw it as a high-stakes undertaking meriting focused leadership-level attention.

Across the Pacific, perspectives and narratives diverged diagonally. Anxious to stop further erosion of US pre-eminence vis-à-vis China, the Trump Administration launched a series of 'national security'-relevant reports reflecting and reinforcing China-focused angst and framing the USA's newly formalised grand-strategic discourse.[30] Resource allocation, military procurement, reorganisation, training and operational deployments would follow. Congress collaboratively acted 'to develop a long-term strategic vision and a comprehensive, multifaceted and principled US policy for the Indo-Pacific region', enacting the Asia Reassurance Initiative Act.[31] The Trump Administration's 'tariff war' against China, a part of Trump's 'all-of-government' anti-China campaign, given the two largest national economies' interdependence and their linkages to the remainder of the global economy, threatened a significant downturn. The newly ensconced Managing Director of the International Monetary Fund (IMF), Kristalina Georgieva, warned the world in her inaugural address, 'In 2019, we expect slower growth in nearly 90% of the world. **The global**

---

[24] Pi et al. (2017).
[25] Report (2018), Editorial (2018) and Shen and Chan (2018).
[26] Peng and Jia (2017).
[27] Xinhua (2017).
[28] Zhou (2017) and Report (2017a).
[29] Pi et al. (2017).
[30] Coats (2019, pp. 4–5, 7, 9, 14–17, 20–22, 24–26, 28, 35), OSD (2018b, pp. i, v–vi, 2–3, 6–7, 11, 24, 31–32, 34–37, 2019), Coats (2018, pp 4–7, 12–13, 15, 18), Mattis (2018, pp. 1–4, 6, 9) and NSC (2017, pp. 2–3, 25–27, 35, 45–47).
[31] US Senate (2018).

economy is now in a synchronised slowdown. This widespread deceleration means that growth this year will fall to its lowest rate since the beginning of the decade'. Georgieva emphasised the toll trade disputes were taking: 'Global trade growth has come to a near standstill. In part because of the trade tensions, worldwide manufacturing activity and investment have weakened substantially. There is a serious risk that services and consumption could soon be affected'.[32]

Given the resonances in mutually aimed US and Chinese pursuit of competing objectives, the lethality of their destructive prowess and the degree of economic integration and linkages interweaving them into the world economy, the two powers' apparently zero-sum approaches, taken to their logical extreme, could thrust the planet into arguably catastrophic conflagrations. It is against that forbidding backdrop—China's determined pursuit of its national-regenerative 'dream', reactive and countervailing responses of US and allied critics in Japan, India, Australia and elsewhere, and the dangerous dialectic dynamic thus unleashed—that BRI merits scrutiny. To restore stability, at least some of the many questions that present themselves must be addressed. Key Questions: Does China seek to build an economy-focused network linking Asia, Europe, Africa and Latin America, using comparative advantages and Chinese finance, to fashion a new geoeconomic order elevating prospects in an unprecedented push for developing countries? Or is the BRI vision, as critics contend, a clandestine geopolitical enterprise with the grand-strategic objective of supplanting *pax Americana* undergirding the US-led liberal-capitalist order with a Sino-centric illiberal, statist and mercantilist successor?

Or could the truth reside elsewhere? What *does* the record show? What were the origins of the BRI vision, and what are its likely consequences? Can the US-led status quo forces defeat allegedly revisionist ones manifest in China's *national rejuvenation* in a non-lethal, geoeconomic, competition and lead the planet peacefully towards a new geopolitical *systemic* equilibrium? Does the evidence offer reasonable ground for optimism? In short, is there any hope, or is BRI a death knell for the post-Cold War world order?

In the context of an apparent juxtaposition of the USA's relative decline and internally focused 'America First' perspective to China's proactive 'going global' policy-framework, these topical, and important questions, having acquired critically elevated salience for the current international relations' discourse, especially on questions focused on global power shifts, at least deserve purposive attempts to provide answers.[33] This book, using primary and secondary evidence from China, the USA, Japan, India, European Union member-states, and elsewhere, seeks to address that crucial need.

---

[32] Georgieva (2019). Emphases in original.
[33] Gu and Ohnesorge (2019), Dervis (2018) and Hoge (2004).

References

Akbar M (2017) Question on BRI of China by Selvaraj A to be answered by the Minister of State in the MEA. Rajya Sabha, New Delhi
Blackwill R (2018) Indo-Pacific strategy in an era of geoeconomics. Japan Forum on International Relations (JFIR), Tokyo
Blackwill R, Harris J (2016) War by other means: geoeconomics and statecraft. Belknap, Cambridge, p 1
Blanchard J, Flint C (2017) The geopolitics of China's MSR initiative. Geopolitics 22(2):223–245
Cavanna T (2018) What does China's BRI mean for US grand strategy? Diplomat, 5 June 2018
Chaudhury D (2017) India, Japan come up with AAGC to counter China's OBOR. Economic Times, 26 May 2017
Chin H, He W (2016) The BRI: 65 Countries and beyond. Fung Business Intelligence, Hong Kong
Clarke M (2017) The BRI: China's new grand strategy? Asia Policy 24:71–79
Coats D (2018) Worldwide threat assessment of the US intelligence community. DNI, Washington
Coats D (2019) Worldwide threat assessment of the US intelligence community. DNI, Washington
Dervis K (2018) Global power is shifting: is it the end of multilateralism? WEF/PS, Geneva, 24 July 2018
Editorial (2018) Xi v Marshall: will China's BRI outdo the Marshall plan? Economist, 8 Mar 2018
FMPRC (2017) Joint communique of the leaders roundtable of the BRF for International Cooperation. Beijing, 16 May 2017
Georgieva K (2019) Decelerating growth calls for accelerating action: 2019 annual meetings curtain-raiser speech. IMF, Washington
Government of the PRC (2017) Belt and Road portal: profiles. https://eng.yidaiyilu.gov.cn/info/iList.jsp?cat_id=10076. Accessed 26 Oct 2017
Gu X, Ohnesorge H (eds) (2019) Global power shift. Springer, Heidelberg
Han M, Sink J (2017) China says Trump open to cooperating on silk road projects. Bloomberg, 23 June 2017
Hoge J (2004) A global power shift in the making. Foreign Affairs, July/Aug 2004
Hopkirk P (1984) Foreign devils on the silk road. University of Massachusetts, Amherst
Hurley J, Morris S, Portelance G (2018) Examining the debt implications of the BRI from a policy perspective. Centre for Global Development, Washington
Kono T, Iwaya T (2019) Defense of Japan 2019. MoD, Tokyo, Preface, pp i–ii
Liu X (2012) The silk roads: a brief history with documents. St. Martin's, Bedford
Matsui N, Nishimura D, Yamagishi K (2017) Abe urges Xi to visit Japan and meet regularly to further ties. Asahi Shimbun, 17 May 2017
Mattis J (2018) Summary of the 2018 National Defense Strategy of the United States of America. DoD, Washington
NSC (2017) National Security Strategy of the United States of America. White House, Washington
Nye J (2017) Xi Jinping's Marco Polo strategy. PS, 12 June 2017
OSD (2018a) Assessment on US Defense implications of China's expanding global access. DoD, Washington, p 12
OSD (2018b) Nuclear Posture Review. DoD, Washington, p 2018
OSD (2019) Missile Defense Review. DoD, Washington, pp vi–vii, xviii, 4, 6–9, 13–15, 19–21, 23
Pence M (2018) Remarks on the administration's policy towards China. White House/Hudson Institute, Washington
Peng Q, Jia D (2017) China state banks provide over $400bn of credits to B&R projects. Caixin, 12 May 2017
Phillips T (2017) China's Xi lays out $900bn silk road vision amid claims of empire-building. Guardian, 14 May 2017
Pi X, Han M, Hong C, Dai M, Dormido H (2017) China's silk road cuts through some of the world's Riskiest countries. Bloomberg, 26 Oct 2017
Ploberger C (2017) OBOR—China's new grand strategy. JCEBS 15(3):289–305

References

Pompeo M (2018) Interview with BBC. DCI, Langley
Pottinger M (2018) Remarks at Chinese embassy national day reception. NSC, Washington
PTI (2017) CPEC concerns keep India away from China's Belt and Road summit. ToI, 13 May 2017
Report (2015) Industrial cooperation between countries along the Belt and Road. China International Trade Institute, Beijing, p 2015
Report (2017a) China to pour trillions into B&R projects. PD, 14 July 2017
Report (2017b) India-Japan 'Freedom Corridor' set to counter China's OBOR. News Bharati, 23 May 2017
Report (2018) China's Belt and Road: bigger than the Marshall plan? Economic and Financial Analysis, ING Bank, Amsterdam
Report (2019) Financial links between China and America deepen, despite the trade war. Economist, 5 July 2019
US Senate (2018) The Asia Reassurance Initiative Act of 2018. 115th congress, 2nd session, Washington
Sevastopulo D (2019) Why Trump's America is rethinking engagement with China. FT, 15 Jan 2019
Shen S, Chan W (2018) A comparative study of the BRI and the Marshall Plan. Nature, Palgrave Commun 4(32):1–11
Tellis A (2017) Protecting American primacy in the Indo-Pacific: SASC testimony. NBR, Washington
Trump D (2018) State of the Union address to Congress. White House, Washington
Voisin J (2017) Fortresses of the silk roads: from the Hindu Kush to the Mediterranean. Nazar Art, Paris
White H (2017) China's BRI to challenge to US-led order. EAF, 8 May 2017
White House (2017a) Donald Trump and Nikki Haley chalk up a victory on North Korea and China. Washington
White House (2017b) Presidential memorandum for the USTR. Washington
White House (2017c) Readout of President Trump's call with President Xi Jinping of China. Washington
Wu G (2017) China touts more than 270 Belt and Road agreements. Caixin, 15 May 2017
Xi J (2018) Address to 2018 Boao Forum for Asia. State Council, Hainan, 11 Apr 2018
Xinhua (2017) CDB to issue B&R project loans in three years. CD, 2 June 2017
Yoshino N (2017) Abe takes a shine to China's Belt and Road plan. Nikkei, 8 June 2017
Zhou X (2017) Remarks at Chinese Firms Going Abroad 50 Forum. NDRC, Beijing

Chapter 2
Fear Factor: Strategists Versus Bankers

2.1 Xi Jinping's 'New Silk Road'

In November 2018, after US Vice President Mike Pence excoriated President Xi Jinping's Belt and Road Initiative (BRI) at the APEC summit in Port Moresby, Papua New Guinea, APEC leaders failed to issue a joint statement. This was the first time since APEC's establishment in 1989 that superpower discord disrupted global diplomacy.[1] Pence, making clear Washington's outrage at Chinese conduct, vowed to respond robustly: 'China has taken advantage of the US for many, many years. And those days are over'. He told fellow summiteers that OBOR/BRI was a poor choice: 'The US offers a better option. We don't drown our partners in a sea of debt. We don't coerce or compromise your independence. The US deals openly, fairly. We do not offer a constricting belt or a one-way road'. The US alternative was truly 'win-win': 'When you partner with us, we partner with you, and we all prosper'.[2]

President Xi Jinping's response sounded sober and circumspect: 'We should reject arrogance and prejudice, be respectful of and inclusive toward others, and embrace the diversity of our world. We should seek common ground while putting aside differences, draw upon each other's strengths and pursue co-existence in harmony and win-win cooperation'.[3]

Although Pence posed an unprecedentedly direct and high-level challenge to Beijing's landmark initiative, tensions had simmered for years. The latest phase began in November 2012. Assuming leadership of the Communist Party of China (CPC), Xi 'articulated a vision for the nation's future' framed as the 'China Dream'.[4] He apparently borrowed the phrase from the work by a PLA political commissar, Col. Liu Mingfu, published during the *Great Recession*.[5] Four months later, ascending

---

[1] Banyan (2018).
[2] Pence (2018a).
[3] Xi (2018).
[4] Report (2014).
[5] Liu (2009).

to China's presidency, Xi told the National People's Congress (NPC): 'the great rejuvenation is a dream of the whole nation, as well as of every individual. The Chinese dream, after all, is the dream of the people'.[6]

Since then, Xi's policy initiatives—e.g. giving market forces 'decisive' influence, launching a sustained and expansive anti-corruption campaign targeting both 'tigers and flies', restricting dissent and media freedom, demanding foreign investors' transfer technology to joint venture partners, constructing artificial islands with military-basing potential on disputed South China Sea (SCS) reefs, or initiating the deepest military reforms since 1949—made him the most assertive and, to critics, concerning, Chinese leader since Deng Xiaoping.[7] Xi abandoned Deng's 'hide our strength and bide our time' dictum, disturbing many. Conflicting perspectives jostled in October 2017 as the CPC reappointed Xi its General Secretary, and he chose its new Political Bureau Standing Committee (PBSC) at its 19th Congress for the next five years.

Xi asserted, 'The Chinese nation now stands tall and firm in the east'. Envisioning China becoming a 'moderately prosperous' socialist country able to realise 'socialist modernisation' by 2035 and turn 'China into one of the world's richest and most powerful' states by mid-century, he ended quiet accumulation of 'national comprehensive power'. Xi suggested, 'socialism with Chinese characteristics offered a new option for other countries and nations who want to speed up their development while preserving their independence'. Beijing explained, 'the Chinese experiment showed the world that socialism could achieve its full potential in a developing country'.[8] The NPC incorporated Xi's 'thought' and his BRI vision into the CPC Constitution, placing him theoretically next to Mao Zedong and Deng Xiaoping. This, Xi's offer of China's experience as a template for the developing world, and the end of presidential term limits, troubled critics. Some detected imperious hubris in Xi's three-and-a-half-hour speech and others highlighted the pushback his assertive policies triggered.[9]

Even before Xi's ascent, as the 2008–2010 *Great Recession* challenged the liberal-capitalist order's stability, Western scholars debated theoretical and practical ramifications of 'power shift', a formulation seeking to explain primarily Chinese economic growth eroding the US-led post-Cold War *international system*, and US *primacy* underpinning it.[10] At the depth of the *Great Recession*, US academic Robert Sutter pointed to 'a time of testing' in US–China relations as Beijing, reflecting 'deep mutual suspicions', robustly diverged from US policies and interests. While both powers sought to manage the public discourse and optics, US arms sales to Taiwan proved particularly insulting and threatening to China, which warned it would be less helpful in addressing issues such as Iran's nuclear programme. Sutter averred that increasing Chinese assertiveness indicated growing confidence from having emerged 'stronger than other major powers, including the USA', from the 2008–2009 economic crisis.

---

[6] Xi (2013a).
[7] Keck (2013), Pollock (2016) and Ching (2017).
[8] Xi (2017c), Xu (2017) and Wang (2017b).
[9] Fan (2017), Mitchell and Clover (2017) and Perlez (2017).
[10] Mayer and Kremer (2011).

2.1 Xi Jinping's 'New Silk Road'

In short, power was, and was seen to be, shifting from the established superpower to an emerging peer.[11] The trend was apparent not just among US observers.

Hugh White, an Australian academic and former-practitioner, warned that Americans, used to exercising *primacy* and unwilling to give it up, believed, 'acceptable international conduct' was 'defined as the acceptance of US primacy'. But 'the story of Asia's power shift' was not about the USA, 'It is about China. This is not a story of American weakness, but of Chinese strength'. With the potential to overtake the USA economically, Beijing was deploying its capacity to deter, or impose severe costs on, US force projection in China's periphery. White cautioned that even if the US economy were 'in perfect shape, China's economic transformation would still pose the biggest threat to America's place' as the *systemic* primate.[12]

Within a decade from the onset of the *Great Recession*, the 'power-shift' school had acquired marked salience within the Western-rooted China discourse. Noted analyst Christopher Layne posited that the unusual foreign policy optics presented by Donald Trump's Administration reflected not just the 'US–Chinese power shift' but effectively, 'the end of the Pax Americana'. Layne reviewed aspects of the USA's relative decline in the realms of economic, military, scientific-technological and diplomatic prowess, and the contrasting ascent of Chinese capabilities, ambition and assertiveness, notably in its maritime periphery, but increasingly, across the wider world. Layne underscored the emergence of an 'institutional challenge to the Pax Americana' arising from Beijing's ability to leverage its economic-financial muscle into a growing list of multilateral initiatives paralleling, if not eroding, the institutional framework undergirding the US-led post-war liberal democratic order. With the balance of power shifting, or appearing to shift, China's ability to supplant the hegemon looked evident.[13]

The 'power-shift' school did not, however, win the debate outright. A large body of opinion, prominently embedded within the national security establishment, academia and other key elements of the US cognoscenti, remained 'committed to a grand strategy of "primacy"'. The core of *primacist* beliefs 'strives for military preponderance, dominance in key regions, the containment and reassurance of allies, nuclear counterproliferation, and the economic "Open Door". The interaction of power and habit' engendered a central belief in the need for, indeed the vital purpose of, American leadership in all international affairs for all time.[14] However, over time, arguably mirroring the external reality, the balance of weight between these two schools shifted to the extent that aside from a shrinking pool of influential believers, the debate increasingly appeared to be focused more on seeking ways to maintain US dominance while managing the transition than on denying the transition itself.[15]

Mirroring stark divergences between Chinese and foreign views of Xi's initiatives, the style and substance of his presidential drive deepened Western anxiety.

---

[11] Sutter (2010).
[12] White (2010).
[13] Layne (2018).
[14] Lieber (2016), Brands and Feaver (2016) and Paul (2016).
[15] Porter (2018b).

If 'China's Dream' was the leitmotif of Xi's early leadership years, the 'Belt and Road Initiative' (BRI), i.e. the Silk Road Economic Belt (SREB) and the twenty-first Century Maritime Silk Road (MSR), became his signature policy-rubric. This potentially transformative framework[16] could shape China's economic–diplomatic perspectives for decades. Xi broached the initiative before regional audiences first in Astana (renamed Nursultan in 2019), the capital of Kazakhstan, and then in Jakarta, Indonesia, in 2013.[17]

Pointing to the Western Han Dynasty (206 BC–AD 24) imperial envoy Zhang Qian's ancient Silk Road mission 2100 years prior, to open 'friendly contacts' with Central Asian polities, Xi told his Nazarbayev University audience about the contributions Silk Road exchanges made to cooperation among China, Central and West Asia, and Europe: 'over the 20-plus years, the ancient Silk Road is becoming full of new vitality with the rapid development of China's relations with Asian and European countries'. Offering to build 'a community of interests' linking the China-inspired Shanghai Cooperation Organisation (SCO) and the Russian-led Eurasian Economic Community (EEC), Xi proposed the SREB to share developmental strategies, open up 'the transportation channel from the Pacific to the Baltic Sea', connect East Asia, West Asia and South Asia, facilitate trade and investment, promote financial 'circulation' and risk-prevention and advance people-to-people exchanges.[18] His audience welcomed BRI.

Three weeks later, addressing the Indonesian parliament, Xi recalled the seven early-fifteenth-century voyages of the Ming Dynasty (1368–1644) Admiral, Zheng He, to the 'Western Seas', Southeast Asia, South Asia and East Africa. That benign narrative, reinforced with recent collaboration, encouraged multi-tiered cooperation envisaged along a twenty-first century MSR linking China, Association of Southeast Asian Nations (ASEAN)-states and other maritime polities. China and ASEAN could 'build trust and develop good neighbourliness', seek win-win cooperation, 'stand together and assist each other', enhance 'mutual understanding and friendship' and 'stick to openness and inclusiveness'. Xi noted, 'the China-ASEAN community of shared destiny is closely linked with the ASEAN community and the East Asia community'.[19] The MSR would weave shared prosperity into the maritime landscape.

These speeches partly marked a charm offensive aimed at mitigating anxiety triggered by Beijing's fast-accreting economic heft, rising military capacity, and Xi's application of these in pursuing China's interests. Three weeks after his Jakarta peroration, Xi addressed the PBSC and China's diplomatic corps on his regional vision underscoring the urgency with which he wished to improve Beijing's engagements. Indicating the import Xi attributed to his guidance to China's envoys, Premier Li Keqiang presided over the proceedings in presence of the entire PBSC. Xi outlined

---

[16] Kovrig (2017, p. 1).

[17] For more information, you can find the 'Official BRI map reproduced by Australian Broadcasting Corporation (ABC)' here: http://www.abc.net.au/radionational/programs/breakfast/the-new-silk-road/7864514 (last accessed October 16th, 2019).

[18] Xi (2013b).

[19] Xi (2013c).

2.1 Xi Jinping's 'New Silk Road'

his blueprint for a collective, mutually beneficial, regional future. Diplomacy pursued the 'strategic goal' of realising the 'two centenary goals and the great rejuvenation of the Chinese nation', i.e. building 'a moderately prosperous society in all respects' before the CPC's 2021 centenary, and building 'a modern socialist country that is prosperous, strong, democratic, culturally advanced and harmonious', by the PRC's 2049 centenary.[20]

Focusing on BRI/OBOR as the organising framework of Beijing's outreach, Xi stressed, 'the basic principle of diplomacy with neighbours is to treat them as friends and partners, to make them feel safe and to help them develop. The concepts of friendship, sincerity, benefit and inclusiveness should be highlighted. Developing good-neighbourly relations is the consistent principle of China's diplomacy with its neighbours'. Xi noted, 'China and neighbouring countries should treat each other as equals and value the friendship'. Stressing 'mutual benefits and win-win results', Xi urged 'joint efforts with relevant countries to accelerate infrastructure connectivity, to build SREB and MSR of the twenty-first Century. We should continuously deepen regional financial cooperation, actively prepare for the establishment of the AIIB (Asian Infrastructure Investment Bank), and improve the regional financial safety net'. Previewing the SREB's networked blueprint, Xi advised, 'We should speed up the opening-up of border areas, and deepen mutually beneficial cooperation between border provinces and areas with neighbouring countries'.[21] The responsibility for forging positive ties rested with China.

Critics, however, discerned threats to the US-led post-Cold War order in most Chinese action. Vice President Pence led the 'New Cold War' charge with a late-2018 speech, then challenging China at regional summits, while the USA deployed a two-CSG flotilla to the South China Sea (SCS).[22] Then-Secretary of Defence, James Mattis, countered 'Chinese threats', allegedly extending to space, with the formation of a Space Command,[23] and by renaming the US Pacific Command (USPACOM) the Indo-Pacific Command (INDOPACOM), extending its area of responsibility to the Indian Ocean, thus formalising the Indo-US counter-China coalition:[24] 'For every state, sovereignty is respected, no matter its size and it is a region open to investment and free, fair and reciprocal trade not bound by any nation's predatory economics or threat of coercion, *for the Indo-Pacific has many belts and many roads*'.[25]

Mattis stressed Washington's determination to defend the USA's *primacy*: 'China's policy in the SCS stands in stark contrast to the openness our strategy

---

[20] Report (2017b).

[21] Xi (2017b) and Perry (2015).

[22] Pence (2018b), Hufbauer (2018), Editorial (2018c), 7th Fleet Public Affairs (2018), Greenert (2018a) and O'Rourke (2018a).

[23] Mattis (2018b).

[24] Ali (2018).

[25] Mattis (2018f); emphasis added.

promotes. It calls into question China's broader goals'. He warned: 'Make no mistake; America is in the Indo-Pacific to stay. This is our priority theatre'.[26] Then-Chairman of the US Joint Chiefs of Staff (CJCS), General Joseph Dunford, similarly underscored growing Chinese and Russian capability to erode the USA's military *primacy*.[27] Secretary of State, Mike Pompeo, asserted, 'the Indo-Pacific must be free and open'.[28] US-aligned statesmen endorsed this strategic vision, best seen in the European Union's (EU) proclamation that China was 'an economic competitor in the pursuit of technological leadership, and systemic rival promoting alternative models of governance'.[29] Such adversarial identification elicited Chinese accusations of the West's 'Cold War mentality' and 'zero-sum approaches' to interstate relations: 'Western sceptics have failed to see the fact that the Chinese don't harbour a zero-sum mentality, but encourage win-win thinking. As Confucius once said, "He who wants success should enable others to succeed". The (Belt and Road) initiative was offered by China for the good of all'.[30]

Beijing's FDI record offered some evidence, with non-financial 2018 FDI in BRI-economies reaching $15.64bn, up 8.9% year-on-year, and 'completed turnover' of 'contractual projects' totalling $89.33bn, but few noticed.[31] That year, some saw the USA was 'defeating' China in a trade war Donald Trump launched in July. China watchers believed Beijing's 'egregious' BRI investments and debt-financing explained its 'defeat' in 'the New Cold War'.[32] Others, sceptical of the efficacy of Trump's tariff-tactics, described these as 'the opening skirmish in a geopolitical Cold War' which, they feared, the USA would not win.[33] As Washington accused Huawei, China's largest telecoms-manufacturer, of violating US sanctions against Iran, and of stealing data, and Chinese hackers of running a 'Technology Theft Campaign', rhetoric mirrored dialectic-dynamics.[34] Against that backdrop, China critics shaped the BRI discourse.

2.2 Belt and Road Initiative's Basic Design

In March 2015, the NDRC, China's apex planning organ and the BRI's intellectual fountainhead, published its blueprint. Describing the 'Silk Road spirit' as a synthesis of 'peace and cooperation, openness and inclusiveness, mutual learning and

---

[26] Mattis (2018h).

[27] Garamone (2018b).

[28] Pompeo (2018).

[29] High Representative for Foreign Affairs and Security Policy (2019).

[30] Zhang (2017).

[31] Department of Outward Investment and Economic Cooperation (2019), Liu and Liu (2018), Ng and Wei (2017), FMPRC (2017) and Chang (2015).

[32] Pei (2018), Report (2018g) and Mourdoukoutas (2018).

[33] Kaletsky (2018), Dixon (2018), Garrett (2018) and Capri (2018).

[34] Public Affairs (2018), Reuters (2018) and Kate (2018).

2.2 Belt and Road Initiative's Basic Design

mutual benefit', it urged the spirit's revival to counteract the global economy's 'weak recovery' and address 'complex international and regional situations'. Recalling Premier Li Keqiang's remarks at the 2013 China-ASEAN Exposition, it stressed MSR's ASEAN orientation, designed 'to create strategic propellers for hinterland development'. It posited that BRI could 'promote the economic prosperity of the countries along the Belt and Road, and regional economic cooperation, strengthen exchanges and mutual learning between different civilisations, and promote world peace and development'.[35]

Connectivity projects would 'help align and coordinate the development strategies' of partner-states, 'tap market potential in this region, promote investment and consumption, create demands and job opportunities, enhance people-to-people and cultural exchanges, and mutual learning'.[36] BRI would mirror 'the purposes and principles of the UN Charter', and China's 'Five Principles of Peaceful Coexistence: mutual respect for each other's sovereignty and territorial integrity, mutual non-aggression, mutual non-interference in each other's internal affairs, equality and mutual benefit, and peaceful coexistence'. Advocating 'tolerance among civilisations', BRI was 'open to all countries, and international and regional organisations'.[37]

A bridge linking Asia, Europe and Africa, BRI would connect 'the vibrant East Asia economic circle' with the 'developed European Economic circle'. The SREB would bring 'together China, Central Asia, Russia and Europe (the Baltic); linking China with the Persian Gulf and the Mediterranean Sea through Central Asia and West Asia; and connecting China with Southeast Asia, South Asia and the Indian Ocean'. The MSR, comprising ports and new infrastructure, would link 'China's coast to Europe through the South China Sea (SCS) and the Indian Ocean in one route, and from China's coast through the SCS to the South Pacific in the other'.[38] China would thus be joined to Southeast Asia, South Asia, Africa, the Middle East and Europe to the south and west, and to Oceania and Latin America to the south-east.[39]

The SREB, connecting cities and industrial parks along trans-Eurasian transport routes, fashioned a 'Eurasian Land Bridge' via the China–Mongolia–Russia, China–Central Asia–West Asia and China–Indochina Peninsula Economic Corridors. The China–Pakistan Economic Corridor (CPEC) and the Bangladesh–China–India–Myanmar (BCIM) Economic Corridor were 'closely related' to BRI. These would improve regional infrastructure, establish 'a secure and efficient network of land, sea and air passages' with industrial clusters joined via railways, roads, waterways, aviation, pipelines, information- and utility networks, to generate shared

---

[35] NDRC (2015), Preface.

[36] Ibid., Chap. 1.

[37] Ibid., Chap. 2.

[38] Ibid., Chap. 3.

[39] For more information, you can find China's 2014 'Official BRI map reproduced by China Central TV (CCTV)' here: http://files.chinagoabroad.com/Public/uploads/v2/uploaded/pictures/1504/new_0.jpg (last accessed October 16th, 2019).

gains.[40] The blueprint's 'major goals' were to forge a Eurasian economic/trading bloc[41] by

- Promoting policy coordination
- Advancing facilities connectivity
- Reinvigorating unimpeded trade
- Enhancing financial integration and
- Boosting people-to-people bonds.

BRI utilised national and multilateral financial institutions e.g. the AIIB, NDB and SRF, while benefiting from the SCO, ASEAN + China (ASEAN + 1), APEC, ASEM and other track-1 and track-1.5 groupings active along BRI alignments.[42] Beijing earmarked $50bn for financing the AIIB, $40bn for SREB, $40bn for SRF and $25bn for MSR.[43] Hu Huaibang, Chairman of China Development Bank (CDB), revealed that by the end of 2017, CDB had lent $110bn to BRI projects and would lend a total of 250bn. Hu noted, during 2014–2016, Beijing invested $50bn in partner-states, with aggregate trade volumes reaching $3tn. He expected China to invest 150bn and import goods worth $2tn from these countries, during 2018–2022. Controlled by the State Council, backed by China's Ministry of Finance and endowed with assets worth $2.4tn, CDB was a leading financier powering BRI implementation.[44]

SREB would modernise China's underdeveloped inland, western and border provinces/regions, e.g. Xinjiang, Tibet, Ningxia, Qinghai, Yunnan, Guangxi, Inner Mongolia, Heilongjiang, Jilin and Liaoning. New links between developed and developing zones would accelerate industrialisation, spreading its gains nationally. Focused investments tying 'hubs' like Xian, Zhengzhou, Wuhan, Changsha, Nanchang and Hefei would reduce inequality and connect China to the rest of Eurasia. MSR, launching from coastal China, would expand free-trade and industrial zones in Shanghai, Fujian, Shenzhen, Guangzhou and Zhuhai.[45] A spur running south from the SCS would link the South Pacific to BRI's catchment area.

The NDRC, coordinating among Ministries of Foreign Affairs (MOFA) and Commerce (MOFCOM), and other State Council agencies, planned for 'every province, autonomous region and municipality' to 'participate in (developing the BRI) blueprint'. Local governments, guided by the framework, would formulate 'specific plans' and 'progress will be pushed by the cooperation among different regions'.[46] Land-corridors linking China to ports in littoral and peninsular Asia, Africa and Europe would serve as SREB/MSR interstices. Infrastructure projects would drive BRI implementation by enterprises, not the central authorities in Beijing.

---

[40] NDRC (2015, Chap. 3) and Tian (2016).
[41] NDRC (2015, Chap. 4).
[42] Ibid., Chap. 5.
[43] Ghiasy and Zhou (2017, p. 51).
[44] Sun (2018).
[45] NDRC (2015, Chap. 6).
[46] Chen (2015).

2.2 Belt and Road Initiative's Basic Design

Together, SREB and MSR would integrate Halford Mackinder's 'World Island', joining it and China to regions farther afield, although Chinese commentary did not emphasise this. Actors driving BRI's diverse elements focused on its economic/commercial, financial and geoeconomic outcomes. The hope was, 'the people of countries along the Belt and Road can all benefit from this initiative'.[47] Proceedings of Beijing's 2017 and 2019 BRFs indicated leaders representing several dozen partner-states agreed. At the inaugural BRF, Xi assured his guests, and others: 'We have no intention to form a small group detrimental to stability. What we hope to create is a big family of harmonious coexistence'.[48] Hosting the second BRF, he said, 'We hope to work with all parties to improve the cooperation concept for high-quality development of Belt and Road cooperation. We should fully implement the principle of extensive consultation, joint contribution and shared benefits'.[49] China's many detractors paid little attention.

2.3 Conflicting Perspectives

US-based and aligned critics, staggered by BRI's scale, ambition and imagination, questioned Beijing's benign postulates.[50] Profoundly sceptical, they insisted it was a 'vast geopolitical project aimed at cementing China's political and trade role over that of the USA, not an economic one in the sense that each project will generate a return'.[51] Economists warned against China's 'predatory' subversion of the US-designed-and-defended liberal trade regime.[52] These questionable views[53] coloured perceptions of the geoeconomic and strategic landscape among those examining cost–benefit commercial calculi and medium-to-long-term investment values.

Resentment united trans-Atlantic alliance-members. Twenty-seven of the European Union's (EU) 28 envoys in Beijing—except Hungary—complained that BRI ran 'counter to the EU agenda for liberalising trade and pushes the balance of power in favour of subsidised Chinese companies'. Their report, foreshadowing a EU-China summit, claimed Beijing sought to 'shape globalisation to suit its own interests' and warned, unless China was 'pushed into adhering to the European principles of transparency in public procurement, as well as environmental and social standards', EU firms could fail to 'clinch good contracts'. They alleged, 'the initiative is pursuing domestic political goals like the reduction of surplus capacity, the creation of new

---

[47] Ibid.; NDRC (2015, Chap. 8).
[48] Xi (2017a).
[49] Xi (2019).
[50] Xinhua (2018d) and Mifune (2018).
[51] Tellis et al. (2019, Chap. 1), Pence (2018a), Suzuki (2018) and Pi et al. (2017).
[52] Friedberg (2018b), Report (2017f), Richman et al. (2017) and Lawder (2017).
[53] Beinart (2019).

export markets and safeguarding access to raw materials'.[54] This view was widely shared.

G7 foreign ministers critiqued China's maritime conduct, human rights records and purported challenge to the 'rules-based order'; Beijing decried it as improper and irrelevant.[55] British, French and Australian officials announced naval collaboration challenging Beijing's SCS operations, reinforcing the Washington-Beijing dichotomy.[56] US monitors of China's 'strategic goals', i.e. boosting national power, its innovation-driven growth-model and military modernisation, insisted that violating a 2015 agreement restraining cyber-targeting of commercial proprietary data, Beijing still acquired 'US technology to include sensitive trade secrets and proprietary information' using both 'licit and illicit methods'.[57] Western economists accused BRI particularly and Chinese lending generally of being state-funded, opaque and risky in the form of 'collateralised loans', allegations Beijing refuted as being 'biased'.[58] Betraying reactive dynamics,[59] BRI attracted very few Western defenders.[60]

2.3.1 China-Rooted Angst

BRI was not the sole object of resentment, but it emerged as a key factor within a broader China-insecurity discourse. Although Donald Trump exploited and intensified the salience of the 'China threat' theory, the postulate and its policy manifestations preceded his ascent to power.[61] The rapidity with which Beijing refined policies, boosting its economic, commercial, scientific-technological, diplomatic and military capabilities, aroused profound anxiety.[62] The fact that the PRC's non-democratic authoritarianism appeared to present a model of success which could not be directly matched by liberal democratic capitalist states deepened spreading angst.[63] The State Council's May 2015 roadmap for advancing China's manufacturing prowess to the world's top tiers—Made in China 2025 (MiC2025)—focusing R&D investment in ten sectors to attain autarky and secure China's national comprehensive power-potential, was in that context deeply troubling[64]:

---

[54] Heide et al. (2018).
[55] G7 (2018) and Hongyi (2018).
[56] Joshi and Graham (2018) and Scimia (2018); France challenges Beijing in SCS. Straits Times, 12 June 2018.
[57] Pompeo and Nielsen (2018) and National Counterintelligence and Security Centre (2018).
[58] Horn et al. (2019) and Hu (2019).
[59] Mattis (2018e), Mu (2018) and Ford (2018).
[60] Jones and Zeng (2019).
[61] Broomfield (2003), Al-Rodhan (2007), Tiezzi (2014), Ai (2016) and Ciovacco (2018).
[62] Politi (2019), Chaguan (2018), Wiseman (2018) and Jacques (2018).
[63] Weiss (2019), Zhou (2019), Jacques (2018) and Runciman (2018).
[64] Information Office (2015), National Advisory Committee on Building a Manufacturing Power (2015).

2.3 Conflicting Perspectives

- Next-generation IT
- High-end numerical control machinery and robotics
- Aerospace and aviation equipment
- Maritime engineering equipment and high-tech vessel manufacturing
- Advanced railways
- Energy-saving and new-energy vehicles
- Electrical equipment
- New materials
- Biomedicine and high-performance medical devices and
- Agricultural machinery and equipment.

The $300bn blueprint envisaged 'basic industrialisation' by 2020, with manufacturing quality, innovative capacity and labour-productivity being 'significantly enhanced' by 2025, taking China to 'the middle level' of global manufacturing by 2035. At the 2049 centenary, its 'comprehensive strength' would be 'among the world's foremost manufacturing powers'. Sectoral plans, with a focus on artificial intelligence (AI), outlined ambitions.[65] A January 2018 update noted China's aim 'to become the world's leading manufacturer of telecommunication, railway and electrical power equipment by 2025'.[66] Despite China's dependence on USA, Japanese and German technology,[67] the USA feared China would overtake it, end Western high-tech exports, and set standards on key technologies, upending scientific hierarchies.[68]

DoD analysts, noting that US *primacy* depended on superior military-technology flowing from innovative enterprises, and seeing 'How Chinese investments in emerging technology enable a strategic competitor to access the crown jewels of US innovation', urged action. USA and Chinese investments shared common foci: AI, autonomous vehicles, augmented/virtual reality, robotics and blockchain technology, which promised a 'significant impact on the future of warfare'. The USA must prevent Chinese investors from 'stealing' these 'crown jewels'.[69] USCC analysts urged legal defences.[70] Identifying seven geoeconomic tools China allegedly used for geopolitical gain, strategists sought comparable counteraction[71]:

- Trade agreements to improve geopolitical ties or 'cutting off trade to coerce states into changing their geopolitical behaviour'
- Investment, 'loans and debt to finance infrastructure projects for geopolitical benefits'
- Economic and financial sanctions to 'force geopolitical change'

---

[65] Lynch and Rauhala (2018) and Kou (2018).

[66] Ma (2018).

[67] Report (2019a), Steinberg (2019) and Woetzel et al. (2018).

[68] Porter (2018a), Cleveland and Bartholomew (2018a) and Report (2017e).

[69] Brown and Singh (2018).

[70] Beeny (2018).

[71] Blackwill (2018).

- Cyber tools to 'steal intellectual property, siphon funds, or disrupt economic activity' as part of a wider 'geopolitical agenda'
- Economic assistance to influence the recipient state's geopolitical conduct
- Fiscal and monetary policy to establish the Yuan/RMB as a global reserve, and to secure 'favourable exchange rates to influence another state's geopolitical behaviour'
- Energy and commodities policies shaping 'supplies and sales to other countries to influence their geopolitical decisions'.

Donald Trump's campaign rhetoric accusing China of one of the 'greatest thefts in the history of the world'[72] partly mirrored anxiety afflicting white Christian Americans whose 'China threat' fears contributed to his election.[73] Early in 2018, NSC staff warned the White House of Chinese threats to US technological pre-eminence posed especially by Huawei.[74] Trump ordered the USTR to determine under Section 301 of the 1974 Trade Act whether to investigate China's possibly discriminatory 'law, policies, practices, or actions' that could 'harm American IPR, innovation, or technology development'.[75] Following Robert Blackwill's analysis, the USTR's arguments for new duties conflated geopolitics and geoeconomics:[76]

- Trump's December 2017 National Security Strategy (NSS) report identified China as a 'revisionist power' determined to erode the US-led order.
- The 1962 Trade Expansion Act (Section 232) granted Washington the right to levy tariffs because China's actions posed a 'threat' to US 'national security'.
- The 1974 Trade Act (Section 301) justified tariffs because Beijing's 'unreasonable' practices 'restrict US commerce'.

That rationale justified blocking Alibaba's acquisition of MoneyGram, sale of ZTE and Huawei ICT equipment, and a proposed Broadcom–Qualcomm merger, slashing Chinese investment. The trade war impacted more visibly. In July 2018, after Trump's Beijing-aided 'historic' summit with Kim Jong-un in Singapore, Trump launched a tariff war with China.[77] In 2016, the USA' imports, at $2251bn, made up 17.6% of the world total, while its exports, at $1455bn, stood at 11.6%. Chinese figures were $1586bn (12.4%) and $2098bn (16.8%), respectively.[78] With total trade at $3706bn and $3685bn, respectively, the two states dominated global commerce. US–China trade tensions inevitably impacted systemically. By late-2018, the 'war' became so serious that Christine Lagarde, then Managing Director of the IMF, warned, the 'uncertainty and lack of confidence already produced by the threats against trade, even

---

[72] Stracqualursi (2017) and Ni (2016).
[73] Mutz (2018).
[74] NSC (2018).
[75] USTR (2018).
[76] Garrett (2018).
[77] Paquette and Rauhala (2018).
[78] Koopman and Maurer (2017).

2.3 Conflicting Perspectives

before it materialises', could 'shock' emerging markets; crises afflicting Argentina and Turkey, for instance, could 'spread' across the developing world.[79]

Trump first imposed 25% duties on steel and 10% on aluminium imported from several countries, later exempting most, but not China. The USTR recommended 25% duties on $50bn of Chinese products. Beijing reciprocated with duties on $50bn of US merchandise, largely affecting Trump's rural support base. Trump asked the USTR to consider tariffs on another $100bn of Chinese goods, but insisted, 'We are not in a trade war with China'.[80] In September, Trump imposed 10% tariff on another $200bn of Chinese goods, threatening to raise it to 25% if Beijing did not concede. China imposed tariffs on $60bn of US-made merchandise. Trump threatened additional tariffs on all imports from China.[81] Many of the Chinese items subjected to early tariffs used MiC2025-listed technologies.[82] After 11 rounds of talks failed to restore peace, in May 2019, threatening to tax all imports from China, Trump raised tariffs on $200bn of imports to 25%. In late-2019, Trump raised existing tariffs on $250bn of Chinese imports from 25 to 30%, and those on remaining $300bn of Chinese goods from the previously announced 10–15%. Beijing responded with new levies on $75bn of US products.[83]

Washington asserted, Beijing was 'seeking to acquire the intellectual property and technologies of the world and to capture the emerging high-technology industries that will drive future economic growth'. This was unacceptable: 'Given the size of China's economy, the demonstrable extent of its market-distorting policies, and China's stated intent to dominate the industries of the future', its 'acts, policies and practices of economic aggression now targeting the technologies and IP of the world threaten not only the US economy but also the global innovation system as a whole'.[84] Washington barred exports of US products to eight Chinese SOEs and their 36 subsidiaries on national security grounds.[85] The White House insisted it would end 'years of unfair trade practices'.[86] Sanctions on the Chinese IT behemoth and 5G pioneer Huawei and its many affiliates followed.[87] Trade turbulence and critique of the BRI blueprint exposed Beijing's determination to challenge slights.[88] Xi urged Politburo colleagues to 'foster stronger confidence in the path, theory, system and culture of socialism with Chinese characteristics' and to 'fear no risks and never be confused by any interference'. A year later, he told Party cadres China faced a period of 'concentrated risks' which they must fight resolutely. As long as the challenges lasted, 'we must carry out a resolute struggle, and we must achieve victory. Cadres in

---

[79] Lagarde (2018b).
[80] Chang (2018).
[81] Report (Report 2018f, 2018j), Paquette and Rauhala (2018) and Xinhua (2018c).
[82] Office of Trade and Manufacturing Policy (2018) and Hamlin et al. (2018).
[83] Politi et al. (2019).
[84] Office of Trade and Manufacturing Policy (2018, p. 20).
[85] Bureau of Industry and Security, Commerce (2018).
[86] Report (2018i) and Trump (2018b).
[87] Cuthbertson (2019) and Bajak and Liedtke (2019).
[88] Report (2018a), Shen (2018), Editorial (2018e) and Partington (2018).

leadership positions must be warriors who dare to struggle and are good at struggle'.[89] There were no indications of compromise.

In Washington, Trump's September 2018 National Cyber Strategy document, like his NSS, National Defence Strategy, Nuclear Posture Review and intelligence community assessments, identified Beijing as a rival. China, Russia, Iran and North Korea, the USA's 'competitors and adversaries' profited from 'the open Internet, while constricting and controlling their own people's access to it'. China hid 'behind notions of sovereignty while recklessly violating the laws of other states by engaging in pernicious economic espionage and malicious cyber activities'.[90] Slamming Beijing for 'cyber-enabled economic espionage and trillions of dollars of intellectual property theft',[91] Trump authorised NSA/Cyber Command offensives to deter, prevent and punish cyber-attacks. China rapidly emerged as the target of an inter-agency counteroffensive.

Trump unprecedentedly sanctioned the CMC's Equipment Development Department, its *materiel* procurement arm, for buying Russian Su-35 combat aircraft and S-400 missile-defence systems.[92] Washington sold F-16 spare-parts worth $330m, sold F-16 upgrades worth $500m and authorised the sale of 66 F-16 V fighters worth $8bn to Taiwan,[93] underscoring its China focus in its regional military build-up and FONOPS. Beijing's responses, alongside trade tensions and BRI implementation, heightened uncertainty.[94] The Admiral responsible for maintaining US *primacy* told Congress, China represented 'the greatest long-term strategic threat to a Free and Open Indo-Pacific and to the USA'.[95] Both rivals struggled to bring each other around to its own views. With hardliners determined to 'win', the risk of escalation from a gathering 'New Cold War' to a shooting one rose.[96]

A US think tank, negating Chinese rebuttals, reignited debate over Beijing's purported string of potential Indo-Pacific bases concealed underneath MSR port-building.[97] US policy documents, reinforcing China's alleged rise as a strategic rival, boosted counteraction.[98] BRI became enmeshed in disputatious Sino-US geopolitical narratives. Resonant economic and strategic threat perceptions reinforced both dynamics, deepening distrust and challenging stability. Xi Jinping's pledges to further open up China's economy and protracted negotiations produced a lull.[99] After Vice President Pence's fulminations in Washington, Singapore and Port Moresby, Trump and Xi called a truce. Meeting at consecutive G20 summits in Argentina and

---

[89] Blanchard (2019) and Report (2018k).
[90] Trump (2018a, p. 1).
[91] Trump (2018a, p. 2).
[92] Bianji (2018a), Bureau of Public Affairs (2018) and Stanway and Wroughton (2018).
[93] AP (2019) and Ihara (2019).
[94] Chen (2018), Report (2018c) and Reuters/AP/NBC (2018).
[95] Davidson (2019).
[96] Liu (2018), Li and Liang (2018), Li (2018c) and Rachman (2018).
[97] Thorne and Spevak (2018) and Report (2018b).
[98] Coats (2019), Public Affairs Office (2018) and Pry (2017).
[99] Xinhua (2018b), Borak (2018) and Lynch and Rauhala (2018).

Japan, they agreed, on the bases of China implementing promised reforms, to partly restore the *status quo ante*. However, as the rivals pursued status quo versus incrementally revisionist interests, structural challenges haunted relations generally and BRI more specifically.

2.3.2 A Darkening Landscape

Even Western *realists* painted a grim picture: BRI aimed 'to redefine the global economy of the twenty-first century' by integrating Europe, Asia and Africa via an unprecedented network of transport-and-communications infrastructure. BRI was central to realising China's *dream* to become a 'rich and strong' country, a vision reinforcing Beijing's parallel ambition to advance key technologies and set global standards. Xi Jinping's articulation of China's centenary goals deepened US and allied angst.[100] Signs that for the first time since early eighteenth century, a non-Western power organised along principles challenging centuries of liberal-capitalist Euro-Atlantic dominance, was assuming planetary prominence, precipitated profound fears.[101]

The doyen of the US' China policy practitioners, Henry Kissinger, noted conflicting perspectives: US analysts believed China sought to displace 'the USA as the preeminent power in the western Pacific' and consolidate 'Asia into an exclusionary bloc deferring to China's economic and foreign policy interests'. Beijing's 'absolute military capabilities' were not 'formally equal' to Washington's, but China possessed 'the ability to pose unacceptable risks in a conflict' and had been 'developing increasingly sophisticated means to negate traditional US advantage'.[102] To a power dedicated to perpetual *primacy*, Beijing's perceived ambivalence towards post-Cold War *unipolarity* was intolerable.

Australia's China-savvy former prime minister, Kevin Rudd, saw Xi's proposed 'new model of major-power relations' as a demand for parity in power, prestige and influence. He thought Xi's leadership, mirrored in the 19th CPC Congress proceedings, manifested a belief that 'Western democracy is corrupt, hypocritical and fails to meet the needs of the poor'. Under Xi, the CPC 'senses that the global spread of liberal democratic ideas has ground to a halt, leaving the West's geopolitical power and prestige ripe for challenge'.[103] Beijing sought an equivalence among US built bodies e.g. the UN and Bretton Woods funds, and Chinese initiatives, e.g. BRI, the AIIB and the NDB. Western *realists* 'simply roll their eyes' when they heard Xi's concept of a 'global community of common destiny for all humankind', but for many Chinese, this was a profound contribution to civilisation's advance.

---

[100] Allison (2017c), White (2017b) and Campbell (2017).
[101] Denmark (2019) and Gehrke (2019).
[102] Kissinger (2012).
[103] Rudd (2017).

Rudd's compatriot, Gareth Evans, warned, China 'wants strategic space in East Asia and is no longer prepared to play second fiddle to the USA, either there or as a global rule-maker'. China was 'parlaying its economic strength into geopolitical influence through the BRI, modernising and expanding its military capability and pursuing expansionist territorial claims in the SCS'. However, unlike 'dragon-slayers' dominating the China *threat* discourse, Evans believed, 'China will not seek to usurp America in the global order but take its place alongside it'; China could live 'quite comfortably' in a milieu defined by 'cooperative security, in which states primarily find their security with others rather than against them'.[104] That was rare optimism.

With debates dividing Australia's Departments of Defence, Foreign Affairs, Trade and Immigration, Prime Minister Malcolm Turnbull decided against joining BRI. The enthusiasm of foreign trade officials was challenged by everyone else. Canberra's Foreign Policy White Paper highlighted China-rooted anxiety, and Australia's intensifying collaboration in China focused naval activities with Quad-partners the USA, Japan and India made this a foregone decision.[105] Even a 'moderate' strategist wrote, 'If the USA and its allies are really determined to resist China's challenge to the old US-led liberal global order, they have to counter Beijing's powerful vision of a future global economy centred on China' with 'an equally powerful and ambitious global economic vision of their own'.[106] None was apparent.

Instead, US *primacists* posited, 'Multiple US administrations have tried and failed, to focus attention on a rising China'. They blamed George Bush and Barack Obama: 'Too easily diverted by near-term distractions, American leaders have failed to act as China has become a military and economic powerhouse capable of contesting US leadership not only in Asia, but also increasingly around the world'. Accusing China of 'closing what was once a seemingly unbridgeable military gap with the US', perpetual-*primacists* warned, BRI would intolerably 'further reinforce Beijing's ability to compete with the USA on a more global footing'.[107]

Western practitioners and analysts built a corpus on China's purportedly unstated geopolitical objectives of 'peacefully' eroding *Pax Americana* and supplanting the USA at the *systemic core* without offering Washington a *casus belli*. Then Secretary of State Rex Tillerson explained: 'China, while rising alongside India, has done so less responsibly, at times undermining the international rules-based order even as countries like India operate within a framework that protects other nations' sovereignty'. Praising India as a partner sharing US values, interests and vision, Tillerson described China, and BRI, as sources of 'disorder, conflict and predatory economics'.[108] He drew India tighter into a grand-strategic embrace: 'China's provocative actions in the SCS directly challenge the international law and norms that the USA and India both stand for. The US seeks constructive relations with China, but we will not shrink from

---

[104] Evans (2017a).
[105] Greene and Probyn (2017), Bishop (2017, Chaps. 2–3) and Singh (2018).
[106] White (2017a).
[107] Kliman and Cooper (2017).
[108] Tillerson (2017a).

China's challenges to the rules-based order and where China subverts the sovereignty of neighbouring countries and disadvantages the USA and our friends'.[109]

Chairman of the Joint Chiefs of Staff (CJCS), General Joseph Dunford, having completed a trip preceding Trump's first presidential visit to the region, noted, 'China's path of capability development and their efforts to address our power projection capability, our ability to deploy when and where necessary to advance our interests, is very much the long-term challenge in the region'. Dunford believed, 'China is the challenge'.[110] Washington's unusual decision to drill three carrier strike group (CSG) in the Western Pacific during Trump's first Asian trip, including a stop in Beijing, was 'a demonstration that we can do something that no one else in the world can'. The display of military prowess undergirding *primacy* could not be any clearer as a warning to challengers.[111]

Dunford told Congress, 'As China's military modernisation continues, the US and its allies and partners will continue to be challenged to balance China's influence'. DoD was deploying 60% of its naval and air forces, especially advanced systems, to the Asia-Pacific by 2020, building up allied-and-partner militaries, and boosting joint combat capabilities. He acknowledged Beijing's defensive drivers: 'China is developing capabilities to deter or defeat third-party intervention in a regional conflict and is focusing on asymmetric capabilities to target key US advantages'. Still, *primacy* drove policy: 'The US will continue to develop a security network through multilateral partnerships and build interoperability and partner capacity…Unilaterally, we will continue to develop capabilities to counter China's improving military capabilities'.[112]

A former CNO, Jonathan Greenert, reported, 'the capacity for accumulating power and President Xi Jinping's somewhat revisionist "China Dream" goals make China the greatest long-term security challenge'.[113] BRI reinforced threat perceptions: 'The Chinese have shown a willingness to use economic leverage as a way to advance the CCP's regional political objectives'. As Beijing's 'international economic presence has expanded through initiatives such as OBOR, China is increasing its ability to respond to global contingencies and protect its interests, such as sea lanes, abroad'.[114] Intelligence assessments reinforced military fears. Then DCI Mike Pompeo noted, 'We can watch very focused efforts to steal American information, to infiltrate the US with spies, with people who are going to work on behalf of the Chinese government against America'. He saw 'infiltration' in 'our schools…our hospitals and medical systems. We see it throughout corporate America. It's also true in other parts of the world'.[115]

---

[109] Ibid.
[110] Kube (2017).
[111] Lendon (2017) and Report (2017g).
[112] Dunford (2017).
[113] Greenert (2018b).
[114] Dunford (2017).
[115] Correra (2018).

Displacement fear coloured the US policy discourse. The USCC, charged by Congress to examine the impact of US–Chinese economic transactions on US 'national' security, and monitoring China's military strength and behaviour, advised, 'Although China claims the (BRI) mega-project is primarily economic in nature, strategic imperatives are at the heart of the initiative'. BRI served 'Beijing's goals to revise the global political and economic order to align with China's geopolitical interests and authoritarian political system'.[116] Legislators boosted Beltway fears of a 'rising' China elementally threatening the US 'national' security.

The Chairwoman of a House Committee, welcoming experts on 'China's pursuit of emerging and exponential technologies', for instance, noted that China's military-technology pursuits and 'the resulting impact on US national security' were 'a critically important topic'. The fact that, 'China continues to increase their research and development investments at an alarming pace and is rapidly closing many of their technology gaps', was particularly worrying: 'More and more, we see China using only domestic Chinese items and creating high market-access barriers to support domestic capacity. The effect is to replace any and all dependency on foreign companies, investments and technologies'.[117] Clearly, a terrifying prospect for the USA.

2.3.3 Warriors to the Fore

Think tanks reinforced policy circularity.[118] Xi Jinping's authoritarian centralisation and assertive proto-revisionism, which BRI allegedly both reflected and reinforced, triggered the latest 'China debate'.[119] An economic commentary urged US-led response via the 'anti-PRC front' against Beijing's 'predatory' BRI activities.[120] Identifying China as the first among 'top 10 threats to the world order', a prominent group warned, Beijing offered 'a viable alternative' to Western liberal democracy: 'For most of the West, China is not an appealing substitute. But for most everybody else, it is a plausible alternative'.[121] Commentators ignored Beijing's rebuttals and reiteration of its long-standing position on China's locus within the dynamic power-hierarchy virtually autonomously recasting the geopolitical landscape: 'China has no leadership ambitions nor any desire to replace any other power'.[122]

Critics also ignored Beijing's emphasis on the continuity in its view from Mao to Xi of China's place within the *international security system*.[123] Some observers

---

[116] Cleveland and Bartholomew (2018b, pp. 259–262) and Bartholomew and Shea (2017, p. 8).
[117] Stefanik (2018).
[118] Mastro (2019), Yan (2019), Nouwens and Legarda (2018) and Ellings and Sutter (2018).
[119] Economy (2019), Bush and Hass (2019) and Chatzky and McBride (2019).
[120] Goodman (2017).
[121] Bremmer and Kupchan (2018).
[122] Li (2018b).
[123] Li (2018b) and Stone (2018a).

2.3 Conflicting Perspectives

examined the economic and sociopolitical risks, e.g. domestic disequilibrium and external debt burdens which BRI allegedly inflicted on partners, offering them few benefits, while Beijing secured its goals.[124] Conspiracy theories, animating the BRI and 'China threat' discourses, generated a dialogue of the deaf. The US military had for years been warning about China threatening the USA' *primacy*. At the 2012, US Army War College (AWC) 'PLA Conference' analysts concluded, the PLAN's incipient carrier-led capabilities were 'posing great challenges to US power projection calculations'. China's emerging carrier capability 'is arguably turning the question of "whether" the balance of power in the Western Pacific established and maintained by the USA since the end of World War II will be altered into a question of "when" and "to what extent" the shifting of power will take place'.[125] AWC scholars, analysing the USA's 'post-primacy' 'hypercompetition' challenges, proposed new 'strategic fundamentals of shifting rules of competition, strategic signalling and strategic manoeuver', to thwart China.[126]

Western analysts insisted Beijing's competitive pursuit of 'global resurgence' drove BRI. Some worried that BRI was not 'sufficiently multilateral and serves to expand China's strategic political and economic influence among participating states'. Others visualised BRI as a tool for deploying Beijing's 'soft power' in a drive to establish China as a major global actor.[127] Unlike BRI-promoters, few non-Chinese observers saw compelling merit in the arguments animating China.[128] BRI thus struck as a stratagem advancing Beijing's geopolitical goals. Official fear of China's 'threats' mirrored technical assessments of the shadow Beijing allegedly cast on US *primacy*. The DoD's 2017 review of 'Ballistic and Cruise Missile Threat' to the USA identified North Korea and Iran as the sources of immediate concerns. However, China and Russia, actors with much bigger arsenals, presented 'strategic' threats.[129]

The DoD's 2018 Nuclear Posture Review (NPR) underscored how China's modest nuclear stockpile shaped the US decision to modernise and expand its large thermonuclear armoury. Then Secretary of Defence Mattis asserted that Beijing was modernising and expanding its 'already considerable nuclear forces'. Notably, 'China is pursuing entirely new nuclear capabilities tailored to achieve particular national security objectives while also modernising its conventional military, challenging traditional US military superiority in the Western Pacific'.[130] China responded to Washington's 'Cold War mentality' that 'plays the zero-sum game and deviates from peace and development', with restraint.[131] Undaunted, Mattis told Congress, 'Regarding

---

[124] Report (2018e), Menon (2018) and Li (2015).
[125] Kamphausen et al. (2014).
[126] Freier et al. (2018).
[127] Clover and Hornby (2015), Johnson (2016), Ghiasy and Zhou (2017, p. ix) and Albert (2017).
[128] Gupta (2017).
[129] NASIC (2017).
[130] Mattis (2018d).
[131] Li (2018a) and Geng and Xinhua (2018).

OBOR, in a globalised world, there are many belts and many roads, and no one nation should put itself into a position of dictating "One Belt, One Road".'[132]

Following a DoD-led review of the US' defence-industrial base, Mattis urged checks on Beijing's investments 'to better combat Chinese industrial policies'.[133] Officials explained 'the threat context' to Sino-US interactions, including over BRI. Undersecretary for Defence Policy, John Rood, noted, 'The Russian government and the Chinese government are competitors who seek a revisionist way to change the international order in a way that favours them and to challenge the rules-based international order that the USA has put in place since World War II'.[134] Susan Thornton, Assistant Secretary of State for East Asia, echoed Rood: 'The Chinese are starting to think about, talk about leadership of the international system. That they want to make the rules…So, this is a huge challenge'.[135]

Then Director of National Intelligence (DNI), Daniel Coats, noted 'China wants to be a global power, and you see them spreading their influence, this OBOR. You see them spreading their influence, whether it is the coral islands or whether it is strategic ports, and so forth'. Coats warned Beijing: 'Look, you cannot steal our secrets. Fine. If you want to innovate, innovate, but do not send your kids here. Do not put your people in our labs. Do not take cyber stealing of our innovation and so forth. We are not going to allow that'.[136] Christopher Wray, FBI Director, said, China presented 'the broadest, most challenging, most significant threat we face as a country'. Beijing's 'traditional espionage' and 'economic espionage' used both traditional and non-traditional tools: 'China is trying to position itself as the sole dominant superpower, the sole-dominant economic power. They are trying to replace the USA in that role'.[137]

Michael Collins, Deputy Director of the CIA's East Asia Mission Centre, posited, 'the Chinese fundamentally seek to replace the USA as the leading power in the world'. But, 'the threat that China poses to US national security, economic interests, political well-being and the international order we stand behind, is not necessarily coming from the country itself, its rise, its contribution to international well-being, nor from the diaspora or the Chinese citizenry in general'. China's 'threat' originated in Xi Jinping's CPC leadership which 'increasingly has been aspiring, expanding its ambitions, its interests, its activities around the globe to compete with the USA, and at the end of the day, to undermine our influence relative to their influence'. In short, 'what they are waging against us is fundamentally a Cold War'. He thought, 'The only problem I had on the South China Sea (SCS) issue, when you think more

---

[132] Mattis (2017).
[133] Mattis (2018a).
[134] Rood (2018).
[135] Thornton (2018).
[136] Coats (2018).
[137] Wray (2018).

2.3 Conflicting Perspectives

broadly what SCS means, I'd argue it is the Crimea of the East'.[138] Congressional analysts agreed.[139]

Academics, reinforcing the official line, pressed Washington to robustly defend its *primacy* from China's challenges.[140] Few recommended objective analyses of *systemic* dynamics, the USA's 'vital' national interests and ways to advance these through *transitional fluidity*.[141] US diplomacy also targeted Beijing's engagement with Latin America, a major MSR destination. Tillerson fumed, 'China's offer always comes at a price...The China model extracts precious resources to feed its own economy, often with disregard for the laws of the land or human rights. Today, China is gaining a foothold in Latin America. It is using economic statecraft to pull the region into its orbit. The question is: At what price?' His conclusion: 'Latin America does not need new imperial powers that seek only to benefit their own people'.[142]

Some BRI critics got the details wrong. For instance, 'The "belt" part of the initiative would link dual-use infrastructure across maritime Asia in a series of ports that would support Chinese Navy power projection and potentially complicate Indian, US, and Japanese transit of the Indian Ocean'.[143] The author, mistaking the terrestrial 'belt' (SREB) for the 'road', i.e. MSR, posited, 'That Mahanian maritime strategy has enormous merit and highlights one American advantage in Asia, which is the natural counterbalancing against China's expansion by major US allies and partners in the region'. BRI's 'threats': 'China's OBOR infrastructure initiative, which was announced in 2013 and consists of ports, railways, roads, and airfields linking China to Southeast Asia, Central Asia, the Middle East, and Europe—a 'New Silk Road' that, if it succeeds, will greatly expand China's economic and diplomatic influence'.[144]

Delhi agreed. The Ministry of External Affairs (MEA) explained India's absence from the 2017 BRF: 'Connectivity initiatives must be based on universally recognised international norms, good governance, rule of law, openness, transparency and equality. Connectivity initiatives must follow principles of financial responsibility to avoid projects that would create unsustainable debt burden for communities, balanced ecological and environmental protection and preservation standards; transparent assessment of project costs; and skill and technology transfer to help long-term running and maintenance of the assets created by local communities. Connectivity projects must be pursued in a manner that respects sovereignty and territorial integrity'.[145] Delhi's indictment accused Beijing of flouting global norms, with little evidence.

---

[138] Collins (2018).

[139] O'Rourke (2018b).

[140] Hass et al. (2019), Blackwill (2019), Blumenthal (2017), Allison (2017b), Wright (2017) and Haddick (2014).

[141] Freeman (2016) and Freeman (2010).

[142] Tillerson (2018).

[143] Green (2017).

[144] Nathan (2017).

[145] Spokesperson (2017).

Prime Minister Narendra Modi, addressing a gathering of national and foreign dignitaries representing national security establishments at India's foremost security conclave, emphasised Delhi's adversarial focus on China generally, and BRI/OBOR in particular: 'In the management of our relationship, and for peace and progress in the region, both our countries need to show sensitivity and respect for each other's core concerns and interests'. Given this perspective, it was not surprising that Modi expressed ill-concealed outrage at the alignment of the China–Pakistan Economic Corridor (CPEC), an enterprise described in both China and Pakistan as a 'landmark project' within the BRI endeavour, as it traversed Pakistani-administered sections of the disputed state of Jammu and Kashmir: 'We appreciate the compelling logic of regional connectivity for peace, progress and prosperity. However, equally, connectivity in itself cannot override or undermine the sovereignty of other nations'.[146] Such absolutist rejection of BRI reflected and reinforced growing 'China-vs.-the rest' tendencies.

This formed a pattern. Think tanks waxed bellicose: 'Just as European imperial powers employed gunboat diplomacy, China is using its sovereign debt to bend other states to its will'. How? 'By wielding its financial clout in this manner, China seeks to kill two birds with one stone'. One, to reduce overcapacity by boosting exports; and second, 'to advance its strategic interests, including expanding its diplomatic influence', and securing resources, promoting RMB internationalisation and 'gaining a relative advantage over other powers'.[147] Such pursuits were purportedly deliberate, uniquely Chinese, and threatened current norms and universal values. This outburst against Beijing's debt-for-equity swap over Sri Lanka's Chinese-financed-and-built Hambantota Port betrayed analytical inconsistency; the author neglected to mention India's purchase of a long-term lease to the Chinese-built Hambantota Airport in the vicinity of above Hambantota Port, with no expectation of economic returns.[148]

2.3.4 The Butter Versus Guns Argument

Western perspectives on China generally and BRI specifically were not uniformly negative or even homogenous. In fact, a clear distinction emerged between two large groups of commentators whose views challenged each other. Many political leaders, their national security advisers and acolytes, notably those from the defence- and intelligence services, various think tank strategists, and academic analysts recorded myriad challenges flowing from China's 'national rejuvenation' generally, and from BRI in particular. They focused on BRI's future ability to build partner-state support for and alignment with China's authoritarian political system and statist economy. If allowed to succeed, BRI would, they feared, enable China to secure the endorsement of its illiberal political–economic model, thereby eroding the influence of the

---

[146] Modi (2017).
[147] Chellany (2017).
[148] Chan (2017b) and Chandran (2017).

2.3 Conflicting Perspectives

US-led liberal paradigm. In contrast, many Western economists, financial experts, Western-led multilateral financial institutions and multinational corporations identified considerable benefits resulting from both China's growing capabilities and from the implementation of its BRI blueprint. This 'strategists vs. bankers' dichotomy dividing the West's China discourse, as projected by the media, similarly coloured perceptions of BRI.

Detractors discerned BRI's political and geoeconomic drivers: 'Beijing hopes to recycle some of its accumulated foreign reserves, utilise its overcapacity in construction materials and basic industries and boost the fortunes of its SOEs by opening new markets'. *Zhongnanhai* allegedly promoted development 'as a means of strengthening and stabilising existing authoritarian regimes around China'. Economic dependence on and 'tighter political ties to Beijing' served the latter's strategic interests. Terrestrial infrastructure built into trans-Eurasia transport networks, knitting East Asia, Central Asia, Southern Asia, Russia and Europe, helped 'hedge against possible disruptions to maritime supply' should conflict break out. Expanding China's 'strategic space will help counter alleged US-led efforts to contain the country's rise'.[149]

Critics complained, 'China's financial, political and diplomatic investments in BRI do not come out of a heartfelt Chinese commitment to serve the common good'. China sought 'to get some concrete geopolitical benefits for itself'.[150] As a strategic instrument, BRI would serve Beijing's 'broader regional ambition of building a Sinocentric Eurasian order'. BRI reflected China's 'newfound willingness' to play a 'leading role in reshaping the world'.[151] Analysts warned Trump, 'China has a grand strategy of its own, and Beijing arguably has the better hand of cards. OBOR is without doubt a strong hint of what China has in mind'.[152] The BRI discourse was built on the belief that geopolitical objectives dressed up as geoeconomics motivated China's vision.[153] Undergirding that belief loomed the question: Would the geopolitical consequences of China's 'national rejuvenation' so threaten the US *primacy* that the *system-manager* felt forced to fight to defend the order it had built and led?[154]

Critics treated Chinese leadership views as though these meant nothing or, worse, were deceptive. PLA General Peng Guangqian explained how foolish it would be to ignore the globalised transformation of economic linkages, especially among 'major powers', and China's absence from the dynamics causing 'US decline'. He noted, 'War is useless' in pursuing national goals: notwithstanding 'the considerable gap' between US and Chinese military strength, China's 'defence forces and strategic counterstrike capabilities suffice for equivalent destructive counterattacks against

---

[149] Roland (2017).
[150] Roland (2018) and Editorial (2018a).
[151] Roland (2017).
[152] Vatanka (2017) and Editorial (2018b).
[153] Miller (2017), Dobell (2017) and Shepard (2017).
[154] Allison (2017b, Chap. 1), Wright (2017, pp. 5–8), Jackson (2017).

any aggressor'. Peng reminded 'dragon-slayers': 'It is safe to say there will be no winner in an all-round war between China and the USA'.[155]

Foreign Minister Wang Yi questioned Western comparison of BRI to the USA's post-War European reconstruction project, refuting the allegation that the blueprint concealed 'a geopolitical strategy': BRI was 'neither a "Marshall Plan" nor a geopolitical strategy'. Wang stressed that Beijing was 'dedicated to constructing high-quality and high-standard projects and focuses on financial sustainability'.[156] On BRI's fifth anniversary, Xi Jinping renewed his pledge to 'benefit people in countries' partnering with China, based on 'dialogue and consultation, joint contribution, shared benefits, win-win cooperation, exchange and mutual learning'. He said BRI would build neither an alliance, nor 'a China club'.[157] Few took notice.

By then, China's trade with BRI-partners exceeded $5.5tn, FDI flows to them reached $80bn, establishing 82 economic/trade cooperation zones creating around 244,000 jobs for locals. Beijing and its partners completed 265 of the 279 items on the BRF list issued in May 2017; 14 others were in progress.[158] In January–July 2018, Chinese firms, signing new contracts worth $57.11bn, invested $8.55bn in 54 BRI states, marking an 11.8% rise year-on-year. In this period, Beijing spent $39.3bn on 'Silk Road scholarships' to students from partner-states studying in China.[159] Western policy-makers signalled diplomatic nuance, but non-official US, Indian, Japanese, European and Australian critics showed disdain for BRI-related activities, and resentment of China's 'plot' to attain what presumably should only be acquired by force.

Five weeks after the first BRF, Narendra Modi visited Donald Trump in Washington and together, they decried China's non-democratic dispensation while describing themselves as 'Democratic Stalwarts in the Indo-Pacific'. Slamming BRI's 'failings', they pledged to 'support bolstering regional economic connectivity through the transparent development of infrastructure' and use 'responsible debt-financing practices, while ensuring respect for sovereignty and territorial integrity, the rule of law, and the environment'. They stressed 'growing strategic convergence' and joint production of 'advanced defence equipment and technology at a level commensurate with that of the closest allies and partners of the USA'.[160]

Pre-emptively counterbalancing Beijing's anticipated BRI gains, in November 2016, Modi and Shinzo Abe initiated a public–private 'Asia–Africa Growth Corridor' (AAGC) to combine 'quality infrastructure and institutional connectivity', thereby 'enhancing capacities and skills and people-to-people partnership'. AAGC would 'improve growth and interconnectedness between and within Asia and Africa for realising a free and open Indo-Pacific region'. AAGC projects, overlaying MSR, linked Japan, ASEAN states, India and African polities along the Indian Ocean

---

[155] Peng (2000).
[156] Mo (2018).
[157] Xinhua (2018d) and Bianji (2018).
[158] Xinhua (2018d) and Hongyu (2018c).
[159] Hongyu (2018c).
[160] Indian Embassy (2017).

2.3 Conflicting Perspectives

littoral. Planners stressed, 'The AAGC will give priority to development projects in health and pharmaceuticals, agriculture and agro-processing, disaster management and skill enhancement'. Pledging, 'The connectivity aspects of the AAGC will be supplemented with quality infrastructure', they designed their challenge by mirroring BRI.[161]

The USA, Japan and Australia expanded these efforts into a 'trilateral partnership to mobilise investments in projects that drive economic growth, create opportunities, and foster a free, open, inclusive and prosperous Indo-Pacific'.[162] This formed the geoeconomic arm of twin-pronged countermoves. US, Indian, Japanese and Australian officials met during a November 2017 regional summit in Manila, reviving the 2007 vintage Quadrilateral Security Dialogue (Quad). Consultations, covering 'issues of common interest in the Indo-Pacific region', focused on 'cooperation based on their converging vision and values for promotion of peace, stability and prosperity in an increasingly interconnected region'. They agreed, 'a free, open, prosperous and inclusive Indo-Pacific region serves the long-term interests of all countries in the region and of the world at large'.[163] Although China was unnamed, the endeavour's Beijing focus was unmistakable.[164] China warned that regional cooperation 'should neither be politicised nor exclusionary'.[165] Polarisation, colouring the geopolitical landscape on which Beijing implemented BRI, was self-evident.

2.4 The Bankers' View

China critics viewed BRI through their monochromatic, geopolitical, lens. Outside the originally 65-strong BRI partnership, opinion varied. The world of high finance and multilateral institutions brought a contrasting perspective to bear on BRI's content and likely outcomes.[166] Those examining BRI's progress from countries arrayed against China's 'rise' appeared split between strategic pessimists and economic realists. Multilateral financial institutions and multinational corporations, making nuanced and granular assessments, welcomed BRI. Since China had established a framework for building BRI projects and forged financial institutions to fund these, they posited, 'the rest of the world needs to go along and use that as a way to get past a lot of the things that have inhibited progress in the past'.[167]

The doyen of the private capital confederacy, founder-Chairman of the World Economic Forum (WEF), Klaus Schwab, described BRI as 'a unique contribution to international cooperation and economic development'. BRI was 'based on the

---

[161] MEA (2017a).
[162] Bishop (2018).
[163] MEA (2017b).
[164] Pant and Bommakanti (2018), Grossman (2018) and Carr (2018).
[165] Jiangtao and Zhou (2017).
[166] Rana (2018).
[167] Wong et al. (2017) and Sneader and Ngai (2016).

stakeholder approach; it seeks to leverage market forces in best ways; it prepares best for the age of the Fourth Industrial Revolution; it is based on the open platform concept; and most importantly', BRI was 'the positive narrative the world needs'.[168] Schwab summarised results of WEF-commissioned research into BRI's designed goals, mechanisms, processes, funding systems and project portfolio. WEF researchers found BRI designers had built-in eight factors promising sustainable success:[169]

- Shared Vision: achieving transnational/international support for strategic infrastructure development promoting well-being and a sense of purpose to trans-border communities
- Multilateralism: curating a multilateral policy dialogue addressing public–private–civic cooperation, early stage project financing, standardised and transparent procurement, levelling the playing field and building trust
- Project Preparation Facility: ensuring a pipeline of bankable projects and a centre of excellence delivering the largest-ever project portfolio
- Risk Mitigation: proactively mitigating political/regulatory risks to boost stakeholder confidence among investors, off-takers and operators
- Sustainable Development: committing to sustainability and affordability to generate value for future generations beyond 2030
- Innovation: enhancing technological- and business-model innovation for sustained value-creation, ensuring future readiness
- People-to-People Exchange: governing domestic and cross-border migration and promoting mutual learning to enable/deepen popular exchanges and transnational cultural cooperation ensuring transcendent growth and universal purpose and
- Human-centric and Future-ready: delivering a new humane infrastructure heritage beyond current usage for the next millennium.

Multinational banks, too, shone a positive light on BRI. An HSBC executive noted, 'Belt and Road is a marathon, not a sprint. It is not a detailed list of imminent spending plans and due-by dates that should by now have delivered a flurry of construction starts and ribbon-cutting ceremonies'. Instead, BRI was a 'multi-year, perhaps even multi-decade vision that will evolve over time, adapting to circumstances and local priorities'. Major projects took 'years to plan, let alone implement', and BRI was 'still at the beginning of a long journey'. In addition to building ports and highways, BRI was also about 'building out the financial building blocks that will oil the wheels of trade and investment and help fund the physical aspects of the initiative'.[170] HSBC commissioned the Economist Intelligence Unit to examine BRI projects' environmental and fiscal viability. EIU economists concluded that China had ensured both ecological and financial sustainability for BRI implementation.[171]

---

[168] Jelinek et al. (2017).
[169] Ibid., p. 3.
[170] Wong (2017).
[171] EIU (2018).

2.4 The Bankers' View

Jim Yong Kim, then President of the World Bank Group (WBG), said, 'the WBG very proudly supports the Government of China's ambitious, unprecedented efforts to light up (the) night sky. The Belt and Road will improve trade, infrastructure, investment and people-to-people connectivity—not just across borders, but on a transcontinental scale'. BRI had the 'potential to lower trade costs, increase competitiveness, improve infrastructure and provide greater connectivity for Asia and its neighbouring regions'. Kim stressed the import of

- Effective governance, which was 'critical to ensuring developmental impact'
- 'Careful attention' to conditions 'at the border' e.g. customs procedures, and 'regulations beyond the border—like non-tariff measures' and
- Access to 'ancillary services for efficient supply chains', e.g. transportation, financing, insurance and telecommunications. Provision was 'uneven across the Belt and Road countries, so reforms like national regulatory policies' were needed.[172]

Kim also underscored BRI's challenges: its scale demanded massive investment; partner-states differed in their state of development, capabilities, constraints and risk-profiles, warranting support-mechanisms to 'define and meet consistent, satisfactory standards'. BRI's benefits, exceeding the sum-total of individual projects, would require 'innovative financing mechanisms—a mix of public and concessional finance and commercial capital'. Success demanded 'project preparation and appropriate risk allocation'.[173]

WBG commitment for BRI infrastructure, trade, power generation and connectivity projects with $24bn for transport-infrastructure, e.g. Afghanistan's Trans-Hindukush Road, Kazakhstan's East–West Roads, Pakistan's Karachi Ports and Uzbekistan's Pap-Angren Railway, which were 'already reinforcing connections', reached $86.8bn in May 2017. International Finance Corporation (IFC), WBG's private-financing arm, collaborated with China's SRF, Exim Bank and other organs to develop hydropower in Pakistan and infrastructure elsewhere. WBG funded $12.9bn to boost intra-BRI trade; its insurance arm invested $1.8bn in BRI lands and provided $1bn in guarantees.[174] WBG economists estimated BRI-triggered trade expansion, depending on partner-state capacities, transport efficacy and regulatory regimes, at 2.5%–4.1%.[175]

However, BRI faced potential-actual differentials. Diverse physical, political–economic and regulatory circumstances meant the undertaking demanded much more than designing, funding and building major intra-state and interstate connectivity. WBG analysts cautioned:

- Infrastructure and policy gaps in B&R corridor economies hindered trade and foreign investment. New infrastructure could help close the gaps but 'it is costly—and investments are occurring in the context of rising public debt'.

---

[172] Kim (2017).
[173] Kim (2017).
[174] Kim (2017).
[175] Baniya et al. (2019).

- BRI transport projects could enhance trade and foreign investment and reduce poverty by lowering trade costs but for some countries, 'the costs of new infrastructure could outweigh the gains'.
- Complimentary policy reforms could maximise the shared benefits of BRI transport projects, and for some states, 'reforms are a precondition to having net gains' from these projects.
- Risks common to large-scale infrastructure projects would be exacerbated by BRI's 'limited transparency and openness' and some partners' 'weak economic fundamentals and governance'.

WBG analysts concurred that trade among BRI partner-states was 30% below potential and FDI, 70% below potential. If they reduced border delays and eased trade restrictions, their real economies could be 'an estimated two to four times larger'. If BRI transport projects were fully implemented, and if the above issues were addressed, world trade could increase by 1.7%–6.2%, raising global real income by 0.7%–2.9%.[176]

Christine Lagarde, then Managing Director of the International Monetary Fund (IMF), believed BRI would connect 'cultures, communities, economies and people' by 'rejuvenating ancient trade-routes and building new ones'. BRI was 'about adding new economic flavours by creating infrastructure projects that are based on 21st-century expertise and governance standards'. Lagarde enumerated BRI's benefits:

- High-quality infrastructure could reduce the developing-world infrastructure gap by $1.5tn annually, lift growth and raise income.
- New ports, roads and power grids would better connect developing economies with global supply chains, boost rural productivity, school enrolment and access to health services, with broader sharing of investment and knowledge, making growth 'stronger, more durable and more inclusive'.
- Stronger economic, business and regulatory cooperation would 'boost global trade, investment and financial cooperation', spreading 'the benefits of growth more widely. All this is good for consumers, good for productivity and good for poverty reduction'.[177]

The IMF was 'deeply committed to helping (BRI-partners) maximise the benefits of more investment, trade, financial connectivity and people-to-people bonds under the BRI. Together, we can create something new—by combining proven policy ingredients with new economic flavours'.[178] Lagarde later warned that a scheme linking 70-plus countries was challenging: 'there is always a risk of potentially failed projects and the misuse of funds. Even when the right project is picked, difficulties often arise during **implementation**'. She stressed, 'these ventures can also lead to a problematic increase in debt, potentially limiting other spending as debt service

---

[176]Ruta et al. (2019).
[177]Lagarde (2017).
[178]Lagarde (2017).

rises and creating balance of payment challenges. In countries where public debt is already high, careful management of financing terms is critical'.[179]

Lagarde launched the China-IMF Capacity Development Centre (CICDC), offering financial management training to BRI executives. The IMF noted, 'the success of the BRI would be enhanced by an overarching framework, better coordination and oversight, more open procurement, and due attention to debt sustainability in partner countries'.[180] Two years later, addressing the 2nd BRF, Lagarde noted that BRI should not 'only go where it is needed', but also, 'it should only go where it is sustainable, in all aspects'. She praised Beijing's new debt-sustainability framework as 'a significant move in the right direction'. She urged 'increased transparency, open procurement with competitive bidding and better risk assessment in project selection'. Her general approbation of BRI was leavened with an emphasis on 'a spirit of collaboration, transparency and a commitment to sustainability'.[181] So, the WBG and IMF, while strongly supporting BRI's design and goals, sought refinements to its implementation.

Other economists urged Beijing to address several home-grown issues:

- China's financial markets, losing almost $5tn during the 2015 stock-market crash, remained volatile.[182] This disrupted lives and cost public confidence in Beijing's ability to manage China's marketising economy. Without massive private funding, including Chinese private-sector capitalisation, BRI implementation could stall.
- Rapid urbanisation enabled demographic and resource concentration and rapid industrialisation, creating a rationale for constructing the world's largest high-speed railway network, transforming travel and freight-moving capacity. However, speculative asset bubbles and ghost cities clouded economic prospects, threatening to rock China and the world, and menacing BRI implementation.[183]
- Rapid urbanisation and industrialisation caused massive pollution. The effect on public health, need for mitigation within project blueprints, making provisions, ensuring compliance and quality control, for all BRI-partners, would be substantial.[184]
- Fast-evolving fusion of artificial intelligence, big data, cloud computing, deep/machine-learning, robotics, 3D printing, 'human cloud' and other information-and-communication technologies into the '4th Industrial Revolution' could replace traditional work.[185] Transitional-period unemployment for millions and immense socio-economic pressure could impact on BRI's hardware-based projects.

---

[179] Lagarde (2018a). Emphasis in original.
[180] Executive Board (2018).
[181] Lagarde (2019).
[182] Report (2017d).
[183] Lim et al. (2018, pp. 62–71).
[184] Albert and Xu (2016) and Levitt (2015).
[185] Zhou (2017), Fulco (2017). Wang (2017a) and O'Connor (2015).

Some analysts discerned elemental dissonances challenging realisation of Beijing's BRI vision. Chinese society had, over millennia, acquired understanding of experience measured in very long temporal frames of reference, but others used much shorter time spans. Targets and achievement along quarterly and financial-year timelines defined Western calculi; the Chinese approach traversed nonlinear trajectories along extended time-windows. Chinese thinking progressed in general directions while non-Chinese calculations demanded precise frameworks of measureable outcomes. Discordance could jar China's collaboration with BRI-partners.[186]

Still, the London-based European Bank for Reconstruction and Development (EBRD) found common cause with BRI. As 'the multilateral development bank (MDB) with the largest footprint in terms of area of operation along BRI, the EBRD is a natural partner for China's initiative'. BRI presented 'significant potential for infrastructure financing, and the EBRD is uniquely placed to help local economies in the region—some of them relatively poorly integrated into the global economy—benefit from the opportunities these investments can bring'. The EBRD and BRI had 'a common vision of growth, sustainability and investment for the economies where we work'.[187]

Mirroring that assessment, global consultancy Deloitte noted, 'BRI is a journey, not a series of one-off infrastructure projects'. BRI was 'much more than an outbound investment programme'. Indeed, BRI's ambition was to 'improve connectivity between Asia, Europe and Africa, and in that way to increase trade, development and prosperity'. Deloitte saw BRI's 'one axis and two wings' design as 'a much-needed step that is kick-starting a new cycle of infrastructure spending, particularly in Asia'.[188]

2.4.1 The Asian Development Bank Perspective

The Asian Development Bank (ADB) noted that Asia's developing economies must invest $26tn over 2016–2030, or $1.7tn annually, in infrastructure to maintain their 'growth momentum, eradicate poverty, and respond to climate change'. Governments could fund up to 40% of Asia's infrastructure investment gap; private investors must finance the rest. Corporations would have to raise their infrastructure investment from $63bn in 2015–2016 to around '$250bn by 2020'.[189] As the ADB underscored the urgency, scale and scope of Asia's infrastructure needs, in 2015, Beijing requested its technical assistance in assessing BRI's SREB/MSR blueprints. ADB's analyses of detailed planning for BRI projects produced baseline documents for officials, planners, implementers and investors. The ADB reviewed the SREB's economic corridors being built 'in coordination with other concerned countries':

---

[186] Lim et al. (2018, pp. 67–69).
[187] Bastian (2017) and International Cooperation Office, EBRD (2017).
[188] Insight-Team (2018).
[189] Hasan et al. (2016).

2.4 The Bankers' View

- PRC–Mongolia–Russian Federation
- New Asia–European Continental Bridge
- PRC–Central Asia–West Asia
- PRC–Mainland Southeast Asia
- PRC–Pakistan
- Bangladesh–PRC–India–Myanmar.

MSR would focus on 'port construction and multimodal transport and logistics, as well as maritime ecological protection', linking coastal China to Europe through the SCS and the Indian Ocean in one route, and through the SCS to the South Pacific in the other.[190] ADB staff inferred, with its strategic agenda focused on 'inclusive economic growth' and 'regional integration', BRI would realise 'a new market opportunity for stakeholders along BRI areas'.[191]

ADB analysts noted that as higher income levels increased the region's domestic savings, governments could invest these in developmental enterprises; savings gave them 'access to commercial credit that was not available to them as low-income countries'. Also, China-led AIIB and NDB were moving 'onto a stage that until recently in Asia was largely occupied by two multilateral lenders: ADB and the World Bank'. This helped since '$1.7tn a year must be spent on infrastructure for the region to sustain its strong growth. But only half of this (was) being spent: in 2015, developing countries in Asia and the Pacific invested a combined $881bn on infrastructure'.[192] ADB assessed, BRI advanced Regional Cooperation and Integration (RCI), which played 'a critical role in accelerating economic growth, reducing poverty and economic disparity, raising productivity and employment, and strengthening institutions'. RCI narrowed development gaps between ADB's developing member-countries by 'building closer trade integration, intraregional supply chains, and stronger financial links, enabling slow-moving economies to speed their own expansion'.[193]

After the 2nd BRF closed at the end of April 2019, ADB President Takehiko Nakao affirmed the ADB's willingness to work with Chinese partners e.g. the AIIB, on implementing future BRI projects which were 'up to the standards like social and environmental impact'. Describing BRI as a 'very good concept and idea', he noted that the ADB and the AIIB were already co-financing five regional projects and more could follow.[194] During a three-day visit to China a month later, he described BRI as 'reasonable and understandable', pledging, 'If there is a good project, we can surely cooperate'.[195] That generally positive impression was, however, somewhat clouded by the fact that ADB was unable to complete until 2020 the originally one-year assessment of the BRI blueprint requested by China in 2015.[196]

---

[190] An et al. (2016, p. 1).
[191] An et al. (2016, p. 3).
[192] Taylor-Dormond (2017).
[193] An et al. (2016, p. 1).
[194] Xinhua (2019).
[195] Liu (2019).
[196] ADB (2016).

2.4.2 The United Nations Vision

The United Nations (UN) Secretary-General, Antonio Guterres posited, China's BRI blueprint and the UN's 17 Sustainable Development Goals (SDG) shared a single vision: 'Both strive to create opportunities, global public goods and win-win cooperation. And both aim to deepen "connectivity" across countries and regions: connectivity in infrastructure, trade, finance, policies and, perhaps most important of all, among peoples'. He averred, 'In order for the participating countries along the Belt and Road to fully benefit from the potential of enhanced connectivity, it is crucial to strengthen the links between the Initiative and the SDGs'. Guterres suggested the SDGs could guide BRI projects towards 'true sustainable development. With the initiative expected to generate vast investments in infrastructure, let us seize the moment to help countries make the transition to clean energy, low-carbon pathways—instead of locking in unsustainable practices for decades to come'.[197]

UNDP, the organ shaping and monitoring BRI's progress had, based on opinion-surveys in 100-plus UN-member-states, formulated a Global Agenda. In September 2015, the UNGA, taking a comprehensive approach to development mirroring the UN's SDGs, focusing on socio-economic and environmental sustainability, adopted it as 'Agenda 2030'. Analysing BRI's vision, mission, project portfolio-planning, and funding arrangements, the UNDP reported that BRI, 'an economic framework covering more than 70 countries, with the 2030 Agenda and the implementation of the SDGs holds the promise to confer substantial developmental benefits and to position BRI as an accelerator for the SDGs and the expansion of global public goods'.[198]

Guterres himself, recognising China 'for its central role as a pillar of international cooperation and multilateralism' at a time of turbulence flowing from the simultaneity of 'inequality, the climate crisis, and the potential risks of globalisation and the Fourth Industrial Revolution', returned to the sustainability refrain in his 2nd BRF address. He urged the world to 'come together' in mobilising resources to realise the SDGs and 'to stop runaway climate change'. Noting Beijing's $125bn investment in renewable energy in 2017, Guterres said, 'China's leadership on climate action is helping to show the way'. Once again, linking Beijing's BRI blueprint to the UN's SDG ambitions, Guterres added, 'I see the BRI as an important space where green principles can be reflected in green action', in the context of BRI offering 'a unique opportunity to build a new generation of climate resilient and people-centred cities and transport systems, and energy grids that prioritise low emissions and sustainability'.[199]

Those monitoring China's overseas direct investment (ODI) flows painted a nuanced picture. Describing media reports of Beijing's ODI flows as often superficial and biased, one economist posited, Chinese and Western financial institutions both competed and cooperated. While Chinese aid might 'circumvent Western sanctions on "problematic" states, Western state agencies and multinationals are guilty

---

[197] Report (2017a).
[198] Horvath (2016).
[199] Guterres (2019).

2.4 The Bankers' View

of similar unethical behaviour'. Notably, 'Western agencies and multinationals frequently circumvent legal and ethical boundaries'.[200] Others saw debt-entrapment risks confronting several BRI-partners. One study found 35 partner-states were either 'investment grade' or 'near-investment grade' as measured by Standard and Poor's, Moody's, and Fitch rating agencies. Fifteen faced no debt-sustainability challenges, but 23 risked 'debt distress', and eight were vulnerable 'due to future BRI-related financing'.[201]

The latter were Djibouti, Kyrgyzstan, Laos, Maldives, Mongolia, Montenegro, Pakistan and Tajikistan.[202] The authors noted that several BRI-partners and others already suffered from debt-overhangs. Chinese ODI began reaching BRI-partners long before BRI's launch. Many pre-BRI projects were incorporated and rebranded since 2013. BRI-related ODI involved mixes of state-to-state credit, 'project-aid' and foreign-contracted projects funded with discounted loans given to SOEs and other Chinese firms invited by partner agencies to construct connectivity infrastructure, power generation units and/or industrial plants. In 2014–2017, some 50 SOEs participated in around 1700 BRI projects.[203]

While 2016 was a 'bumper year' with Chinese ODI reaching $170bn, in January–October 2017, flows fell by over 40% year-on-year. Non-financial Chinese ODI in BRI-economies fell from $14.8bn in 2015 to $14.5bn in 2016, then declined by 13.7% over January–September 2017.[204] Beijing's restraints on 'irrational' ODI imposed in 2017 explained the slide. One assessment suggested BRI-related ODI mirrored Beijing's general ODI patterns. BRI-partners took a small proportion of Chinese ODI in 2017, with Malaysia climbing to 4th position, Kazakhstan to 12th, Thailand to 18th and Iran to the 19th among top recipients. Pakistan, at the heart of the $62bn 'flagship' China–Pakistan Economic Corridor (CPEC), ranked 41st.[205]

In 2018, Chinese ODI grew by 4.2% to reach $130bn, of which non-financial ODI remained at $120bn. Investment into North America and Europe fell by 73% from $111bn in 2017 to $30bn. North American regulators rejected 14 Chinese acquisitions while their European counterparts cancelled seven. ODI into BRI-states grew by 8.9% to over $15bn. After many African states signed on to BRI in September 2018, BRI-partners accounted for 40% of China's 2018 ODI. With Chinese private firms representing 44% (compared to 31% in 2017) of total ODI value in 2018, investment focused on leasing and business services, manufacturing, and retail- and wholesale trade, rather than energy and commodities. Overseas SEZs grew in importance in attracting Chinese ODI: in 2018, they drew $2.5bn, with ODI stock in these zones exceeding $20bn by the end of 2018.[206] In Jan–June 2019, Chinese ODI reached $50.4bn, up 0.1% year-on-year. Of this, investment into 51 BRI-partner

---

[200] Camba (2018).
[201] Hurley et al. (2017, p. 6).
[202] Hurley et al. (2017, p. 11).
[203] EIU (2017, p. 21).
[204] EIU (2017, p. 22).
[205] EIU (2017, pp. 4–6).
[206] Report (2019c).

states increased by $6.8bn, or 12.6% of total Chinese ODI in that period in which Chinese SOEs/agencies signed 389 projects worth over $50m each. Nearly 70% of these focused on electricity- and transport infrastructure.[207]

Cancelled or stalled BRI projects highlighted challenges: a $5.5bn Indonesian railway project was held up in August 2017 over land acquisition disputes; in November, Nepal cancelled its largest hydropower project contracted to a Chinese firm over alleged bidding irregularities; next, Pakistan cancelled a hydel project over ownership-and-financing terms. Beijing incentivised SOEs to initiate risky BRI projects, but without state-backed insurance protection, private firms remained cautious. In 2017, China's State Council and Finance Ministry framed due diligence, feasibility studies and operational parameters of SOE-run BRI projects. Chinese financial and insurance firms began applying higher project-assessment standards. Although private capital enjoyed fewer benefits in financing BRI projects, after the 19th National Congress wrote in BRI into the CPC Constitution in October 2017, firms like Wanda and HNA evinced interest in aligning their global operations with BRI priorities.

Private finance could sustain investment in an expanding BRI; its absence could hobble it. In late-2017, Chinese private-bankers estimated potential BRI-linked investment in railways, urban-development, logistics and cross-border e-commerce could reach $10tn. Nine Chinese banks, with 62 branches in 26 partner-countries, were involved in BRI. With annual investment likely to exceed $60bn across 1676-plus projects, around 0.5% of China's nominal GDP, Beijing's anxiety to retain control was 'strategic'.[208] Economists identified the developing world's need to raise annual infrastructure investment from $0.8tn–$0.9tn in 2008 to $1.8tn–$2.3tn by 2020, bridging a $1tn annual gap. Some saw BRI-financing as an 'ideal opportunity' to do so with China's large RMB-denominated assets, simultaneously advancing Beijing's interest in internationalising the RMB.

That endeavour made slow progress, however: in September 2009, the RMB was the 5th most-used global payment currency, and 8th in foreign-exchange transactions, accounting for 29% of China's trade settlements and 10% of its ODI.[209] These trends accelerated after the IMF added the RMB to its Special Drawing Rights (SDR) basket in October 2016. However, Beijing's tightened capital outflow controls, US monetary policy changes, very low European Union (EU) interest rates and uncertain prospects for global recovery created a fluid backdrop for Beijing's efforts. The impact on BRI financing remained moot.

Not all economists waxed bullish about BRI. Key concerns: Beijing extended soft-loans to 'foreign governments, who will then use the Chinese funds to pay the Chinese companies'; BRI-partners included 'some of the riskiest developing countries in the world'; facing massive overcapacity, Beijing was seemingly 'expanding the problem to projects overseas'. Given BRI's scale, critics questioned its sustainability. Failure

---

[207]Report (2019b).
[208]Pieraccini (2017).
[209]Liu et al. (2017).

'could be damaging not just for China but for the global financial system'.[210] Former World Bank official, David Dollar, countered that BRI should not cost China more than $100bn annually and total investments would not run into 'trillions of dollars'. With its annual current account surplus of $200bn–$300bn, Beijing could annually invest $100bn in BRI projects easily.[211] Still, potential pitfalls, including the trade war's uncertain outcomes, confronted Beijing and its BRI-partners.[212]

2.5 Belt and Road Initiative's Adversarial Strategic Backdrop

Anxiety about Xi Jinping's assertive China challenging the US-led order troubled economists, too. An influential journal, asking, 'Is China challenging America for global leadership?' advanced a complicated argument. Xi's speeches at the World Economic Forum in Davos, and at the UN in Geneva, in January 2017, highlighted the *systemic* order's frailties and offered lessons from China's experience in escaping the worst effects of capitalist fragility exposed by the 'Great Recession'. Next, Premier Li Keqiang's annual 'Work Report' to the National People's Congress (NPC) atypically discussed diplomacy and mentioned *quanqiu* (global) and *quanqiuhua* (globalisation) 13 times—compared to only five such mentions in 2016.[213]

Xi's inauguration of the 1st BRF in May 2017 juxtaposed to Trump's 'America First' perspective, and Xi's promotion of 'China solutions' to problems, indicated Beijing's shifting approach. Sceptics shaped the trans-Pacific discourse. Anxious to sustain pre-eminence by thwarting all challenges, the USA pushed back as DoD vowed: China's leaders 'need to understand that while we seek cooperation where our interests align, we will compete where we must'.[214] Beijing rejected permanent subordination within a system designed by others. Tensions were inherent in that mismatch of mutual insecurity and response dialectics. That was the context in which China's critics viewed BRI and other Chinese initiatives.

2.5.1 Beltway–Zhongnanhai *Dialectics*

US legislators were advised that during the Cold War, China feared encirclement by Soviet forces deployed to the USSR's Central Asian republics. After the Soviet-collapse, China 'cultivated ties with Central Asian governments, peacefully settled outstanding boundary disputes, and sought to take advantage of the region's vast

---

[210] Gilchrist (2017).
[211] Tsuruoka (2017).
[212] Hongyu (2018b), Tan (2018) and Kovrig (2017).
[213] Editorial (2017a).
[214] Garamone (2018c).

mineral wealth'. In the 2000s, after the USA built military bases there, China again feared encirclement, now by US forces. In response, Beijing boosted ties to Central Asian states. BRI/SREB flowed from goals Beijing began pursuing long before Xi's 2013 Almaty speech; its success would affect US interests[215]:

- Encouraging economic engagement between Xinjiang and Central Asia, to bolster development and stability in China's restive western province
- Eradicating the 'three evils' of extremism, separatism and terrorism from Xinjiang, and
- Expanding China's 'economic and geostrategic influence' across Eurasia.

Displacement anxiety deepened gloom. Australian legislators learnt, 'China has been…using its massive financial assets to dominate smaller economies through long-term control of infrastructure, natural resources and associated land assets, and through offering less than desirable credit terms for infrastructure loans'. Beijing's 'production capacity cooperation' with BRI-partners 'often involves the simple transfer of Chinese-owned production capacity to countries where production is cheaper and markets are closer'. The 'economic nature of the OBOR agenda' notwithstanding, BRI was 'simultaneously a strategic program' designed for 'further validating China's claims to the islands of the SCS, while on the other side of the Indian Ocean, Djibouti is providing China with both a trade port and its first overseas military base'.[216] Legislators heard 'broader concerns' with 'the longer-term aims of China': 'the OBOR agenda is aimed at creating a Eurasia-wide, China-led bloc to counter the USA', shaping 'a post-Westphalian world', posing 'a profound challenge to the current global political and economic status quo'.[217]

Anxiety triggered by *systemic transitional fluidity* threatening to supplant three centuries of Euro-Atlantic dominion with Chinese eminence coloured the policy discourse. Trump and Abe, mirroring the Indo-Japanese AAGC, initiated 'high-quality infrastructure investment alternatives in the Indo-Pacific region', forging a 'Strategic Energy Partnership' promoting 'universal access to affordable and reliable energy in Southeast Asia, South Asia and Sub-Saharan Africa', pledging 'high-quality energy infrastructure solutions to the Indo-Pacific region'.[218] Counter-BRI intent notwithstanding, Abe's then-Foreign Minister, Taro Kono, stated only days later that BRI would boost the global economy.[219] Ambivalence suggested that while keen to secure US support for defence vis-à-vis Beijing, Tokyo was hedging its bets.

An influential former-diplomat, Robert Blackwill, underscoring China's 'subtle challenges' to the USA's *primacy*, urged shifts essential to sustaining US dominance. Noting that the USA 'squanders opportunities and dilutes its own foreign policy outcomes', enabling China 'to coerce its neighbours and lessen their ability to resist',

---

[215] Cleveland and Bartholomew (2018b, pp. 261–263) and Bartholomew and Shea (2017, pp. 391–393).
[216] Wade (2017).
[217] Wade (2017).
[218] White House (2017).
[219] Kono (2017).

2.5 Belt and Road Initiative's Adversarial Strategic Backdrop

Blackwill warned, Washington was neglecting economic instruments in accomplishing geopolitical objectives despite military power's limited effect. Such US policy 'gives China free rein in vulnerable African and Latin American nations'.[220] He recounted the USA's record of deploying geoeconomic leverage to secure strategic goals, the tradition's decline during the Vietnam War, and how China adopted the template. Highlighting alleged Sino-Russian geoeconomic pressure pushing target states to recast their policies, Blackwill urged Washington to follow suit. His school emphasised economic/commercial leverage as a strategic tool, and an inclination to view all Chinese initiatives as threating US *primacy*.[221] BRI fit this bill and Trump's rhetoric resonated.

Washington had, however, long been concerned about post-Soviet competition. Six weeks after the USSR imploded, the Secretary of Defence (SoD) instructed military commanders to secure US *unipolar*-leadership by 'ensuring that a hostile power does not dominate a critical area of the world, including Western Europe, East Asia, Southwest Asia, or Russia, or mount a global challenge'. DoD's 'first objective' was to 'prevent the re-emergence of a new rival, either on the territory of the former Soviet Union or elsewhere, that poses a threat on the order of that posed formerly by the Soviet Union'.[222] US forces would be designed to this end. A revision issued 11 days later carried annotations stressing US interest in 'East Asia (especially NE Asia)' and 'Persian Gulf (espl SWA)' where US forces must prevent any rival's rise.[223]

DoD assumed *primacy* demanded shaping the regional and planetary future, but 'World events repeatedly defy near-term predictions; our ability to predict events over longer periods is even less precise. History is replete with instances of major, unanticipated strategic shifts over multi-year time frames, while sophisticated modern forces take many years to build'.[224] DoD envisaged an expansive and flexible strategy, with forces sized for it. The USA faced no 'peer-rival', but 'we could still face in the more distant future a new antagonistic superpower or some emergent alliance of hostile regional hegemons'. The USA must 'preclude the development of any potentially hostile entity that could pursue regional or global domination in competition with the US and our allies'.[225]

DoD's policy-guidance, repeatedly refined, stressed, 'Working with our allies and friends to preclude hostile, nondemocratic domination of a region critical to our interests, and also thereby to strengthen the barriers against the re-emergence of a global threat to the interests of the US and its allies'.[226] Cold War-era US administrations had initiated major foreign policy changes via presidential endorsement of NSC-led initiatives; following the USA's ascent to unchallenged *unipolarity*, DoD initiated

---

[220] Blackwill and Harris (2016).
[221] Rolland (2019), Lukin (2019), Roy (2019), Friedberg (2018a), Inkster (2018), Kennedy (2018).
[222] Principal Deputy Under Secretary of Defence (1992).
[223] OSD (1992, p. 1).
[224] OSD (1992, p. 3).
[225] OSD (1992, p. 12).
[226] Principal Deputy Under Secretary of Defence (1992).

the shift to ensuring the indefinite extension of the USA's sole-superpower status. Until December 1989, when President George H.W. Bush sent National Security Advisor Brent Scowcroft and Deputy Secretary of State Lawrence Eagleburger for the second time as his personal emissaries to Deng Xiaoping after the Tiananmen Square crackdown, China had been a tacit ally in the US' covert campaigns against Soviet forces and proxies.[227] Now, the possibility of China challenging US *primacy* sometime in the future began colouring iterations of the Guidance.

Washington posited, 'to virtually all of' China's neighbours' except Russia, 'either by its direct involvement in territorial disputes in the ECS and SCS or by the indirect impact on major shipping lanes transiting these waters', China was seen 'as posing a potential threat'.[228] Beijing's anxiety flowing from US-imposed sanctions on high-technology sales to post-Tiananmen China, Washington's 1991 coalition-war against Iraq, Taiwan's procurement of modern US naval and air combat platforms, and concerns vis-à-vis untrammelled US power, transformed China into an insecure, defensive, actor. In 1991, US intelligence detected the first PLA drills involving 'strikes' on US forces deployed across northeast Asia.[229]

Against this backdrop, China planned a nuclear test in October 1993. President Bill Clinton insisted 'there's no reasonable threat to China from any other nuclear power',[230] but Beijing persisted. Around then, CPC Central Committee (CPCCC) and CMC leaders convened an 11-day gathering of Chinese strategists to review post-Soviet security dynamics. The participants' 60 papers forged a consensus that the USA posed the 'greatest threat to China' over the next decade. Its 'open door of personnel exchanges and propaganda for ideological infiltration' would subvert China's 'upper strata'. Washington would also fund hostile forces 'to create turbulence'. CPCCC and CMC leaders circulated the report among China's civil-military elites.[231] Beijing's potentially terminal fear of the US-led *order* was thus formalised shortly after the 'sole superpower' emerged.

With the USA focused on perpetual *unipolarity* shaping the *system* around its grand-strategic goals, and China struggling to acquire strategic autonomy, their national narratives diverged. Events deepened polarisation. President Lee Teng-hui's moves towards Taiwan's sovereign democratisation, the 1995–1996 PLA drills culminating in 'missile tests' in Taiwan's vicinity, US deployment of two CSGs in Taipei's defence, Beijing's failure to even detect these flotillas, and reciprocal rhetorical flourishes, deepened trans-Pacific mutual insecurity. Sino-US collaboration in the wake of the May 1998 Indian and Pakistani nuclear tests, manifest in a Clinton-Jiang Zemin joint statement, recalled past cooperation, but goodwill soon evaporated.

Early on 7 May 1999, at the height of US-led NATO air operations against Serb forces during the Kosovo War, a USAF B-2 stealth bomber targeted the Chinese

---

[227] Scowcroft and Eagleburger first carried Bush's message to Deng in July 1989; former President Ricard Nixon met Deng in October 1989. Sandler (1989) and Kristoff (1989).

[228] Whiting (1995, p. 4).

[229] Whiting (1995, p. 4).

[230] Whiting (1995, p. 8).

[231] Tsung (1994).

2.5 Belt and Road Initiative's Adversarial Strategic Backdrop

chancery in Belgrade, destroying 'sovereign property' and killing or wounding several Chinese civilians. This first US attack on Chinese assets since the Korean War lit a firebomb of outrage. Demonstrators surrounded US missions across the country. Clinton and his DCI, George Tenet, claiming the CIA's outdated maps provided incorrect targeting data, apologised, and Washington proffered compensation. However, intelligence officials, accusing Chinese diplomats of relaying Serb leadership's instructions to field-units after the USA had destroyed Serb military communications facilities, confirmed the strike was deliberate.[232] Mutual mistrust deepened.

US executive and legislative perceptions of Chinese challenges resonated. While revising the National Defence Authorisation Act (NDAA) 2000, Congress instructed DoD to cut all links to the PLA, thus preventing transfers of militarily significant information, submit annual reports on China's military power and activities comparable to the 1980's-era reports on Soviet forces, and establish an NDU institute to study the PLA's doctrine, strategy, operational tactics and order of battle. Congress then set up a United States-China Economic and Security Review Commission (USCC) to annually analyse the security fallout from bilateral economic relations.

The executive branch, too, acted. In mid-1999, DoD's Office of Net Assessment (ONA), under the Advisor to the SoD for Net Assessment, Andrew Marshall, convened analysts at the Naval War College, Newport, for its annual summer study, prognosticating 'Asia 2025'. Conferees inferred that in 2025, Asia's security milieu would be shaped by actors with strategic leverage manifest in coercion, intimidation and denial, enabled by the proliferation of

- Long-range precision strike capability
- Hi-tech/low-skill weaponry
- Long-range reconnaissance and
- Asymmetric capabilities[233]

In that context, ONA analysts surmised that a strong China's goals would be:

- Maintain control of sovereign territory
- Gain control over territory they claim
- Minimise potential regional threats
- Gain veto power over neighbour's policies—a Chinese Monroe Doctrine?
- Sinocentric Asia: Re-establish tributary system? China's only positive historical model
- PRC develops infrastructure links, economic inducements, diplomatic, mil-to-mil ties
- 2025: Effective zone of continental control; clearly dominant in peninsular Southeast Asia, 'Central Asia protectorates', 'extensive penetration of RFE/Siberia'
- Dominance helps PRC manage internal problems—minorities/regions
- PRC 'already on the road to continental dominance'[234]

---

[232] Holsoe and Vulliamy (1999) and Report (1999).
[233] Under Secretary of Defence (1999, p. 15).
[234] Under Secretary of Defence (1999, pp. 101–105).

ONA strategists concluded a Sino-Indian condominium would doom US *primacy* as Asians accommodated themselves to the former order. American cooperation with India, the strategic 'swing-state', could pre-empt Sino-Indian alignment. While China was a 'force for instability and constant competitor', the USA needed 'India's strategic potential', shift resources from the European Command to the Pacific Command, focus on the Asia–Pacific's naval and airspace-relevant features, and defend the USA's power projection assets there. The question was, how to engage with India's power-potential against a resurgent China.[235] An answer soon arrived.

2.5.2 Polarisation Cemented

In 2000, Clinton ordered Israeli Prime Minister Ehud Barak to cancel a $250m contract to sell two *Phalcon* airborne warning and control systems (AWACS) to Beijing.[236] He then travelled to India and Vietnam, reviving security relations with two states that harboured deep post-conflict insecurity vis-à-vis China. Shared anxiety and a determination to internally and externally balance against a 'rising' China fashioned the nucleus of a tacit counter-China coalition. By the time a collision between a US ISR aircraft and a Chinese interceptor killed the Chinese pilot and forced the US crew to land on Hainan Island in April 2001, triggering a crisis early in the George W. Bush Administration, Sino-US competitive lines had been drawn.

Washington's subsequent focus on its 'Global War on Terrorism' (GWoT), especially campaigns in Afghanistan and Iraq, diluted focused rivalry with China, but a 2004 Global Posture Review began the redeployment of substantial *materiel* to the PACOM theatre.[237] The Obama Administration's Asia-Pacific *Pivot/Rebalance* built on it,[238] but failed to overwhelm Beijing's determination to defend its 'core interests', or allay regional insecurity. Beijing's strategic parity-focused formulation of 'A New Type of Major-power Relations' ruffled US feathers.[239] Mutual disappointment challenged stable relations between these two 'major' powers.

China's Shanghai International Energy Exchange, formed to manage future-contracts for hydrocarbon imports, boosted the world's largest energy-importer's energy security, and accelerated RMB internationalisation, exposing tensions between Beijing's economic priorities, and external fears of their outcomes. China imported around 7.6m barrels of crude oil daily in 2016, largely via long-term contracts between energy-SOEs and foreign suppliers. Having for decades chafed under the dollar's domination of global trade and finance and resented the US' unilateral sanctions enabled by it, Beijing sought the RMB's transformation into an interstate

---

[235]Under Secretary of Defence (1999, pp. 135–145).
[236]Perlez (2000).
[237]Rumsfeld et al. (2004).
[238]Obama (2011) and White House (2015).
[239]Tisdall (2016) and Perlez (2014).

2.5 Belt and Road Initiative's Adversarial Strategic Backdrop

commercial currency. China's stature within global energy markets and shared interests binding Beijing and its suppliers in avoiding dollar-based restrictions could erode the dollar's 'exorbitant privilege', unpredictably modifying the post-1945 economic landscape.[240]

Against that backdrop, boosting US forward-deployed forces near China, counter-China coalition expansion, and Beijing-targeted economic policies deepened the divide. Both prepared for conflict. China insisted it had consistently rejected power-politics since the July 1971 meeting between Henry Kissinger and Zhou Enlai. Zhou told Kissinger that Mao Zedong had repeatedly asserted China 'would absolutely not become a superpower', adding, 'What we strive for is that all countries, big or small, be equal. It is not just a question of equality for two countries'.[241] Beijing now stated, 'Xi Jinping's concept of a new type of international relations is designed to bring an end to the old way of international relations of gaining power at the expense of others',[242] precisely the charge Western strategists levelled against China.

Trump Administration's policy-rhetoric reinforced the trend. In his 2017 National Security Strategy (NSS) document, Trump accused China of waging 'economic aggression'. Given the risks of escalation from possible conflict over trade relations, North Korea's nuclear-and-missile programmes, Taiwan, Sino-Japanese maritime/territorial disputes in the ECS, China's island base-building in the SCS, cyberspace competition, or Sino-Indian territorial disputes, the possibility of war breaking out 'was real enough to require prudent policies and effective deterrent measures'. Trump insisted, 'China and Russia challenge American power, influence, and interests, attempting to erode American security and prosperity'. Beijing and Moscow, in US view, were 'determined to make economies less free and less fair, to grow their militaries, and to control information and data to repress their societies and expand their influence'.[243]

Competition stood front and centre. In addition to structural dissonance between the established hegemon and an emerging rival, Sino-US tensions carried elemental, value-based, mutual threats. Competition 'between free and repressive visions of world order' and a contest between 'those who value human dignity and freedom and those who oppress individuals and enforce uniformity' drove rivalry. Trump insisted, 'we are engaged in a new era of competition. We accept that vigorous military, economic and political contests are playing out all around the world…when America does not lead, malign actors fill the void to the disadvantage of the US. When America does lead from a position of strength and confidence and in accordance with our interests and values, all benefit'. He said, 'none should doubt our commitment to defend our interests. An America that successfully competes is the best way to prevent conflict'.[244]

---

[240] Evans (2017b), Pettis (2016), Marsh (2016) and Ryan (2011).

[241] Stone (2018a).

[242] Stone (2018a).

[243] Trump (2017a, pp. i–ii, 2).

[244] Trump (2017a, p. 3).

INDOPACOM's theatre, from 'the west coast of India to the western shores of the US', was roiled by 'a geopolitical competition between free and repressive visions of world order'. The USA sought cooperation, but 'China is using economic inducements and penalties, influence operations, and implied military threats to persuade other states to heed its political and security agenda'. Trump asserted, 'China seeks to displace the USA in the Indo-Pacific region, expand the reaches of its state-driven economic model, and reorder the region in its favour'.[245] Notably, 'China's infrastructure investments and trade strategies reinforce its geopolitical aspirations'.[246] Dynamics defining Beijing's behaviour:

- China's newly built and militarised SCS outposts 'endanger free flow of trade, threaten the sovereignty of other nations, and undermine regional stability'.
- China's rapid military modernisation was 'designed to limit US access to the region and provide China a freer hand there'.
- China presented its ambitions 'as mutually beneficial, but Chinese dominance risks diminishing the sovereignty' of regional states.
- Indo-Pacific states were 'calling for sustained US leadership in a collective response that upholds a regional order respectful of sovereignty and independence'.[247]

To protect US interests, Trump pledged unilateral and collective diplomatic, economic and military responses, boosting force deployments and coalition-building efforts. Rex Tillerson explained Trump's transactional approach to US–China relations: 'A central component of our strategy is persuading China to exert its decisive economic leverage' on Pyongyang. Acknowledging Beijing had applied 'certain import bans and sanctions, but it could and should do more', he warned, Washington would 'continue to pursue American interests' in 'trade imbalances, intellectual property theft and China's troubling military activities in the SCS and elsewhere'.[248] Failure to satisfy US demands explained the application of coercive tools.

Stressing its contrasting and cooperative stance, Beijing regretted that Washington had 'not given up its hegemonic ambitions and will do everything it can do to try to ensure that world powers rise under the USA, and not with it and never above it'. Trump's NSS revealed 'a determination to try to make the global system serve the interests of the USA, rather than the interests of the global community'. Noting that in 2016, Sino-US trade hit $550bn and two-way investment, $200bn, supporting 2.6m US jobs, China complained, putting 'the national interests of some countries above the interests of other countries and the international community' was 'downright selfish'. Rejecting zero-sum 'Cold War mentality', China would 'always remain committed to world peace, global development, and international order'.[249]

---

[245] Trump (2017a, p. 25).
[246] Trump (2017a, pp. 45–46).
[247] Trump (2017a, p. 46).
[248] Tillerson (2017b).
[249] Stone (2017).

2.5 Belt and Road Initiative's Adversarial Strategic Backdrop 53

Underscoring the gulf defining the Pacific and concerned over US dependence on Chinese supply of 'critical mineral commodities', Trump ordered his secretaries of interior and defence to locate alternative sources by exploring US deposits of such minerals. His belief that this 'dependency of the US on foreign sources creates a strategic vulnerability for both its economy and military to adverse foreign government action, natural disaster, and other events that can disrupt supply of these key minerals', reflected fears of Chinese geoeconomic 'threats' to US security.[250] DoD's National Defence Strategy report affirmed a Sino-centric focus: 'Interstate strategic competition, not terrorism, is now the primary concern in US national security. China is a strategic competitor using predatory economics to intimidate its neighbours while militarising features in the SCS…China is leveraging military modernisation, influence operations, and predatory economics to coerce neighbouring countries to reorder the Indo-Pacific region to their advantage'.

Washington feared, 'As China continues its economic and military ascendance, asserting power through an all-of-nation long-term strategy, it will continue to pursue a military modernisation program that seeks Indo-Pacific regional hegemony in the near-term and displacement of the USA to achieve global pre-eminence in the future… we must make difficult choices and prioritise what is most important to field a lethal, resilient, and rapidly adapting Joint Force'. DoD concluded, a 'rapidly innovating Joint Force, combined with a robust constellation of allies and partners, will sustain American influence and ensure favourable balances of power that safeguard the free and open international order'.[251]

2.5.3 Belt and Road Initiative's Hostile Landscape

Adversarial policy-shifts led DoD commanders to reorganise force-packages for combat with China, and Russia.[252] JCS leaders noted that war with China would be concentrated in the air and at sea, other domains playing subsidiary roles. Then CJCS, General Joseph Dunford, emphasised 'multidomain, multiregional conflict and using US forces and their ability to be all over the world to put pressure on adversaries everywhere'. To improve prospects against China (and Russia) by synergising the power of US forces, DoD 'developed 13 initiatives' to 'find a way that the joint force, working together, can hold the initiative, because…the side that wins will be the side that can command the initiative by driving an operational tempo that the other side can't keep up with'.[253] The USAF Air Combat Command, the US Army Training & Doctrine Command, and the USN collaboratively fashioned new operational concepts and manoeuvre frameworks to defeat China in battle.

---

[250] Trump (2017b).
[251] Mattis (2018g, pp. 1–2).
[252] Mehta (2018).
[253] Garamone (2018a).

Accusing Washington of 'Cold War mentality', MND spokesman Ren Guoqiang noted, 'China has steadfastly taken the path of peaceful development and followed defensive national defence policies'. It never sought 'military expansion or sphere of influence and has always been a builder of world peace'.[254] Still, Washington counteracted Beijing's geoeconomic advances with military strategies. To allay US anxiety, Xi Jinping pledged to developing-world leaders to not 'export' Chinese models of political or economic policy-paradigms: 'We will not ask other countries to copy the Chinese practice'. Having suffered protracted war and conscious that only sustained peace could allow development, the CPC knew 'deeply the value of peace and (held) firm resolve in maintaining peace'. Xi noted, over 2500 PLA personnel were engaged in UN peacekeeping missions in eight trouble-spots, 'safeguarding local peace and tranquillity, despite difficulties and dangers',[255] a reminder that China provided more peacekeepers than all other permanent UNSC members combined.

Still, China confronted the US' expanding coalition. Xi's proposed 'community of shared human destiny' as an organising principle, and BRI a framework for realising it, elicited suspicious derision among 'dragon-slayers' who insisted BRI evoked 'imperial nightmares' among China's neighbours.[256] Chinese assessments differed. One commentator noted, Xi's paradigm urged 'due consideration to the legitimate concerns of other countries while pursuing their own interests'. It would be a commonly beneficial replacement of 'the current dominant conception of international relations—namely one of anarchy, power politics and a winner-takes-all dynamic'. Such a shift would reinforce economic liberalisation with 'a new global system that is more equitable, inclusive and fair'.[257]

Chinese analysts advanced accretive perspectives: 'Some Western politicians and media organisations tend to view BRI as a source of political leverage for China to seize the global leadership role, which is unnecessary paranoia'. They insisted, 'China has never had such an intention behind the BRI'.[258] Another reaffirmed, China had 'no ambition to replace the US as the world leader. This is not a job China has ever aspired to'. Beijing sought to realise Xi's 'community with shared future for mankind' by trying 'to add a little to the existing system' with BRI and other initiatives to benefit the developing world—'A community of shared future. This is the future for humanity'.[259]

In response, Washington and Tokyo reinforced adversarial dialectics by deploying their F-35 Lighting stealth fighter aircraft in China's vicinity in 2017–2018, with the commander of USPACAF explaining: 'The F-35A gives the joint warfighter unprecedented global precision attack capability against current and emerging threats'.[260] While the USA had been reinforcing combat-assets near China since its 2004 Global

---

[254] Xinhua (2018a).
[255] Chengsheng (2017).
[256] Daly and Rojansky (2018).
[257] Yong (2018).
[258] Wang (2018).
[259] Lau (2018).
[260] Report (2017h) and Report (2018h).

2.5 Belt and Road Initiative's Adversarial Strategic Backdrop

Posture Review, hastening the PLA's modernisation efforts, Japan's conversion of its 'helicopter carriers' into aircraft carriers posed a more acute challenge to China. Warning Tokyo against such action, Beijing published reports on carrier-borne EW systems capable of neutralising stealth and other hostile advances.[261]

Washington's determination to forcefully sustain *systemic primacy* was explicit: 'the DoD will be prepared to defend the homeland, remain the preeminent military power in the world, ensure the balances of power in the Indo-Pacific, Europe, the Middle East and the Western Hemisphere remain in our favour, and advance an international order that is most conducive to our security and prosperity'.[262] That drive caused friction. Shortly after Trump signed NDAA2018, authorising the first ship visits between the USA and Taiwan since 1979, Beijing accused the USA of 'interfering in China's internal affairs', lodging a 'stern representation'. When a Congressional delegation met Taiwanese leaders to discuss 'security', a Chinese diplomat stationed in Washington warned, 'The day that a US Navy vessel arrives in Kaohsiung, is the day that our People's Liberation Army unites Taiwan with military force'.[263]

Beijing reminded Washington that it had 'never conceded and will not yield an inch on Taiwan being part of China. It is a matter that leaves no room for negotiation. This seems to be something that needs restating…US policymakers would do well to bear in mind there is a line that should not be crossed when it comes to Taiwan'.[264] After an FoN sortie close to the Scarborough Shoal by the US missile-destroyer *USS Hopper* in January 2018, Beijing warned, 'If the US keeps stirring up trouble and creating tension in the SCS, China will be forced to come to the conclusion that it is indeed necessary to strengthen and speed up the building of its capabilities in the SCS so as to safeguard regional peace and stability'.[265] Washington followed up with an extended SCS sortie by the *USS Carl Vinson* CSG, calling at Filipino and Vietnamese ports—the latter for the first time since the Vietnam War,[266] setting precedents.

Xi struggled with a challenging quandary: how to reinforce China's status as a 'major country' on the geopolitical landscape without elementally threatening the other 'major country', hitherto the 'sole superpower', thus obviating a potentially uncontrollable spiral.[267] Ideologically focused and absolutist US policy documents, e.g. the 2017 NSS, 2018 NDS, 2018 NPR, and 2019 NDAA, designated China a strategic and military adversary.[268] The prospects of war became an oft-repeated

---

[261] Report (2017c, 2018d), Shim (2017) and Huang (2018).

[262] Mattis (2018g, p. 4).

[263] Seidel (2017), Martina and Yu (2017) and Chan (2017a).

[264] Editorial (2017b).

[265] Stone (2018b).

[266] Gomez (2018).

[267] Yan (2018) and Brown (2018).

[268] Jost (2017), NSC (2017), Mattis (2018c), SoD (2018) and Comptroller (2018).

refrain.[269] US analysts urged efforts to cleave deepening Sino-Russian collaboration, easing prospects of US 'victory'.[270]

The dynamic accelerated after the March 2018 NPC removed term-limits to China's presidency, enabling Xi to theoretically become 'President for life'. US observers decried Beijing's 'regression' while China braced for a 'New Cold War'.[271] Even some within the USA's national security establishment were troubled by the trajectory of US policy-rhetoric. A key collective of advisers to the US military, persuaded that by 2030, China's nominal GDP could exceed that of the USA, making it potentially a far more capable adversary than either Nazi Germany or the USSR had been, but also convinced that China showed 'no interest in matching US military expenditure, achieving a comparable global reach, or assuming substantial defence commitments beyond its immediate periphery', urged Washington to engage with Beijing to address concerns, rather than allow that dangerously dialectic dynamic to spiral towards avoidable and potentially catastrophic conflict.[272]

These analysts identified more granular causal connections, but US leaders viewed things differently. So, the Trump Administration, paralleling BRI implementation by deploying the spectrum of the USA's power and influence, engaged Beijing in comprehensive competition. Trump's response, to both advice from DoD-funded analyses, and Beijing's own actions, came in the form of a trade war covering much of Sino-US commerce, part of legal, diplomatic, economic and military measures in all-of-government 'neo-containment' strategies. That was the geoeconomic-vs.-geopolitical context in which the US-led coalition viewed BRI, a mirror for the West's 'China threat' *problematique*. Sino-US perceptual divergences in an era defined by *systemic transitional fluidity*, triggered by their schizophrenic competitive–cooperative mutual insecurity, fashioned the landscape on which the BRI discourse evolved.

References

7th Fleet Public Affairs (2018) Ronald Reagan, John C Stennis CSGs Operate in Philippine Sea. Philippine Sea, 15 Nov 2018
ADB (2016) TA 9124-PRC: study of the BRI—Sovereign (Public) Project 50141-001. Technical Assistance Reports, Manila, June 2016
Ai X (2016) An international analysis of the formation of 'China threat theory'. In: Advances in social science, education and humanities research, vol 92, Atlantis Press, pp 450–454
Albert E (2017) China's big bet on soft power. CFR, New York
Albert E, Xu B (2016) China's environmental crisis. CFR, New York
Ali I (2018) In symbolic nod to India, US PACOM changes name. Reuters, Pearl Harbour
Allison G (2017a) China's ready for war against the US if necessary. LAT, 8 Aug 2017

---

[269] Bronson (2018), Farley (2018b), Editorial (2017c, 2018d), Basrur (2018), Majumdar (2018), Ness (2017), Allison (2017a), Farley (2017a) and Buruma (2017).

[270] Sutter (2018) and Lopez (2017).

[271] Beinart (2018), Editorial (2018f), Landler (2018) and Perlez (2018).

[272] Dobbins et al. (2017).

References

Allison G (2017b) Destined for war: can America and China escape Thucydides's trap?. Houghton-Mifflin Harcourt, New York

Allison G (2017c) What Xi Jinping wants. Atlantic, 31 May 2017

Al-Rodhan K (2007) A critique of the China threat theory: a systematic analysis. Asian Perspect 31(3):41–66

An B, Chang P, de Castro M, Lo C, Nacpil E, Wu A, Zhang Y (2016) PRC: study of the BRI. ADB, Manila

AP (2019) Trump administration tells of plans to sell F-16s to Taiwan. VoA, 16 Aug 2019

Bajak F, Liedtke M (2019) Huawei sanctions: who gets hurt in dispute? AP/USA Today, 21 May 2019

Baniya S, Rocha N, Ruta M (2019) Trade effects of the new silk road: a gravity analysis. WBG, Washington, p 4

Banyan (2018) Superpower rivalry blows the APEC summit to smithereens. Economist, 24 Nov 2018

Bartholomew C, Shea D (2017) Report to congress. USCC, Washington

Basrur R (2018) Avoiding nuclear crises in Asia. RSIS, Singapore

Bastian J (2017) The potential for growth through Chinese infrastructure investments in Central and South-Eastern Europe along the 'Balkan Silk Road'. EBRD, London, pp 44–52

Beeny T (2018) Supply chain vulnerabilities from China in US federal ICT. USCC, Washington, pp 19–35

Beinart P (2018) Trump is preparing for a new cold war. Atlantic, 27 Feb 2018

Beinart P (2019) China Isn't Cheating on Trade. Atlantic, 21 Apr 2019

Bianji H (2018) BRI not meant to be 'China club'. PD, 28 Aug 2018

Bishop J (ed) (2017) Foreign policy white paper. Department of Foreign Affairs, Canberra, Nov 2017

Bishop J (2018) Australia, the United States and Japan announce a trilateral partnership. Department of Foreign Affairs and Trade, Canberra, 30 July 2018

Blackwill R (2018) Indo-Pacific Strategy in an era of geoeconomics. Japan Forum on International Relations, Tokyo, pp 2–3

Blackwill R (2019) Trump's foreign policies are better than they seem. CFR, New York, pp 8–17

Blackwill R, Harris J (2016) War by other means: geoeconomics and statecraft. Belknap, Cambridge, p 1

Blanchard B (2019) China's Xi says country facing a period of 'concentrated risks'. Reuters, Beijing

Blumenthal D (2017) Trump needs to show that he is serious about America's Rivalry with China. FP, 2 Nov 2017

Borak D (2018) China 'welcomes' Mnuchin's offer to hold trade talks in Beijing. CNN, 22 Apr 2018

Brands H, Feaver P (2016) Should America retrench? For Aff 95(6):164–172

Bremmer I, Kupchan C (2018) Top risks for 2018: risk 1—China Loves a Vacuum. Eurasia Group, Washington

Bronson R (2018) It is now two minutes to midnight. BAS, 25 Jan 2018

Broomfield E (2003) Perceptions of danger: the China threat theory. J Contemp China 12(35):265–284

Brown K (2018) China's 'major-country diplomacy' is a balancing act. EAF, 12 Aug 2018

Brown M, Singh P (2018) China's technology transfer strategy. DIUx, Washington, pp 2–8

Bureau of Industry and Security, Commerce (2018) Addition of certain entities; and modification of entry on the entity list. Department of Commerce, Washington, 1 Aug 2018

Bureau of Public Affairs (2018) CAATSA section 231: addition of 33 entities and individuals to the list of specified persons and imposition of sanctions on the equipment development department. DoS, Washington, 20 Sept 2018

Buruma I (2017) Are China and the United States headed for war? New Yorker, 19 June 2017

Bush R, Hass R (2019) The China debate is here to stay. Brookings, 4 Mar 2019

Camba A (2018) The grand narratives of Chinese FDI aren't so great. EAF, 9 Jan 2018

Campbell C (2017) Ports, pipelines, and geopolitics: China's new silk road is a challenge for Washington. Time, 23 Oct 2017
Capri A (2018) We may be in the early stages of a new Cold War. Channel News Asia, Singapore, 11 July 2018
Carr A (ed) (2018) Debating the quad. ANU, Canberra
Chaguan (2018) China should worry less about old enemies, more about ex-friends. Economist, 15 Dec 2018
Chan M (2017a) Increased military drills suggest mainland China is preparing to strike against Taiwan, experts say. SCMP, 19 Dec 2017
Chan T (2017b) India is buying world's emptiest airport in its battle for territorial dominance with China. Business Insider, 13 Dec 2017
Chandran N (2017) India and China compete for control of an almost empty Sri Lanka airport. CNBC, 13 Dec 2017
Chang F (2015) China's 'win-win' Development Bargain: China, the AIIB, and the International Order. FPRI, Philadelphia, 24 June 2015
Chang G (2018) Trump's right to say he's not launching a trade war with China. He's doing something bigger. Beast, 6 Apr 2018
Chatzky A, McBride J (2019) China's massive BRI. CFR, New York, 21 Feb 2019
Chellany B (2017) China's creditor imperialism. PS, 20 Dec 2017
Chen J (2015) 'Belt and road' takes new route. CD, 15 Apr 2017
Chen J (2018) What the US has really done to China over the past 25 years. CD, 2 Nov 2018
Chengsheng (2017) China will not 'export' Chinese model: Xi. Xinhua, Beijing, 1 Dec 2017
Ching F (2017) How Xi Jinping will cement himself as China's most powerful leader in 25 years. Forbes, 17 Oct 2017
Ciovacco C (2018) Understanding the China threat. NI, 29 Nov 2018
Cleveland R, Bartholomew C (2018a) Annual report. USCC, Washington, 14 Nov 2018, pp 20–21
Cleveland R, Bartholomew C (2018b) Report to Congress. USCC, Washington, Nov 2018
Clover C, Hornby L (2015) China's great game: road to a new empire. FT, 13 Oct 2015
Coats D (2018) A look over my shoulder: the DNI reflects and foreshadows. DNI/Aspen Institute, Aspen, 19 July 2018
Coats D (2019) National intelligence strategy of the USA. DNI, Washington, p 2019
Collins M (2018) Remarks at ASF. CIA/Aspen Institute, Aspen, 20 July 2018
Comptroller (2018) FY2019 defense Budget. DoD, Washington, pp. 1-1, 2-1, 2-2, 2-3, 2-5, 3-1
Correra J (2018) CIA chief says China 'as big a threat to US' as Russia. BBC, Langley, 30 Jan 2018
Cuthbertson A (2019) Huawei ban: more than 130 US companies blocked from selling to Chinese giant. Independent, 29 Aug 2019
Daly R, Rojansky M (2018) China's global dreams give its neighbors nightmares. FP, 12 Mar 2018
Davidson P (2019) Statement before the SASC on US INDOPACOM posture. DoD, Washington
Denmark A (2019) Problematic thinking on China from the State Department's Head of Policy Planning. WotR, 7 May 2019
Department of Outward Investment and Economic Cooperation (2019) China's outward investment and cooperation in 2018. MofCOM, Beijing
Dixon R (2018) Is the US in a new Cold War with China? How much worse could things get? LAT, 29 Aug 2018
Dobbins J, Scobell A, Burke E, Gompert D, Grossman D, Heginbotham E, Shatz H (2017) Conflict with China revisited. RAND, Santa Monica, pp 2–3
Dobell G (2017) China's B&R promise: Asia's fear and greed. Strategist, 22 May 2017
Dunford J (2017) Answers to policy questions: nominee for reconfirmation as CJCS. DoD, Washington
Economy E (2019) The problem with Xi's China model. Foreign Affairs, 6 Mar 2019
Editorial (2017a) Is China challenging the US for global leadership? Economist, 1 Apr 2017
Editorial (2017b) No interference will be brooked in internal affairs. CD, 15 Dec 2017
Editorial (2017c) Will America and China go to war? Economist, 6 July 2017

References

Editorial (2018a) China, Elbows out, charges ahead. NYT, 5 Feb 2018

Editorial (2018b) China's B&R plans are to be welcomed—and worried about: the 'project of the century' may help some economies, but at a political cost. Economist, 26 July 2018

Editorial (2018c) Hank Paulson and Wang Qishan illustrate a superpower divide. Economist, 17 Nov 2018

Editorial (2018d) The next war. Economist, 25 Jan 2018

Editorial (2018e) Why China cannot concede in trade war. GT, 21 Apr 2018

Editorial (2018f) Xi Jinping dreams of world power for himself and China. NYT, 27 Feb 2018

EIU (2017) China Going Global Investment Index. Economist, London, p 21

EIU (2018) Ensuring sustainability throughout the BRI. HSBC, London, 18 Jan 2018

Ellings R, Sutter R (eds) (2018) Axis of Authoritarians: implications of China-Russia Cooperation. NBR, Seattle

Evans D (2017a) China aims for dollar-free oil trade. Nikkei, Denpasar, 14 Sept 2017

Evans G (2017b) Australia in an age of geopolitical transition. EAF, Canberra

Executive Board (2018) IMF country report no. 18/240: PRC. IMF, Washington, pp 2, 29, 33

Fan J (2017) At the communist party congress, Xi Jinping plays the emperor. New Yorker, 18 Oct 2017

Farley R (2017a) Terrifying tale: why a war between China and America would be all sorts of awful. NI, 1 Aug 2017

Farley R (2018b) How would a war between the US and China End? Diplomat, 25 Jan 2018

FMPRC (2017) New opportunities for win-win cooperation: outcomes of Mar-a-Lago meeting and the BRF for International Cooperation. PRC Consulate-General, Houston, 26 May 2017

Ford L (2018) Was China's RIMPAC exclusion an opening salvo or a wasted shot? EAF, 20 July 2018

Freeman C (2010) China's challenge to American hegemony. Middle-East Policy Council, London

Freeman C (2016) The end of the American empire. WotR, 11 Apr 2016

Freier N, Hayes J, Hatfield M, Lamb L (2018) Game on or game over: hypercompetition and military advantage. AWC, Carlisle, 22 May 2018

Friedberg A (2018a) Competing with China. Survival 60(3):7–64

Friedberg A (2018b) The Trump administration is at a crossroads on China trade policy. WotR, 11 Jan 2018

Fulco M (2017) The WeChat economy. Cheung Kong Graduate School of Business Knowledge, vol 26, Summer 2017, pp 20–24

G7 (2018) G7 Foreign Ministers' Communique. MoFA, Charlevoix, 23 Apr 2018, pp 8–9, 21

Garamone J (2018a) Air force, army developing multidomain doctrine. JCS/DoD News, Washington, 25 Jan 2018

Garamone J (2018b) Great power strategy affects DoD priorities, allocations, Dunford says. DoD News, Halifax, 17 Nov 2018

Garamone J (2018c) US, Allies, aim to maintain free, open Indo-Pacific region. DoD News, Washington, 8 Aug 2018

Garrett G (2018) A new cold war? Why the US and China would both lose. Wharton Knowledge, Philadelphia, 6 Aug 2018

Gehrke J (2019) State Department preparing for a clash of civilizations with China. Washington Examiner, 30 Apr 2019

Geng S, Xinhua (2018) US NPR deviates from peace, development: FMPRC. CD, 6 Feb 2018

Ghiasy R, Zhou J (2017) The SREB: considering security implications and EU-China cooperation prospects. SIPRI, Stockholm

Gilchrist K (2017) China's BRI could be the next risk to the global financial system. CNBC, 24 Aug 2017

Gomez J (2018) Navy says it won't be deterred by Chinese-built islands. AP, Aboard USS Carl Vinson, 17 Feb 2018

Goodman M (2017) Predatory economics and the China challenge. Global Econ Month CSIS VI(11):1–2

Green M (2017) Asia awaits Trump's visit with trepidation. FP, 27 Oct 2017
Greene A, Probyn A (2017) OBOR: Australian 'strategic' concerns over Beijing's bid for global trade dominance. ABC, 23 Oct 2017
Greenert J (2018a) Tenets of a regional defense strategy: considerations for the Indo-Pacific. NBR, Seattle, Aug 2018, pp 1–8, 17–31
Greenert J (2018b) Tenets of a regional defense strategy: considerations for the Indo-Pacific. NBR, Seattle, Executive Summary, pp 2–6
Grossman D (2018) The quad needs broadening to balance China—and now is the time. RAND, Santa Monica, 22 Oct 2018
Gupta S (2017) The BRI should learn from paths already travelled. EAF, 16 May 2017
Guterres A (2019) Remarks at the opening ceremony of the BRF for International Cooperation. UN Secretariat, Beijing, 26 Apr 2019
Haddick R (2014) America must face up to the China challenge. NI, 17 Oct 2017
Hamlin K, Pi X, Zhao Y (2018) How 'MiC 2025' frames Trump's trade threats. Bloomberg, 11 Apr 2018
Hasan R, Yi J, Zhigang L (eds) (2016) Meeting Asia's infrastructure needs. ADB, Manila, pp xi–xix
Hass R, Jones B, Mason J (eds) (2019) China's B&R: the new geopolitics of global infrastructure development. Brookings, Washington
Heide D, Hoppe T, Scheuer S, Stratmann K (2018) EU ambassadors band together against Silk road. Handelsblatt, 17 Apr 2018
High Representative for Foreign Affairs and Security Policy (2019) EU-China: a strategic outlook. European Commission, Strasbourg, 12 Mar 2019, p 1
Holsoe J, Vulliamy E (1999) Nato bombed Chinese deliberately. Guardian, 17 Oct 1999
Hongyi B (2018) Irresponsible practices of the G7 reveal its hypocrisy. PD, 28 Apr 2018
Hongyu B (2018a) China urges US to cancel sanctions. CD, 22 Sept 2018
Hongyu B (2018b) China will not surrender to US threatening tactic. PD, 6 Aug 2018
Hongyu B (2018c) Factbox: BRI in five years. Xinhua, Beijing, 26 Aug 2018
Horn S, Reinhart C, Trebesch C (2019) China's overseas lending. Kiel Institute for the World Economy, Kiel, pp 1–5
Horvath B (2016) Identifying development dividends along the BRI. UNDP, New York, p 3
Hu W (2019) Biased views deceive world about BRI-related loans. GT, 1 July 2018
Huang K (2018) Why America's stealth jet forces should fear China's new unarmed eye in the sky. SCMP, 24 Jan 2018
Hufbauer G (2018) The unfolding of a new Cold War. EAF, 18 Nov 2018
Hurley J, Morris S, Portelance G (2017) Examining the Debt implications of the BRI from a policy perspective. Centre for Global Development, Washington, Mar 2017
Ihara N (2019) Taiwan to buy up to 66 F-16 fighter jets from US. Nikkei, Taipei, 7 Mar 2019
Indian Embassy (2017) United States and India: prosperity through partnership. MEA, Washington
Information Office (2015) Made in China 2025. State Council, Beijing, 8 May 2015. http://www.gov.cn/zhengce/content/2015-05/19/content_9784.htm. Accessed 6 Mar 2018
Inkster N (2018) Chinese culture and soft power. Survival 60(3):65–70
Insight-Team (2018) Embracing the BRI ecosystem in 2018. Deloitte, London, 12 Feb 2018
International Cooperation Office, EBRD (2017) The EBRD and BRI. https://www.ebrd.com/what-we-do/belt-and-road/ebrd-and-bri.html. Accessed 6 Aug 2018
Jackson V (2017) To war or not to war? US-Chinese relations as the central question of our times. WotR, 1 Nov 2017
Jacques M (2018) Can the West's democracy survive China's rise to dominance? Economist, 14 June 2018
Jelinek T, Schwab O, Buehler M, Sucholdolski S, de Almeida P (2017) Eight transformation factors for driving the new silk road towards sustainability and shared prosperity. WEF, Geneva, p 1
Jiangtao S, Zhou L (2017) Wary China on 'quad' bloc watch after officials from US, Japan, India and Australia meet on Asean sidelines. SCMP, 13 Nov 2017

References

Johnson C (2016) President Xi Jinping's BRI: a practical assessment of the CCP's roadmap for China's global resurgence. CSIS, Washington

Jones L, Zeng J (2019) Understanding China's BRI: beyond 'grand strategy' to a state transformation analysis. Third World Quart 20:1–2

Joshi S, Graham E (2018) Joint FoN patrols in the SCS. Lowy Interpreter, 21 Feb 2018

Jost T (2017) The new pessimism of US strategy towards China. WotR, 28 Dec 2017

Kaletsky A (2018) The US will lose its trade war with China. PS, London, 21 Sept 2018

Kamphausen R, Lai D, Tanner T (eds) (2014) Assessing the PLA in the Hu Jintao Era. AWC, Carlisle, p 8

Kate D (2018) Huawei arrest reignites US-China tensions. Bloomberg, 6 Dec 2018

Keck Z (2013) Xi Jinping: China's most powerful leader since Deng and Mao? Diplomat, 5 Aug 2013

Kennedy A (2018) China's innovation trajectories. Survival 60(3):71–86

Kim J (2017) Remarks at the BRF for international cooperation. WBG, Beijing

Kissinger H (2012) The future of US-Chinese relations. Foreign Affairs, Mar/Apr 2012

Kliman D, Cooper Z (2017) Washington has a bad case of China ADHD. FP, 27 Oct 2017

Kono T (2017) Belt and Road to benefit the global economy. MoFA/ONTV, Tokyo

Koopman R, Maurer A (eds) (2017) World trade statistical review 2017. WTO, Geneva, p 103

Kou J (2018) China's AI talent strategy: domestic boom, international lure. PD, 12 Apr 2018

Kovrig M (2017) The twists and turns along China's belt and road. ICG, Beijing, Brussels

Kristoff N (1989) Better relations depend on US, Deng tells Nixon. NYT, 1 Nov 1989

Kube C (2017) Can China beat the US without a fight? NBC, 1 Nov 2017

Lagarde C (2017) BRI: Proven policies and new economic links. IMF, Beijing

Lagarde C (2018a) BRI: strategies to deliver in the next phase. In: IMF-PBC conference, Beijing, 12 Apr 2018

Lagarde C (2018b) US-China trade war could inflame emerging market crisis. IMF/FT, 11 Sept 2018

Lagarde C (2019) BRI 2.0: stronger frameworks in the new phase of belt and road. Address at BRF2. IMF, Beijing

Landler M (2018) As Xi tightens his grip on China, US Sees Conflict Ahead. NYT, 27 Feb 2018

Lau N (2018) The destiny of humanity is shared future. CD, 2 Feb 2018

Lawder D (2017) US business groups say WTO unable to curb many Chinese trade practices. Reuters, Washington

Layne C (2018) The US-Chinese power shift and the end of the Pax Americana. Int Aff RIIA 94(1):89–111

Lendon B (2017) US flexes naval muscle in Asia ahead of Trump's visit. CNN, 25 Oct 2017

Levitt T (2015) Ma Jun: China has reached its environmental tipping point. Guardian, 19 May 2015

Li S (2015) One belt, one road, many risks. Reuters/Asian Legal Business, 18 Dec 2015

Li J (2018a) China rejects Pentagon report for false speculation of its nuclear development program. MND, Beijing

Li L (2018b) China has no intention of leading, replacing any country: FM. PD, 4 Jan 2018

Li X (2018c) A trade war's impact wide, official says. CD, 10 Apr 2018

Li C, Liang D (2018) Why is it hard for China and the US to cut a deal on trade? Brookings, Washington

Lieber R (2016) Retreat and its consequences: American foreign policy and the problem of world order. Cambridge University Press, Cambridge, pp 89–139

Lim C, Mack V, Zhi W (2018) 'Chinese dream', global ambition: Beijing's BRI. Global Asia 13(3)

Liu M (2009) Zhongguo Meng (the China dream). Friendship Press, Beijing

Liu N (2018) Confrontation, whether in the form of a cold war, a hot war, or trade war, will produce no winners: Xi. PD, 17 Nov 2018

Liu H (2019) ADB president willing to engage in BRI projects, sees AIIB as partner. Xinhua, Beijing

Liu J, Liu Y (2018) Inquiry into FDI pattern of China and the economies along the B&R. Technology and investment. Sci Res Publ 9:161–177

Liu D, Gao H, Xu Q, Li Y, Song S (2017) The BRI and the London market: the next steps in Renminbi internationalization. Chatham House/RIIA, London, pp1–2

Lopez L (2017) China and Russia are totally playing Trump together. Business Insider, 6 July 2017

Lukin A (2019) The US-China trade war and China's strategic future. Survival 61(1):23–50

Lynch D, Rauhala E (2018) With tariffs, Trump starts unravelling a quarter century of US-China economic ties. WP, 15 June 2018

Ma S (2018) MiC 2025 roadmap updated. State Council News, Beijing, 27 Jan 2018

Majumdar D (2018) The US military's mission is clear: crush Russia or China in a war. NI, 22 Jan 2018

Marsh D (2016) China SDR announcement enshrines move away from its dollar focus. Market Watch, 11 Apr 2016

Martina M, Yu J (2017) China angered as US considers navy visits to Taiwan. Reuters, Beijing, Taipei

Mastro O (2019) The stealth superpower. Foreign Affairs, Jan/Feb 2019

Mattis J (2017) Testimony before the SASC. DoD, Washington

Mattis J (2018) Assessing and strengthening the manufacturing and defense industrial base and supply chain resiliency. DoD, Washington, p 4

Mattis J (2018b) Final report on organizational and management structure for the national security space components. DoD, Washington, p 4

Mattis J (2018c) NDS of the USA. DoD, Washington, pp 1–3, 9

Mattis J (2018d) Nuclear Posture Review. DoD, Washington, pp I, v–vi, 5, 7, 11, 31–32

Mattis J (2018e) Remarks at the Shangri-La Dialogue. DoD, Singapore

Mattis J (2018f) Remarks at US Indo-Pacific Command change of command ceremony. DoD, Pearl Harbour

Mattis J (2018g) Summary of the 2018 National Defense Strategy: sharpening the American military's competitive edge. DoD, Washington

Mattis J (2018h) US leadership and the challenges of Indo-Pacific security. DoD/IISS, Singapore

Mayer M, Kremer J (eds) (2011) Global power shift I: theoretical perspectives and approaches. In: Annual convention proceedings, International Studies Association, Montreal

MEA (2017a) AAGC: a vision document. GoI, Delhi, pp 3–12

MEA (2017b) India-Australia-Japan-US consultations on Indo-Pacific. GoI, Delhi

Mehta A (2018) The Pentagon is planning for war with China and Russia: can it handle both? Defense News, 31 Jan 2018

Menon S (2018) The unprecedented promises—and Threats—of the BRI. Brookings, Washington

Mifune E (2018) What Japan needs to be careful of Japan-China Cooperation on BRI. Global Forum of Japan (GFJ) Commentary, Tokyo

Miller T (2017) China's Asian dream. Zed, London, pp 7–9

Mitchell T, Clover C (2017) Xi Jinping's ambitions on show during epic congress address. FT, 18 Oct 2017

Mo J (2018) BRI no Marshall Plan or political strategy. CD, 25 Aug 2018

Modi N (2017) Inaugural address at 2nd Raisina Dialogue. PMO/ORF, New Delhi

Mourdoukoutas P (2018) China will lose the trade war with America, and that's good for its citizens. Forbes, 5 Aug 2018

Mu X (2018) Xi meets with US Secretary of Defense. Xinhua, Beijing

Mutz D (2018) Status threat, not economic hardship, explains the 2016 presidential vote. Proc Natl Acad Sci, 2–3

NASIC (2017) Ballistic and cruise missile threat report. DIA, Washington

Nathan A (2017) The Chinese world order. New York Review of Books, 12 Oct 2017

National Advisory Committee on Building a Manufacturing Power (2015) MiC 2025 major technical roadmap. State Council, Beijing, 29 Oct 2015. http://www.cae.cn/cae/html/files/2015-10/29/20151029105822561730637.pdf. Accessed 24 Apr 2018

References

National Counterintelligence and Security Centre (2018) Foreign economic espionage in cyberspace. DNI, Washington, pp 5–7
NDRC (2015) Vision and actions on jointly building SREB and the 21st century MSR. State Council, Beijing
Ness P (2017) Is a war between China and America simply unstoppable? NI, 17 Aug 2017
Ng A, Wei L (2017) Strategic collaboration: how inclusive management helps Chinese enterprises win overseas. China Mergers & Acquisitions Association, Beijing
Ni V (2016) Trump accuses China of 'raping' US with unfair trade policy. BBC, 2 May 2016
Nouwens M, Legarda H (2018) China's pursuit of advanced dual-use technologies. IISS, London
NSC (2017) NSS of the USA. White House, Washington, pp 1–2, 25–28, 45–46
NSC (2018) Secure 5G: the Eisenhower National Highway system for the information age. White House, Washington, pp 2–3, 11–12
O'Connor S (2015) The human cloud: a new world of work. FT, 9 Oct 2015
O'Rourke R (2018a) A shift in the international security environment: potential implications for defense. CRS, Washington, pp 2–10
O'Rourke R (2018b) China's actions in South and East China Seas: implications for US interests. CRS, Washington, pp 2–27
Obama B (2011) Remarks to the Australian parliament. White House, Canberra
Office of Trade and Manufacturing Policy (2018) How China's economic aggression threatens the technologies and intellectual property of the US and the world. White House, Washington
OSD (1992) Defense Planning Guidance, FY1994–1999: Revised Draft for Scooter Libby. DoD, Washington
Pant H, Bommakanti K (2018) Can the quad deal with China? ORF, Delhi
Paquette D, Rauhala E (2018) As Trump's trade war starts, China retaliates with comparable tariffs of its own. WP, 6 July 2018
Partington R (2018) Trump tariffs: China retaliates with new levies on US products. Guardian, 4 Apr 2018
Paul T (2016) (ed) Accommodating rising powers: past, present, and future. Cambridge University Press, Cambridge, pp 20–24
Pei M (2018) China is losing the new Cold War. PS, 5 Sept 2018
Pence M (2018a) Remarks at the 2018 APEC CEO summit. White House, Port Moresby
Pence M (2018b) Remarks at the Hudson Institute on the administration's policy toward China. White House, Washington
Peng G (2000) A Chinese general's view on whether China is a challenge to America. Huffington Post. https://www.huffingtonpost.com/peng-guangqian-/china-america-challenge_b_6107744.html. Accessed 3 Nov 2017
Perlez J (2000) Israel drops plan to sell air radar to China military. NYT, 13 July 2000
Perlez J (2014) China's 'new type' of ties fails to Sway Obama. NYT, 9 Nov 2014
Perlez J (2017) Xi Jinping pushes China's global rise despite friction and fear. NYT, 22 Oct 2017
Perlez J (2018) As Xi Jinping extends power, China braces for a new Cold War. NYT, 27 Feb 2018
Perry S (2015) China's new FYP and the two centenary goals. PD, 11 Nov 2015
Pettis M (2016) How China's 'currency manipulation' enhances the global role of the US dollar. FP, 22 July 2016
Pi X, Han M, Hong C, Dai M, Dormido H (2017) China's silk road cuts through some of the World's riskiest countries. Bloomberg, 26 Oct 2017
Pieraccini F (2017) The $10 trillion investment plan to integrate the Eurasian supercontinent. Strategic Culture, 8 Dec 2017
Politi J (2019) US and China race for technologies of the future. FT, 21 Jan 2019
Politi J, Smith C, Greeley B (2019) Trump raises tariffs on Chinese goods after stocks tumble. FT, 24 Aug 2019
Pollock J (2016) Xi Jinping is now China's most powerful leader since Deng Xiaoping. NATO Association, 19 Sept 2016

Pompeo M (2018) America's Indo-Pacific economic vision: remarks at the Indo-Pacific business forum. DoS, Washington

Pompeo M, Nielsen K (2018) Joint statement: Chinese actors compromise global managed service providers. DoS/DHS, Washington

Porter L (2018a) DoD's AI structures, investments, and applications: HASC testimony. DoD, Washington

Porter P (2018b) Why America's grand strategy has not changed: power, habit, and the US foreign policy establishment. Int Secur 42(4):9–46

Principal Deputy Under Secretary of Defence (1992) FY 94-99 defense planning guidance sections for comment. DoD, Washington, pp 2, 31

Principal Deputy Under Secretary of Defense (1992) Memorandum for the Secretary: Defense Planning Guidance. DoD, Washington, p 3

Pry P (2017) Nuclear EMP attack scenarios and combined-arms cyber warfare. Commission to Assess the Threat to the United States from EMP Attack, Washington

Public Affairs (2018) Two Chinese hackers associated with the MSS charged with global computer intrusion campaigns targeting IP and confidential business information. DoJ, Washington

Public Affairs Office (2018) Competing in space. NASIC, Washington, p 2018

Rachman G (2018) America v China: how trade wars become real wars. FT, 12 Mar 2018

Rana P (2018) Healthy competition and cooperation for Asian development finance. EAF, 22 Dec 2018

Report (1999) Truth behind America's raid on Belgrade. Observer, 28 Nov 1999

Report (2014) Potential of the Chinese Dream. CD USA, 26 Mar 2014, p 11

Report (2017a) At China's BRF, UN chief Guterres stresses shared development goals. UN Information Centre, Beijing, New York

Report (2017b) China a step closer to centenary goals. Xinhua, Beijing

Report (2017c) China urges Japan to follow the road of peaceful development. Xinhua, Beijing

Report (2017d) Chinese tycoon 'in contact' with family and business after vanishing from Hong Kong. SCMP, 6 Feb 2017

Report (2017e) MiC 2025: Global ambitions built on local protections. US Chamber of Commerce, Washington, pp 6–8, 40–41

Report (2017f) Official: trump to press China on 'predatory' trade practices during visit. World Trade, 26 Oct 2017

Report (2017g) US plans three-carrier drill as Trump visits Asia. Maritime Executive, 27 Oct 2017

Report (2017h) USAF F-35 fighters fly to Okinawa, 1st deployment in Asia. Kyodo, Naha

Report (2018a) Beijing vows firm answer to new US tariff threat. CD, 20 June 2018

Report (2018b) China refutes US research report on BRI. Xinhua, Beijing

Report (2018c) China warns against $330m US arms sale to Taiwan. Asia Times, 26 Sept 2018

Report (2018d) China's new EW aircraft enhances navy's combat capability. PD, 24 Jan 2018

Report (2018e) CPEC: Opportunities and Risks. ICG, Brussels

Report (2018f) Donald Trump is fighting trade wars on several fronts. Economist, 20 July 2018

Report (2018g) Is China losing the trade war against America? Economist, 11 Aug 2018

Report (2018h) Japan mulls deploying F-35B fighters on helicopter carrier. Asia Times, 3 Jan 2018

Report (2018i) President Donald J. Trump is confronting China's unfair trade policies. White House, Washington

Report (2018j) Trump threatens to tax virtually all Chinese imports to US. VoA, 7 Sept 2018

Report (2018k) Xi stresses importance of The Communist Manifesto. Xinhua, Beijing

Report (2019a) America still leads in technology, but China is catching up fast. Economist, 16 May 2019

Report (2019b) China's ODI increases in H1 2019, 13% goes to BRI countries. CGTN, Beijing

Report (2019c) China's outward investment in 2018: 5 point summary. Belt and Road Ventures, Institute for the World Economy, Kiel, 20 Jan 2019. https://beltandroad.ventures/beltandroadblog/china-2018-overseas-investment-odi. Accessed 31 Aug 201

Reuters (2018) Canada arrests Huawei CFO. She faces US extradition for allegedly violating Iran sanctions. CNBC, 6 Dec 2018
Reuters/AP/NBC (2018) China won't let US warship dock in Hong Kong amid trade tensions. EuroNews, 25 Sept 2018
Richman H, Richman J, Richman R (2017) China's predatory economics and how to stop it. American Thinker, 25 Oct 2017
Roland N (2017) China's Eurasian century?. NBR, Seattle, pp 2–3
Roland N (2018) USCC testimony: China's BRI—five years later. NBR, Washington
Rolland N (2019) A China-Russia condominium over Eurasia. Survival 61(1):5–22
Rood J (2018) Remarks at the Aspen security forum (ASF). DoD/Aspen Institute, Aspen
Roy D (2019) Assertive China: irredentism or expansionism? Survival 61(1):51–74
Rudd K (2017) When China leads. PS, 27 Oct 2017
Rumsfeld D, Fargo T, LaPorte L (2004) The global posture review of US military forces stationed overseas: SASC testimony. DoD, Washington
Runciman D (2018) China's challenge to democracy. WSJ, 26 Apr 2018
Ruta M, Dappe M, Lall S, Zhang C, Constantinescu C, Lebrand M, Mulabdic A, Churchill E (2019) Belt and Road economics: opportunities and risks of transport corridors. IBRD/WBG, Washington, pp 9, 11–18
Ryan J (2011) The decline of the dollar: the loss of the exorbitant privilege. LSE Exchanging Ideas on Europe, London
Sandler N (1989) Second secret trip to China by US envoys disclosed. UPI, Washington
Scimia E (2018) French and British navies draw closer in the Pacific. Should China worry? SCMP, 4 June 2018
Seidel J (2017) China to attack Taiwan 'the day a US warship visits', diplomat. AP, Washington, 10 Dec 2017
Shen Y (2018) Talk of US cyber war on China ridiculous. GT, 20 June 2018
Shepard W (2017) China's challenges abroad: why the BRI will succeed. Forbes, 17 Oct 2017
Shim E (2017) China warns Japan against modifying helicopter carriers. UPI, Beijing
Singh A (2018) India wants a Quad to counteract China's expanding influence. ORF, Delhi
Sneader K, Ngai J (2016) China's OBOR: will it reshape global trade?. McKinsey, Hong Kong
SoD (2018) NPR. DoD, Washington, pp I–VI, 2, 6–8, 11, 31–32
Spokesperson (2017) Response to a query on participation of India in OBOR/BRI Forum, MEA, New Delhi
Stanway D, Wroughton L (2018) China cancels military talks with US in protest at sanctions over Russia military equipment. Reuters, Shanghai, Washington
Stefanik E (2018) Opening remarks: China's pursuit of emerging and exponential technologies. House Subcommittee on Emerging Threats and Capabilities, Washington
Steinberg J (2019) US vs. China: a technology cold war? Asian Review, 19 Mar 2019
Stone C (2017) Trump plays up competition in national security while China lays out cooperation strategy. GT, 19 Dec 2017
Stone C (2018a) The real enemy is not China but anti-China hysteria. PD, 9 Jan 2018
Stone C (2018b) USN's reckless operations in SCS will only hit a brick wall. PD, 22 Jan 2018
Stracqualursi V (2017) 10 times Trump attacked China and its trade relations with the US. ABC, 9 Nov 2017
Sun N (2018) CDB commits $250bn to B&R. Nikkei, 15 Jan 2018
Sutter R (2010) China, the United States and a 'Power Shift' in Asia. Discussion paper no. 24, UNISCI, Madrid, pp 10–21
Sutter R (2018) How the US influences Russia-China relations. NBR, Washington, p 2018
Suzuki Y (2018) The presence of Britain the maritime empire and China's ambition. GFJ 11(5):7
Tan W (2018) The trade war is complicating China's efforts to fix its economy. CNBC, 18 July 2018
Taylor-Dormond M (2017) Think bigger on Asia development. ADB, Manila, 20 June 2017

Tellis A, Szalwinski A, Wills M (eds) (2019) Pursuing global reach: China's not so Long March toward preeminence. Strategic Asia: China's Expanding Strategic Ambitions, NBR, Washington

Thorne D, Spevak B (2018) Harbored ambitions: how China's port investments are strategically reshaping the Indo-Pacific. C4ADS, Washington

Thornton S (2018) Remarks at ASF. DoS/Aspen Institute, Aspen

Tian J (2016) OBOR: connecting China and the world. McKinsey, Beijing, p 2016

Tiezzi S (2014) Beijing's 'China threat' theory. Diplomat, 3 June 2014

Tillerson R (2017a) Defining our relationship with India for the next century. DoS/CSIS, Washington

Tillerson R (2017b) I am proud of our diplomacy. NYT, 27 Dec 2017

Tillerson R (2018) US engagement in the Western hemisphere. DoS/University of Texas, Austin

Tisdall S (2016) Barack Obama's 'Asian Pivot' failed. China is in the ascendancy. Guardian, 25 Sept 2016

Trump D (2017a) National Security Strategy of the USA. White House, Washington

Trump D (2017b) Presidential executive order on a Federal Strategy to ensure secure and reliable supplies of critical minerals. White House, Washington

Trump D (2018a) National Cyber Strategy of the USA. White House, Washington

Trump D (2018b) Remarks at signing of a presidential memorandum targeting China's economic aggression. White House, Washington

Tsung L (1994) CPC decides on its international archenemy. Cheng Ming, 1 Jan 1994

Tsuruoka D (2017) OBOR affordable, India participation vital: Brookings fellow. Asia Times, 14 June 2017

Under Secretary of Defense (1999) Asia 2025. ONA/DoD, Washington

USTR (2018) Findings of the investigation into China's acts, policies, and practices related to technology transfer, intellectual property, and innovation under Section 301 of the Trade Act of 1974. Executive Office of the President, Washington, 22 March 2018, pp 4–5

Vatanka A (2017) China courts Iran: why OBOR will run through Tehran. Foreign Affairs, 1 Nov 2017

Wade G (2017) China's 'OBOR' initiative. Australian Parliament, Canberra. https://www.aph.gov.au/About_Parliament/Parliamentary_Departments/Parliamentary_Library/pubs/BriefingBook45p/ChinasRoad. Accessed 11 Dec 2017

Wang W (2017a) China and the intelligent future, vol 26. Cheung Kong Graduate School of Business Knowledge, pp 25–29

Wang X (2017b) What China's leadership reshuffle means for Xi Jinping's New Era. SCMP, 22 Oct 2017

Wang J (2018) Endorsement of BRI looms over China visit by UK's May. GT, 28 Jan 2018

Weiss J (2019) A world safe for autocracy? China's rise and the future of global politics. Foreign Affairs, July/Aug 2019

White H (2010) Power shift: Australia's future between Washington and Beijing. Quart Essay 39:27–30

White H (2017a) China's BRI to challenge US-led order. EAF, 8 May 2017

White H (2017b) China's OBOR to challenge US-led order. Straits Times, 25 Apr 2017

White House (2015) Fact sheet: advancing the rebalance to Asia and the Pacific. Washington, 16 Nov 2015

White House (2017) President Donald J. Trump's visit to Japan strengthens the US-Japan alliance and economic partnership. Tokyo, 6 Nov 2017

Whiting A (1995) East Asian military security dynamics. Asia/Pacific Research Centre, Stanford

Wiseman P (2018) Why they fight: US and China brawl over high technology. AP, Washington

Woetzel J, Seong J, Leung N, Ngai J, Manyika J, Madgavkar A, Lund S, Mironenko A (2018) Reimagining global ties: how China and the world can win together. McKinsey Global Institute, New York

Wong P (2017) More than just a 'road'. HSBC, Beijing

Wong G, Booker S, Barthe-Dejean G (2017) China and Belt and Road infrastructure. PWC, Hong Kong

References

Wray C (2018) Remarks at ASF. FBI/Aspen Institute, Aspen

Wright T (2017) All measures short of war: the contest for the 21st century and the future of American power. Yale University, New Haven

Xi J (2013a) Address to the closing session of the NPC. CPCCC, Beijing

Xi J (2013b) Promote people-to-people friendship and create a better future. FMPRC, Astana

Xi J (2013c) Speech to Indonesian parliament. FMPRC/China-ASEAN Centre, Jakarta, 2 Oct 2013

Xi J (2017a) Inaugural address at the BRF. MoFA, Beijing

Xi J (2017b) Let the sense of community of common destiny take deep root in neighbouring countries: conference on the diplomatic work with neighbouring countries. State Council, Beijing

Xi J (2017c) Work report to the CPC 19th Congress. CPCCC, Beijing

Xi J (2018) In report. President Xi calls for inclusive, rule-based world economy. Xinhua, Port Moresby, 18 Nov 2018

Xi J (2019) Address at the leaders' roundtable of the second BRF. MoFA, Beijing

Xinhua (2018a) 'Cold War' mentality for US to play up 'Chinese military threat': spokesperson. PD, 22 Jan 2018

Xinhua (2018b) Highlights of Xi's keynote speech at Boao Forum. CD, 10 Apr 2018

Xinhua (2018c) US has ignited largest trade war in economic history: China MofCOM. PD, 6 July 2018

Xinhua (2018d) Xi gives new impetus to BRI. CD, 28 Aug 2018

Xinhua (2019) ADB says to continue cooperating with China on BRI. CD, 26 Apr 2019

Xu L (2017) Socialism is Great—CPC Congress sets eyes on 2049 and beyond. Xinhua, Beijing

Yan (2018) Xi urges breaking new ground in major country diplomacy with Chinese characteristics. Xinhua, Beijing

Yan X (2019) The age of uneasy peace. Foreign Affairs, Jan/Feb 2019

Yong W (2018) China's vision for a new world order. EAF, 25 Jan 2018

Zhang R (2017) China's BRI not a threat. Xinhua, 4 May 2017

Zhou L (2017) China and the Fourth Industrial Revolution (4IR), vol 26. Cheung Kong Graduate School of Business Knowledge, pp 5–7

Zhou L (2019) China's authoritarian way can rival liberal democracy if it doesn't tear itself apart, says End of History author. SCMP, 27 Mar 2019

Chapter 3
China's Belt and Road: An Evolving Network

3.1 A 'New Silk Road' Emerges

2017–2018 was a seminal period for Beijing's BRI vision. In October 2017, the CPC's 19th Congress, 'following the principle of achieving shared growth through discussion and collaboration and pursuing the Belt and Road Initiative', incorporated Xi Jinping's 2013 blueprint into the CPC constitution. This transformed BRI, an economic/commercial/transportation/communications and infrastructure-building endeavour, into a central tenet of China's national policy.[1] That decision formalised BRI's locus within the Xi regime's worldview, which the inaugural BRF had underscored in May. Parallel events throughout 2017, fashioning a palimpsest of economic/commercial, transportation/communications, energy and infrastructure, industrial, financial and regulatory convergences, some evolving over decades, began bearing fruit.

Six weeks before the 2017 BRF, Beijing launched seven free-trade zones (FTZ) in Chongqing, Henan, Hubei, Liaoning, Shaanxi, Sichuan and Zhejiang, adding to older FTZs in Shanghai, Fujian, Guangdong and Tianjin.[2] Five of the new FTZs, lying within China's Western Development Strategy catchment area, directly contributed to overland trade along BRI alignments. A month before the BRF, the 1454 km Kazakh–China gas pipeline, linking southern Kazakh gas fields with Chinese consumers, became operational. With an installed capacity of six billion cubic metres (bcm), the pipeline would annually transfer 5bcm to China National Petroleum Corporation (CNPC), the joint venture's Chinese partner. This was a 'significant' section of the longer Central Asia China gas pipeline, which began at Gedaim on the Turkmen–Uzbek border, crossed Central Uzbekistan and southern Kazakhstan before reaching the Khorgos 'dry-port' rising astride the Kazakh–Chinese border, to enter China's Xinjiang Uyghur Autonomous Region.[3]

---

[1] Ying (2017).
[2] Report (2017c) and Bureau of Economic & Business Affairs (2017).
[3] Zheng (2017).

This new source of gas complemented oil pipelines bringing crude oil from Russia, Kazakhstan and Myanmar, securing and diversifying Beijing's energy supplies. Two parallel pipelines conveying Russian oil from Skovorodino to Daqing in Heilongjiang Province via Mohe on the Inner Mongolian border—the first operational from 1 January 2011, and the second, seven years later—doubled the annual flow of Russian oil from 15m tons to 30m tons, reinforcing 'the China-proposed BRI'.[4] In the late 2017, the China Petroleum Engineering & Construction Corporation contracted Russia's Gazprom to jointly build the Amur gas processing plant, further advancing China's energy security.

SREB's transport connectivity component, too, progressed. On New Year's Day, 2017, Beijing launched a China–Britain freight rail service. Trains traversed 7456 miles across Kazakhstan, Russia, Belarus, Poland, Germany, Belgium and France, before reaching the UK. This was part of a venture running nearly 1000 freight trains from Chengdu in Sichuan Province alone to European cities in 2017, almost doubling the 73,000 tons of goods worth $1.56bn shipped from Chengdu by rail in 2016, adding to the rail freight service linking 25 Chinese hubs with European cities, and carrying two-way cargo shipments worth $17bn in the first half of 2016. The network grew in August 2017 with a new service, linking Zhengzhou in Henan Province to Munich, Germany, via Central Asia, the Czech Republic, Hungary and Austria, also linking Spain, Italy and France.[5] By the end of 2017, 35 Chinese cities were connected to 34 points in 12 European states with 3271 freight trains running along 57 routes.[6]

At the end of January 2017, a container-loaded 'Silk Road train' left China's Yeiu city, traversed 7908 km of Chinese and Kazakh territory before reaching Serkhyatak on the Turkmen–Kazakhstan border. It then travelled 1156 km to Serakhs on the Turkmen–Iranian frontier, marking the test launch of the BRI's China–Kazakhstan–Turkmenistan–Iran (CKTI) freight route. The train reached Tehran, 10,399 km from Yeiu, in mid-February.[7] CKTI was formalised 10 months later when the four partners harmonised customs, transportation and other regulations and railway hardware, regularising freight traffic along the route.[8] BRI's terrestrial network, the 'Eurasian Land Bridge', thus rapidly crystallised.

BRI's financial arrangements similarly advanced. Weeks after the 2017 inaugural BRF, China and Russia established joint financial instruments and mechanisms for trans-border trade and infrastructure development spanning not just their territory, but also other members of the Russian-led Eurasian Economic Union (EEU)—Armenia, Belarus, Kazakhstan and Kyrgyzstan. Moscow's sovereign wealth fund, the Russian Direct Investment Fund (RDIF) and the China Development Bank (CDB) set up a $10bn fund to finance BRI and EEU projects, enhancing China–EEU trade and development. RDIF also boosted the Russia–China Investment Fund (RCIF), launched in

---

[4]Xinhua (2018b).
[5]Webb (2017) and Xinhua (2017d).
[6]Zhang (2018).
[7]Xinhua (2017a).
[8]Report (2017a).

2012, with another $1bn from Beijing's sovereign wealth fund, the China Investment Corporation (CIC).

RCIF, committed to funding 19 manufacturing, transport infrastructure, finance and technology projects in Russia, Central Asia and Central and Eastern European states, expanded its operations.[9] Sino-Russian financial cooperation cemented the SREB's political–economic foundations, knitting the Eurasian landmass into a cooperative development zone; RMB-denominated financial vehicles insulated the SREB from US-led sanctions against Russia, enhancing project predictability. Bypassing US control, the accord accelerated RMB internationalisation. In 2017, 35 of 68 BRI partner states signed currency deals with China to conduct transactions in RMB; by year end, half of them had begun doing so. These agreements, the SREB's financial core, enabled the fashioning of the terrestrial New Silk Road linking China to the remainder of Eurasia. BRI's resource base was thus secured.[10]

Still, 2018 brought mixed prospects. Celebrating BRI's 5th anniversary, Beijing pointed to advances in implementing the SREB and MRI blueprints.[11] Days later, at the Beijing Forum on China–Africa Cooperation (FOCAC), Xi Jinping received all African heads of state/governments bar one, further expanding China's trade, aid and investment via the MSR framework and demonstrating its popularity among Africa's rulers.[12] As for Washington, then-head of the US Overseas Private Investment Corporation (OPIC), Ray Washburne, marked BRI's 5th anniversary by insisting China used BRI to 'debt-trap' recipients and then secure 'their rare earths and minerals and things like that as collateral for their loans'. He believed China was not 'in it to help countries out, they are in it to grab their assets'.[13]

Clouding BRI's prospects, elections in Malaysia, Maldives and Pakistan brought in leaders who either cancelled several projects, as Malaysia's Mahathir Mohamad did, or reviewed those being implemented, as Imran Khan did in Pakistan.[14] In Europe, Hungary—whose officials described their country as 'a key state on the Silk Road', was severely criticised by fellow EU members for eroding democracy and liberal values.[15] That trend continued into 2019, reflecting the mixed prospects for China's globe-girdling infrastructure connectivity and trade blueprint.

---

[9]McGrath (2017).
[10]Zhang (2018).
[11]ZX (2018).
[12]Mo (2018).
[13]Washburne and Churchill (2018).
[14]Banyan (2018).
[15]Report (2018d) and Maraczi (2017).

3.2 The Silk Road Economic Belt's Accretive Evolution

BRI incorporated myriad initiatives, including many long preceding Xi Jinping's twin proclamations. Policy-pronouncements and financial/investment endeavours publicised in 2017 affirmed Xi's vision, but BRI was rooted in history. When the Soviet Union fissioned into 15 'republics', their immediate concerns focused on legitimising their revived sovereignty while stabilising the transition via links securing respective interests in a transformed milieu. Somnolent Central Asian *Stans*—Kazakhstan, Kyrgyzstan, Tajikistan, Turkmenistan and Uzbekistan—reappeared as independent Eurasian polities, confronting severe dislocation as their traditional roles within the centralised Soviet system suddenly came to an end.

Kazakhstan, Kyrgyzstan and Tajikistan shared 4000 km of borders inheriting ethno-cultural spillover into, and disputed sections with, China's poorly assimilated, marginalised, poverty-stricken and restive Muslim majority Xinjiang, which also abutted Southern Asia, making it central to both the ancient and revived Silk Road. Around 3500 miles of the ancient Silk Road, weaving between the Taklamakan Desert to the south and a prairie to the north separated by a mountain range, traversed Xinjiang.[16] Since 1964, Xinjiang provided Beijing with its Lop Nor nuclear test site and missile test ranges, but later became an emotive fault line in Sino-Soviet animosity.

Bitterly fought military clashes in 1969 along disputed Western Xinjiang borders triggered Mao Zedong's instructions to four of the PLA's ten Marshals to review China's strategic insecurity ecology for initiating appropriate defensive and deterrent action. That, and President Richard Nixon's coincidentally resonant initiative to build diplomatic bridges in Beijing to increase American leverage vis-à-vis Moscow and Hanoi in negotiating an end to the Indochina War, engendered a tacit anti-Soviet US-Chinese alliance which engaged in myriad covert proxy campaigns against Soviet clients on three continents.[17] These drained the USSR, accelerating the Soviet state's evisceration during the late 1970s and the 1980s.

The most spectacular Sino-US campaign targeted Soviet-occupied Afghanistan, reinforcing Central Asia's role in great power contestation and deepening Chinese anxiety. Xinjiang and Central Asia were a theatre of clandestine operations during the Cold War's final stages. After Iran's 1979 Islamic Revolution decimated US signals intelligence (SIGINT) assets monitoring Soviet nuclear and ballistic missile (BM) tests in Central Asia, the USA sought new facilities. When Sino-US diplomatic relations were formalised, DCI Stansfield Turner visited China to craft joint building of such stations at Qitai and Korla in Xinjiang. Manned by US-trained Chinese staff, their output, shared by partner intelligence agencies, gained Beijing and Washington insights into Soviet missile forces. It also helped them to grasp Moscow's strategic thinking generally.

In late 1979, Moscow's Afghan invasion, implementing the 'Brezhnev doctrine' in defence of Kabul's Soviet-allied socialist regime, deepened Chinese angst

---

[16]Campbell (2017).

[17]Ali (2005).

over Xinjiang's fate. Beijing partnered with the US–Saudi-led coalition sponsoring *Mujahideen* militias, eventually forcing a Soviet retreat. The campaign reinforced China's anxious determination to secure Xinjiang from turbulence precipitated by intensifying religiosity triggered and exploited by the coalition in deploying *Salafist* sentiments sweeping the Muslim world, whose volunteers flocked to fight Soviet 'infidels'. Sino-US covert collaboration continued after Soviet withdrawal but Beijing's June 1989 Tienanmen Square crackdown embittered ties, and the USSR's collapse eroded the anti-Soviet adhesive.[18] These dynamics shaped China's post-Soviet Central Asia policy.

Having helped the USA to undermine Soviet power, China was now disturbed by Soviet collapse precipitating US *unipolarity*. Anxious about the impact of Central Asian renascence amid growing Islamist-nationalist tendencies on Xinjiang's stability, Beijing sought to prevent spillover instability by consolidating Xinjiang's integration into Han-China.[19] In late 1990, while the USSR struggled to survive gathering domestic and external turbulence, east-west tracks of the future national railways in the five Central Asian *Stans* were connected to PRC railways.[20] Viewing interactions with these states as complementary to improved relations with Russia, China recognised the *Stans* in December 1991, later establishing diplomatic relations. China-Soviet railway links, operationalised just months before the USSR collapsed, enabled the notional 'New Silk Road' to realise China's westward progression.

The foundations had, however, been laid even earlier. Soon after the PRC arose, Sino-Soviet alliance leaders initiated travel along a version of the 'Eurasian Land Bridge', now the SREB's spine. To establish treaty relations between the Kremlin and *Zhongnanhai*, Mao Zedong travelled to Moscow by rail, arriving on 16 December 1949 and demonstrating the feasibility of such passenger services. On 16 January 1954, the first of three railway routes linking Soviet and Chinese networks opened. Connecting Moscow to Beijing via Harbin in Heilongjiang Province, and using a section of the Soviet Trans-Siberian Express route, it carried Communist bloc elites. The Moscow–Beijing Express turned south at Chita, crossing the border into China at Zabaikalsk/Manzhouli, on to Harbin, and then to Beijing. Gauge differences forced passengers to change coaches at the border.

The need to change rolling stock limited interest in cargo shipments. However, the initiative focused attention on routes capable of comprehensive service between the allies. They agreed on two other alignments. With coaching stock built in East Germany in 1955–1959, the shorter, Trans-Mongolian route opened in 1959. Chinese- and Soviet-built locomotives hauled the Moscow–Beijing trains on respective sides of the border. As with the first route, gauge variation required trans-shipment, first at Tsining in China and then, since 1966, at Erlian on the Sino-Mongolian border.[21] But the longest route, linking China's east coast ports to those in the Soviet Baltics

---

[18] Cooley (2002).
[19] Karrar (2006, p. 28).
[20] Shigeru (2001).
[21] Gardner and Kries (2014) and Report (2007c).

via Kazakhstan, held the greatest promise, presaging the SREB's central transport infrastructure alignment.

China's pre-existing rail tracks from the Yellow Sea port of Lianyungang crossed Xian, Shaanxi province, terminating at Lanzhou, Gansu province. Soviet-era tracks in Kazakhstan ended at Aktogay, 340 km west of the Sino-Kazakh border, and 320 km west of Druzhba, Kazakhstan's eastern most township. Fashioning a rail link via Kazakhstan required new tracks on both sides. In 1954, Moscow and Beijing agreed to lay these. China offered to build the Lanzhou–Urumqi–Alashankou line, bringing its railhead to the Kazakh border. The allies decided to construct the Urumqi–Aktogay section in 1956, and Moscow completed the Aktogay–Druzhba section in 1959. When completed, these linked sections would enable rail transport from China's Pacific coast to Baltic ports. However, Sino-Soviet relations chilled in 1959 and the alliance turned adversarial. China completed the Lanzhou–Urumqi extension in 1962 but suspended work on the Urumqi–Alashankou section.

On the Soviet side, the new Druzhba–Aktogay line lay dormant for three decades. When Sino-Soviet relations thawed in 1985, Beijing revived the Urumqi–Alashankou section, completing construction in 1990. In mid-September, the Chinese and Soviet rail networks were linked up at Druzhba. Freight trains began running in July 1991 and, in June 1992, after the Soviet collapse, China and Kazakhstan launched an Urumqi–Almaty passenger service. As Central Asian states expanded their rail networks around Soviet-era tracks, it became feasible to start from China's Lianyungang harbour, travel 11,000 km westward and reach Rotterdam, Europe's largest port. For several years, trains ran between Urumqi and Tashkent, the Uzbek capital, via Almaty. However, by late 2000, given low demand, services were reduced to just two weekly return trips between Urumqi and Almaty. Chinese and Kazakh railways shared responsibility for running these *Zhibek Zholy,* or Silk Road, trains.

3.3 Eurasia's Post-Soviet Coalescence

Diplomacy followed the railways. In 1992, the Presidents of Kazakhstan, Kyrgyzstan and Uzbekistan visited Beijing; in November 1992, China's Foreign Minister Qian Qichen travelled to these states and Russia. In 1993, the Presidents of Tajikistan and Turkmenistan visited China. These exchanges laid the foundations for China-Central Asian engagement. Deepening concerns over economic divides fracturing China reinforced Beijing's diplomacy. In late 1993, the State Council announced plans to develop 'the regions along the Eurasian Continental Bridge', aiming to narrow the gulf between China's fast-growing coastal belt and the underdeveloped hinterland, as parallel plans to connect China to the rest of Eurasia gelled.[22]

In April 1994, Premier Li Peng visited Kazakhstan, Kyrgyzstan, Turkmenistan and Uzbekistan, propounding principles, objectives and processes guiding Sino-Central

---

[22] Apel and Gallagher (2001).

Asian relations: mutually beneficial economic and diversified cooperation, prosperity via shared use of natural resources, improvement of communications and transport infrastructure, offers of Chinese aid, peaceful coexistence, non-interference in internal affairs and respect for territorial integrity and sovereignty.[23] Li's commercial focus was manifested in the delegation of SOE executives accompanying him. While addressing the Uzbek parliament in Tashkent on 19 April, Li proposed to found Sino-Central Asian relations on stable ties, economic cooperation and non-interference, symbolised by mutually beneficial collaboration along a 'New Silk Road'. He was the first Chinese leader in the modern era to proclaim, 'the People's Republic of China is now willing to work to build a new Silk Road'.[24]

SOE leaders signed four agreements, two contracts and around 20 letters of intent on cooperation in petroleum, natural gas, metallurgy, electronics, textiles, construction and other industries.[25] Through the 1990s, as Eurasia evolved, security, stability and economic cooperation remained the *leitmotif* of Beijing's diplomacy. A NATO-affiliated assessment identified two motivations driving it: the quest for new energy sources to fuel China's booming economy, and expanding land access westward 'in order to reduce the economic dependence on coastal maritime commerce'.[26] Beijing engaged with all five *Stans* but, given its political–economic stature, the length of its border with China, its scientific–technical progress and policy-proximity to Moscow, Kazakhstan received most attention, shaping China's policy.

Astana shared Beijing's interest in forging the Eurasian Land Bridge with new roads and pipelines, although railways remained central to the enterprise. In 1990, the cross-border line reached Aktogay, joining Soviet-era tracks running southwest to Kazakhstan's old capital, Almaty. From there, it ran westward for about 1000 km before crossing into Uzbekistan, reaching its capital, Tashkent, Central Asia's largest city and a trading hub on the ancient Silk Road. From Tashkent, the line ran westward for 350 km, to another fabled Silk Road market city, Samarkand. From there, 1,000 km west, it reached Turkmenistan's capital, Asgabad. Before Asgabad, a spur line branched off at Tedzhen, heading south to Sarakhs on the Iranian border. In 1996, a newly built 100 km line joined Sarakhs to Mashhad, allowing train travel from China to Tehran and then, via Sirjan, to Bandar Abbas on the Persian Gulf, giving Central Asian economies maritime access for the first time. Europe, too, beckoned.

In May 1996, Beijing, with UN support via UNDP and the Department for Development Support and Management Services, hosted a three-day international symposium on 'Economic Development along New Euro-Asia Continental Bridge'. More than 400 officials, experts, representatives of multilateral institutions and business executives attended. Ministers from regional states discussed Eurasian development with senior figures from the UN, EU, WBG, ADB and other institutions. Song Jiang, China's Minister of the Commission for Science and Technology and conference co-chair, stressed the 'urgency of formulating collective actions within the countries and

---

[23] Alam (2014) and Karrar (2006, p. 104).
[24] Li (1994) and BBC (1994).
[25] Report (1994).
[26] Vingoe (2015).

among the people' of Eurasia to boost shared development. He invited groups and individuals, including those from outside Eurasia, to join 'the challenging endeavour' lasting 'decades to come'. Speakers pressed for transportation and telecommunications infrastructure, economic and trade cooperation, sustainable development, environmental protection, population and social affairs and poverty alleviation.[27] Eurasian collaboration gained salience in China's development plans.

Another group coincidentally coalescing in 1996 also drove Eurasian connectivity. Following a November 1994 Singaporean–French joint initiative, 56 Asian and European members and dialogue partners of ASEAN and the EU, established the Asia–Europe Meeting (ASEM). ASEM urged deepening post-Cold War intercontinental ties, reinforcing political, economic, sociocultural and other cooperation, fostering 'open discussions on geopolitical issues of common concern', creating 'a space of peace and shared development', promoting 'poverty eradication, protection of cultural heritage' and advancing 'intellectual endeavours, economic and social development, knowledge and educational resources, science and technology, commerce, investment and enterprise'.[28] Members hoped economic cooperation would stabilise Eurasia through transformational development.

Premier Zhu Rongji reminded ASEM science ministers at their first meeting: 'There has been a long history of science-and-technology exchanges between the two continents. Even in the ancient times, our ancestors were engaged in a two-way dissemination and exchanges of advanced scientific and technological inventions through the ancient Silk Road, benefiting the people of both continents and contributing to world civilization'. Meaningful Euro-Asian exchanges required ASEM's recognition that 'science and technology advance by leaps and bounds and all countries are attaching greater importance to the important role of scientific and technical advancement in socio-economic development'. Zhu noted, links between scientific exchanges and socio-economic development had 'become an important component of international relations'.[29] Hence the Silk Road's revival.

ASEM established inter-ministerial commissions to drive sector-specific collaboration. Leaders acknowledged that unequal development between an industrialised Europe and still industrialising Asia must be addressed by exchanging scientific–technological knowledge, information, skills, expertise and data. Ministers, striking complex balances, crafted mechanisms for disseminating scientific knowledge without draining investment and revenue streams from firms which had patented new processes and sought to protect their intellectual property. In 2001, China extended the idea of fashioning mutually beneficial commercial, cultural and 'civilisational' connectivity from the Pacific Ocean to Europe's shores via terrestrial transport linkages along ancient alignments. Welcoming ASEM foreign ministers to Beijing for their third meeting, Jiang Zemin urged: 'ASEM should become a major channel for exchanges between eastern and western civilizations. Both Asia and Europe are cradles of human civilizations and have long been associated with each other'.

---

[27] Information Office (1996).
[28] EU External Action (2016).
[29] Zhu (1999).

Significantly, 'the ancient Silk Road, which used to be an important passage of interflow of Asian and European civilisations, has played a unique role in the exchanges between eastern and western civilisations. ASEM should build up a new Silk Road to actively boost exchanges between these two civilisations in the new century so that countries in Asia and Europe will build on their respective civilisations and respect, learn from, complement and benefit each other'.[30] The fact that China and the USA were recovering from a crisis following the collision between two of their military aircraft near Hainan Island seemingly reinforced the impetus to forging China–Europe links via a new Silk Road.

Recognising this was a long-term vision, China intensified Eurasian diplomacy, especially with Kazakhstan. By mid-2008, Sino-Kazakh trade represented about 70% of China's Central Asian commerce. Kazakh success in halving its poverty level in 15 years and raising *per capita* GDP to $9400, second only to Russia's $12,000 among former Soviet Republics, encouraged Beijing.[31] Astana reciprocated by opening consulates in Hong Kong and Shanghai, boosting economic exchanges. In the 1990s, diplomacy pursued goals of 'good neighbourliness' and stabilisation by settling border disputes. Since Hu Jintao's June 2003 visit and the launch of a five-year cooperative programme, economics predominated. In May 2004, the partners established a Kazakhstan–China Cooperation Committee to implement agreed programmes and drive further growth. Given Kazakhstan's relative wealth, China's 2004 offer of $900m credit for Central Asia, executed from 2006, targeted poorer *Stans*.[32]

In July 2005, Hu Jintao and his Kazakh host, Nursultan Nazarbayev, proclaimed a 'strategic partnership'. Trade expanded by 22% to $8.3bn in 2006.[33] Nazarbayev's return visit in December 2006 produced a 'Cooperation Strategy for the twenty-first Century' and a 'Plan for Economic Cooperation'. The leaders agreed to build a four-lane motorway linking Lianyungang in eastern China to St. Petersburg, Russia, by extending China's Lianhua highway to Khorgos and on to Western Russia. Hu returned to Astana in August 2007, signing nine agreements, stressing mutual determination to diversify trade composition from energy and commodities, and secure balance. Cabinet- and official-level exchanges implemented leadership-level accord. A decade before Xi Jinping hosted the first BRF, Beijing and Astana began discussing knitting together Eurasia with connectivity hardware and software.

In March 2007, Kazakh Prime Minister Karim Massimov inspected the Korgas–Khorgos and Dostyk–Alashankou border crossings to better understand issues relating to frontier inspections regimes, customs procedures and trans-shipment infrastructure. China mostly imported petroleum and ferrous and non-ferrous metals, while Kazakhstan bought consumer goods, e.g. textiles, shoes, appliances, toys, electronics, spare-parts, foodstuffs and pharmaceuticals. Despite a dramatic growth in commerce, however, the trade balance mirrored overall imbalances: in 2006, Kazakhstan

---

[30]Jiang (2001).

[31]Peyrouse (2008, p. 34).

[32]Peyrouse (2008, p. 35).

[33]Report (2007b).

represented just 0.49%, and Central Asia, 0.69%, of China's international trade.[34] Goods crossing the China–Kazakh border nonetheless rose to 13.1m tonnes in 2006, but dropped to 12m tonnes in 2007, indicating the routes' potential.[35]

That April, Beijing began laying double-track sections in northern Xinjiang, while electrifying the entire Chinese section. Stipulating that their countries 'will use the potential of freight transport to the maximum, as well as the consolidation of port capacity and will promote the construction of the international transport corridor for protecting transport between China and Europe through China and Kazakhstan', Hu and Nazarbayev slashed bureaucratic regulatory obstacles, enabling the expansion of both bilateral and transit trade.[36]

Container railway freight service between Lianyungang and Moscow began after the August 2007 Hu-Nazarbayev summit. China's appetite for Central Asian energy and mineral resources grew, but the region's locus within Beijing's commercial firmament was that of a way station, not the final destination. The partners began work on the Lianyungang-St. Petersburg highway in 2008 and opened the 5246-mile motorway to commercial traffic in October 2018, reducing travel time to 10 days. This was the outcome of protracted, complex and challenging diplomatic, regulatory and engineering negotiations. The partners managed a dynamic environment while maintaining focus on a trans-continental vision.[37]

3.4 The 'Shanghai Spirit' Paradigm

Security proved paramount. Soviet collapse transformed the interstate landscape. As China struggled to comprehend its new circumstances, its strategic analyst community devoted the 1990s to examining a fluid, unfriendly, international milieu. Confronting the consequences for China, it generated a series of primarily *Sinic* constructs: the New Security Concept, the New Development Approach, the New Civilisation(al) Outlook and the Harmonious World Concept. Together, these comprised China's post-Cold War 'New Diplomacy'.[38] These historically rooted and culturally derived frameworks informed Beijing's approaches to China's post-Soviet neighbourhood, with the emphasis on proximate Eurasia.

China's diplomatic goals drew upon Deng Xiaoping's 'reform and opening up' initiative, i.e. to 'focus on the central task of economic construction and to ensure a peaceful and favourable international environment' for sustained development.[39] A dynamic ecology demanded agile adaptation. Even as conceptual paradigms evolved, policy-makers confronting unfamiliar challenges needed to act promptly. Still, the

---

[34]Peyrouse (2008, p. 36).
[35]Ilie (2010).
[36]Hu and Nazarbayev (2007), Kazakh MoFA (2007), Hu and Nazarbayev (2009) and Fang (2009).
[37]Tang (2018).
[38]Gao (2010, pp. 1–2, fn. 2).
[39]Gao (2010, p. 2).

urgency of establishing longer-term frameworks for effective diplomacy precipitated 'new' thinking. On 3 March 2004, Hu proposed to the CPCCC his 'Scientific Development Theory' to inform China's 'New Development Approach'. It posited: 'all countries should strive to achieve mutual benefit and win-win situations in their pursuit of development. They are encouraged to open up rather than close themselves, to enjoy fair play instead of profiting oneself at the expense of others'.[40] The formulation being primarily directed at the CPC elite, rather than at external audiences, it likely represented beliefs animating the PBSC's response to post-Soviet *unipolarity*.

In contrast to the US 'global leadership' or *systemic primacy*, and the European Union's legal order, Beijing's 'Harmonious World Order' and 'New Civilisation Outlook' constructs challenged deterministically competitive Western and allied notions of interstate relations. The 'Harmonious World Concept' stripped interstate dynamics of elemental antagonism, stipulating that actors pursue peaceful inter-civilisational coexistence, each imbued with inalienable, inviolable, rights to choose its developmental path suited to its own circumstances.[41] Essentially demanding, no external impositions, and cooperation instead of competition.

A 'Harmonious World' would be peaceful, stable, open and tolerant, enabling equal prosperity and enduring peace. Diverse civilisations bearing distinct social organisation and developmental designs, sharing mutual trust and respect, communicated with and learnt from each other; states maintained peace and security by using 'just and efficient' security institutions to address international affairs with dialogue and negotiations on the basis of international law.[42] Resonating with the UN's founding principles and foundational aspirations, the 'New Civilisation Outlook' encouraged 'inter-civilisation dialogue' to build a harmonious world based on sovereign equality among states.

Responding to the West's elation at its 'victory' over Cold War adversaries and the apparent dead-end confronting polities lying outside the expanding capitalist-liberal-democratic circle, specifically post-Tiananmen Square China, these frameworks argued for legitimising normative equality between the victorious pluralist domain led by the USA and the residual authoritarian states, principally China itself. Without explicitly challenging the post-Soviet *unipolar* order, Beijing's mutually consistent paradigms demanded equal representation, adequate strategic space in which to function, and intra-*systemic* autonomous agency.

In a milieu reverberating with aftershocks of the USSR's collapse, strategic insecurity drove conceptual work. The Cold War's end lowered great power tensions, but China's national security environment brought 'new challenges': 'The vicious rise of the "Taiwan independence" forces, the technological gap resulting from RMA, the risks and challenges caused by…economic globalization, and the prolonged existence of unipolarity vis-à-vis multipolarity—all these will have a major impact on China's security'. Beijing was 'determined to safeguard its national sovereignty and security no matter how the international situation may evolve, and what difficulties it

---

[40] Hu (2004c) and Li (2005).
[41] Li (2005).
[42] Zhang (2006).

may encounter'.[43] The CMC ordered the PLA to 'maintain the fundamental principle and system of absolute Party leadership over the armed forces'.[44]

Assuming CMC Chairmanship three months before the MND published its December 2004 Defence White Paper, Hu Jintao expounded his Scientific Development Theory's national defence elements on Christmas Eve. Summarising the PLA's 'new historic mission' as a 'Three-Provides-and-One-Role' framework, he decreed that the PLA

- Provide an important guarantee of strength for the CPC to consolidate its ruling position
- Provide a strong security guarantee for safeguarding the period of important strategic opportunity for national development
- Provide powerful strategic support for safeguarding China's growing national interests, and
- Play an important role in safeguarding world peace and promoting common economic development[45]

The missions, codified in Beijing's 2006 Defence White Paper, and ratified by the CPC's 17th (2007) and 18th (2012) Congresses, underpinned the PLA's newly proclaimed responsibility to 'go global' in defending China's rapidly expanding economic interests.[46] Beijing's decision, following the December 2008 UNSC Resolution-1851, to protect Chinese shipping from pirates in the Gulf of Aden by despatching a naval flotilla and maintaining rotational deployments since reflected this shift.[47] China's launch of an aircraft carrier programme with a view to building a CSG-centric blue water fleet similarly indicated global aspirations.[48] Chinese rhetoric stressed developmental, economic and defensive impetus driving policy, but in its theoretical postulates and policy-praxis, Western critics saw a newly 'assertive' China.

Diplomacy flowered. In 1992, China inherited border disputes with Russia, Kazakhstan, Kyrgyzstan and Tajikistan. With Moscow coordinating, the five governments negotiated frontier disarmament and border demarcation. In April 1996, heads of the five governments, meeting in Shanghai, signed an 'Agreement on Deepening Military Trust in Border Regions', catalysing the 'Shanghai five' process and generating an 'Agreement on Reduction of Military Forces in Border Regions' in April 1997. These agreements, combined into a joint statement, proposed, 'on the principles of equality, trust, consultation and mutual benefit', the parties reduce military deployments, lowering threats to regional and national development.[49] China resolved border disputes with Kazakhstan in 1998, Kyrgyzstan in 1999, Russia in 2004 and Tajikistan in 2006.

---

[43] CMC (2004, Chap. 1).
[44] Ibid., Foreword.
[45] Hu (2004a).
[46] Editorial (2006), Huang (2009) and Cooper (2009).
[47] Lai et al. (2014).
[48] Ibid. pp. 6–7.
[49] Hali (2016).

3.4 The 'Shanghai Spirit' Paradigm

Peaceful negotiations encouraged cooperation across the economic–political security spectrum. Perceived marginalisation, even alienation, vis-à-vis the US-dominated post-bipolar order, stimulated Eurasian collaboration.

The Shanghai Five platform enabled cooperation against Beijing's 'three evils', i.e. terrorism, separatism and extremism, as well as drugs and human trafficking and weapons proliferation. In 1998, the five began joint action on counter-narcotics and narcotic finances, cutting off terrorist funding, thus lending the group an anti-terrorism focus. Consolidating cooperative experience, in June 2001, the 'Shanghai Five' signed the 'Shanghai Convention on Fighting Terrorism, Separatism and Extremism', proclaimed the Shanghai Cooperation Organisation (SCO) and welcomed Uzbekistan as a member. A year later, the six adopted the SCO Charter, establishing a Secretariat and ministerial committees to address shared concerns. Initially focused on intra-regional issues, the group gradually widened the remit of collective action.

Since 2002, when the SCO established the Regional Anti-Terrorist Structure as the group's focal institution, members held progressively larger and more complex drills, enabling shared learning and best practice adoption. The 'Peace Mission' drills evolved from counterterrorism exercises to more conventional combat-oriented combined arms and joint services manoeuvres. Peace Mission 2007, held in Siberia's Chelyabinsk region, marked a departure. It combined armed forces of all six member-states, with the PLA dispatching its first-ever training contingent overseas. It involved about 6000 troops, over 1000 combat vehicles and dozens of aircraft far from home 'fighting' week-long 'operations'.[50] All six SCO presidents, after concluding their annual summit in Bishkek, Kyrgyzstan, travelled to Chebarkul in Western Siberia, to review the drills' final stages. The symbolism reverberated powerfully.

The SCO also looked outward, especially at Afghanistan, the source of myriad concerns, inviting President Hamid Karzai to the 2004 annual summit. In 2005, the SCO and Kabul established a Beijing-based liaison body to be the official channel of communication on cross-border crime, drug trafficking and intelligence sharing. Afghanistan, the first SCO Observer State, was later joined by Belarus, Iran, India, Pakistan and Mongolia. Armenia, Azerbaijan, Cambodia, Nepal, Sri Lanka and Turkey became 'Dialogue Partners'. In March 2009, the SCO hosted an international conference on Afghanistan in Moscow, inviting external stakeholders, including the USA, NATO and the UN. At their June 2009 summit, SCO leaders pledged to work with other interested parties in addressing the Afghan conflict's fallout. In 2017, they welcomed India and Pakistan as full members. The SCO thus accreted influence in managing Eurasian security and economic cooperation.

Insecurity drove coherence. The Shanghai Five's July 2000 Dushanbe Statement underscored a determination to not only counter separatism or 'liberation movements', terrorism and extremism, but also 'oppose intervention in other countries' internal affairs' on the pretexts of 'humanitarianism' and 'protecting human rights' and support the efforts of one another 'in safeguarding the five countries' national

---

[50] Radyuhin (2016).

independence, sovereignty, territorial integrity and social stability.'[51] US observers noted the potential use China or Russia might make of the SCO to counteract the US post-Soviet pre-eminence. Some described Jiang Zemin's remarks: 'We should strengthen mutual support in safeguarding the national unity and sovereignty of our nations and resist all kinds of threat to the security of the region', and the SCO's opposition to the 'use of force or threat of force in international relations without the UNSC's prior approval and…any countries or group of countries' attempt to monopolise global and regional affairs out of selfish interests', as challenging the USA in Asia.[52]

US bases in Kyrgyzstan and Uzbekistan underscored close security ties, but the former's 'colour revolution' and the latter's 'Andijan incident' in 2005 transformed perceptions. That year's SCO summit urged Washington to specify its base withdrawal timeframe while Sino-Russian Peace Mission 2005 drills deepened collaboration. Two years later, Washington's top diplomat for Central Asia described the SCO as 'a subject that seems to make a lot of Americans' blood just boil'.[53] Events partly explained US outrage. Just before the Shanghai Five became the SCO, Sino-US relations plunged over the collision between a US EP-3E *Aries* ISR aircraft and a PLANAF J-8II interceptor 70 miles from China's Hainan Island. The J-8II crashed, killing the pilot. The damaged EP-3E made an 'unauthorised' landing at a PLA airfield on Hainan, its 24-strong crew being detained for 11 days. A bushfire of anti-US anger spreading across China triggered the George W. Bush Administration's first 'China-crisis'.[54]

Contradictory narratives mirroring conflicting interpretations of international maritime law exposed fundamental divergences in US and Chinese perspectives, claims and interests. USN and NSA investigations concluded that Beijing had obtained 'large volumes' of 'technical data' and 'sensitive COMINT (communications intelligence) equipment' from the damaged EP-3E.[55] The crisis, eventually diplomatically resolved, nonetheless exacerbated mutual mistrust. While US anger at Chinese 'assertiveness' was understandable, scrutiny suggested outrage was misplaced. Towards the end of Hu's tenure, with international relations gaining focus in Hu's second term, his platitudinous formulation, 'Major powers are the key; surrounding/peripheral regions are the first priority; developing countries are the foundation, and multilateral fora are the important stage',[56] reflected and explained Beijing's largely defensive Eurasian outreach and shaped the continuity in Chinese diplomacy under Hu's successors.

The 'Shanghai Spirit' manifest in the SCO's formation represented the first instance of China's 'New Security Concept' in practice. SCO members collaborated as partners, without accepting legally binding mutual obligations inherent in

---

[51] Report (2000a) and Gill and Oresman (2003).
[52] Gill (2001).
[53] Feigenbaum (2007).
[54] Kan et al. (2001) and Rosenthal (2001).
[55] USN/NSA (2001, p. v).
[56] Liu et al. (2011).

3.4 The 'Shanghai Spirit' Paradigm

an alliance. SCO members did not endorse Russia's invasion of Georgia in 2008, nor its annexation of Crimea in 2014. Unlike NATO members, who are obliged to treat an attack on one as an attack on all, SCO members took consensually negotiated decisions, each choosing which decisions to implement. This afforded the *Stans* formal equality to Russia and China, giving them a stake in collective approaches to shared concerns, without compromising their recently regained sovereign identities.[57] This loosely egalitarian collective security structure hastened trans-frontier trade-related linkages.

The low bar of mutual obligations limited the SCO's ability to act in unison, but by allowing cooperation among actors pursuing varied goals, the formulation stabilised much of the Eurasian landmass after a traumatic transition. Western analysts, ignoring the occasionally mutually defensive insecurity driving several SCO members, betrayed misplaced anxiety by asking if the SCO was becoming a 'Eurasian NATO' or even a rival to NATO.[58] After the group's 18th summit in Qingdao inducted India and Pakistan in June 2018, one observer averred, 'the SCO remains focused on relationship-building, rather than achieving concrete outcomes'.[59] Contrasting origins, motivations, purposes and frameworks undergirding the NATO alliance and the SCO collective belied the validity of western fears.

The SCO's one challenge to the US-led order resided in the possibility of alternatives it suggested to suddenly self-aware subaltern actors subsisting precariously on a landscape dominated by a *unipolar system* premised on values threatening those on which they were themselves organised. Authoritarian, underdeveloped, and eclectically mixing and matching varied combinations of public, private and foreign capital to grow, these states were outliers in the post-Soviet US-led order. Their ideational preferences, developmental needs, state-building priorities, accretive experience of cooperation and response to persistently perceived external challenges, helped to crystallise a collective Eurasian mindset vital to imagining and then sustaining the realisation of the future BRI/SREB blueprint.

In one of his final formal diplomatic tasks in hosting Central Asian and other regional delegations, Wen Jiabao noted: 'Over 2000 years ago, our ancestors…opened the Silk Road that connected both ends of the Eurasian continent and served as a bridge for interactions between the East and the West…The ancient Silk Road has regained its past vigor and vitality…Today, we should draw strength from our historical heritage and, with greater confidence and in a pioneering spirit, work together for new glory of the Silk Road and a better future of the Eurasian people'.[60] A year later, Xi Jinping proposed the SREB in Kazakhstan's capital, Astana: 'In order to make the economic ties closer, mutual cooperation deeper and space of development broader between the Eurasian countries, we can innovate the mode of

---

[57] Maksutov (2008) and Xing and Sun (2007).
[58] Fillingham (2009) and Darling (2015).
[59] Report (2018f).
[60] Wen (2012).

cooperation and jointly build the 'Silk Road Economic Belt' step by step to gradually form overall regional cooperation'.[61]

Continuity notwithstanding, the vision would have remained a dream without strong endorsement and active participation of Russia and Kazakhstan. Russia's key role has been analysed elsewhere.[62] The changing trajectory of Central Asia's trade relations underscored the pattern. By 2016, Russia's $18bn commerce with the five *Stans* was dwarfed by China's $30bn.[63] Investments followed a similar pattern. Kazakhstan, by far the largest Central Asian economy, reflected the shift in the China–Central Asia–Russia dynamic. During 2013–2018, Astana and Beijing agreed on the construction or completion of 127 BRI projects worth $67bn, although implementation was uneven. The most visible ones included a $1.9bn light railway system in Astana, and a $600m, five-year transport infrastructure plan covering Kazakhstan's other areas. In 2018, Kazakhstan's debt to China crossed $12.3bn.[64]

Another impetus to Beijing's westward drive came from the vulnerability of the explosive expansion of its maritime trade, especially growing energy imports powering its economy, to interdiction by US and allied fleets deployed along Southeast Asian choke points, i.e. China's 'Malacca Dilemma'. Dependence on shipping through the Malacca and Lombok/Makassar straits explained Beijing's concerns. The shortest link between the Indian and Pacific Oceans, stretching north from Singapore between the Indonesian island of Sumatra and the Malay Peninsula, the Malacca Strait was one of the world's busiest 'choke points'. The deeper Lombok/Makassar Strait between Indonesian islands of Lombok and Bali, used by Very Large Crude Carriers, too, mattered to China's crucial energy supplies.

Growth enlarged China's 'national security interests' from conventionally defined territorial integrity to include the maritime domain, space and the electromagnetic spectrum.[65] Middle Eastern supplies comprised 60% of China's crude oil imports in 2006, then projected to rise to 75% by 2015. In fact, this ratio reached 80% by 2012.[66] Despite slowing growth, China imported 440m metric tons of crude and 125bcm of gas, marking year-on-year increases of 11% and 31.7%, respectively, in 2018, with import-dependence of 69.8% and 45.3%, respectively.[67] Beijing feared that in a crisis, hostile US and allied naval fleets, controlling the choke points, could cut off energy supplies to strangulate China, holding its economy to ransom. Beijing lacked the ability to defend its vital sea lines of communications (SLoC), but boosting its forces appropriately could trigger hostile pre-emption; hence the 'dilemma'.

In November 2003, noting that 'certain major powers' were bent upon controlling the Malacca Strait, potentially threatening China's strategic well-being, Hu Jintao asked the party state to prepare imaginative countermeasures to this, in Chinese eyes

---

[61] Xi (2013).

[62] Majumdar (2018), Stronski (2018), Ali (2017b), Wishnick (2017) and Goldstein (2017).

[63] Stronski (2018).

[64] Kurmanov (2018).

[65] General Political Department (2006).

[66] Storey (2006) and Pottter (2012).

[67] Zheng (2019).

3.4 The 'Shanghai Spirit' Paradigm

very real, challenge.[68] Some US analysts acknowledged China's defensive concerns driving its efforts to forge energy connectivity with Central Asia–Eurasia, but insisted China's premise and response were 'misguided'.[69] US critique neither allayed Beijing's anxiety nor slowed the fashioning of alternative supply routes via Russia, Central Asia, Myanmar, Pakistan and possibly Iran, designed to mitigate the 'dilemma'. Economic insecurity drove China's expanding engagement with East Asian, Southeast Asian, Central Asian, South Asian and West Asian—i.e. Eurasian, neighbours. Developmental motivations reinforced a broader 'going out' policy-outlook.

3.5 Beijing's Regional Developmental Drivers

As 2018 ended, the NDRC and line ministries with economic charters outlined 2019 plans at work conferences. They pledged to 'promote western region development, northeastern region revitalisation, the central region's rise and support the eastern region' in pursuing 'optimal development'.[70] Regionally focused perspectives drove BRI, specifically, SREB's evolution. Early in the reform era, labour and capital-intensive foreign-invested assembly and production, growing trade from coastal hubs and investment in scientific–technological education and skills catalysed rapidly rising disposable incomes. Beijing's WTO accession and explosive trade growth accumulated surpluses, raising savings-based investment. Regulatory reforms nurtured entrepreneurial middle classes, concentrated wealth and deepened inequality. Expanding commercial interests caused disruptions.[71] Fearful of inter-regional disparities, in 1999, the Ministry of State Planning and Development drafted a 'Go West'/'Open Up the West'/'poverty-reduction' campaign.

In the 1980s–1990s, China attracted FDI worth over $300bn, fuelling its manufacturing prowess, but the central and western hinterland only received $9.9bn, of which $6.8bn went to just Sichuan. In contrast, Shanghai's Pudong District alone received $11bn.[72] Early in the twenty-first century, 'communist' China stood internally transformed and socio-economically divided.[73] CPC leaders responded with a 'Develop the West' campaign in 2000, pushing investments into Gansu, Guangxi, Guizhou, Inner Mongolia, Ningxia, Qinghai, Shanxi, Sichuan, Tibet, Xinjiang and Yunnan Provinces/autonomous regions, and the province-sized Chongqing Municipality. The focus-area comprised two-thirds of China's territory, and 22.8% of its population. The western region shared 3500 km of land borders with a dozen neighbouring states. Plans to improve transport and communications links between coastal

---

[68] Hu (2004b).
[69] Shea (2014).
[70] Report (2018e).
[71] Haltmaier (2013), Parietti (2016) and Morrison (2017).
[72] Report (2001a, pp. 1–2).
[73] Kujis (2008).

China and the west, and the latter and neighbouring economies, promised to transform the region into a 'second golden area' for reform and opening-up efforts.[74]

In mid-September 2000, Jiang Zemin chaired a PBSC/CPCCC meeting attended by China's non-communist parties,[75] the Federation of Industry and Commerce and several 'eminent persons', presumably including retired leaders. He explained Beijing's new Western Development Strategy. Targeting the gulf separating China's coastal belt and its hinterland and the threats this posed to national cohesion, Jiang cautioned those present that developing the west demanded 'generations of strenuous efforts'. He warned against 'any short-term behaviour'. Beijing would drive the process and expected the participants' engagement,[76] but implementation required financial, technical and scientific support from other sources. To attract foreign investment into the region, China announced a preferential tax regime, demanding only 10–15% after the initial tax-exempted period expired. Provincial governments' project approval authority was raised to $30m, the same as coastal counterparts.

Jiang acknowledged tensions rocking China's developmental experience in his awkwardly translated 'Three Represents' theory in 2000, explaining it on the CPC's 80th anniversary in 2001. His postulate was incorporated into the CPC Constitution at the 16th Congress which transferred party leadership to Hu in November 2002. Jiang recorded 'the development trends of advanced production forces', 'the orientations of an advanced culture' and 'the fundamental interests of the overwhelming majority of the people of China'.[77] He sought to reconcile the nascence of a proto-*bourgeoisie*, aspirations of the emerging middle-classes and continuing CPC support for the toiling masses whose labours sustained growth. Jiang's formulation reflected concerns over imbalances between a rapidly urbanising, prosperous and globally connected coastal belt and China's poor, rural and neglected central and western hinterland.

In March 2001, Premier Zhu Rongji told the 4th Session of the 9th NPC that to reduce disparities, Beijing sought the 'development of science and technology' in the western region, devoting 70% of its FY2000 budget, $396bn, to infrastructure construction there. In 2001–2002, another $45.5bn helped to speed up western development: local governments received $37.1bn in subsidies; $8.4bn went into building infrastructure. Several special economic and technical development zones, piloted by ones in Xian, Shaanxi Province and Yibin, Sichuan Province, were planned. Over the next decade, the region would add more than 2500 miles of rail tracks and 20,000 miles of motorways.[78] To attract foreign investors, the 2001 China International Fair for Investment and Trade (CIFIT), held every September in Fujian Province, launched

---

[74] State Council (undated-a) and Ding and Neilson (2003).

[75] China Revolutionary Committee of the KMT, China Democratic League, China Democratic National Construction Association, China Association for the Promotion of Democracy, Chinese Peasants and Workers' Democratic Party, China Zhi Gong Dang, Jiusan Society, and Taiwan Democratic Self-government League.

[76] Report (2000c).

[77] State Council (undated-b).

[78] Zhu (2001) and Report (2001b).

10,000 projects targeting western development. Regional entities signed more than 200 contracts worth $4.3bn with foreign investors.[79]

Although Beijing planned to mobilise resources and channel managerial and planning skills towards hinterland urbanisation and development, given the coastal belt's experience, environmental protection acquired salience. While seeking eastern provinces' investment support towards western industrialisation and infrastructure building, Zhu warned coastal enterprises against transferring their obsolescent, energy-intensive and polluting operations to the hinterland. To obviate fresh imbalances, western farmland unsuited to agriculture would be reafforested, grasslands protected from over-grazing and scarce water resources afforded improved conservation management.[80] The region received significant leadership attention, but implementation proved challenging.

Western critics questioned Beijing's 'sincerity' in promoting western development: 'the campaign is an excuse for China to exploit the vast natural resources of the western region, while imposing strict security measures that will counter any attempts by ethnic separatists to break away from the PRC'. Although few countries tolerated separatism at home, China's conduct was considered problematic. Still, foreign governments and corporations showed interest in the strategy, seeking benefits in energy, petro-chemicals, transportation and telecommunications sectors, but critics feared, 'building physical infrastructure will attract large numbers of ethnic Chinese and weaken many of the unique minority cultures in the west'.[81] This perspective found Beijing's efforts to integrate underdeveloped regions with developed ones, raise incomes and deepen societal cohesion across ethnicities, disingenuous.

State consolidation may have informed Beijing's Western Development Strategy, but rising inequality and its socio-political consequences did deepen elite anxiety. Deng was aware of income variations inherent in capitalist modes of production resulting from varied levels of risks, uncertainty and individual responses, particularly in societies used to the levelling comforts of statist, egalitarian, distributive regimes,[82] but believed distribution could follow accumulation. Western critique rested on GDP and GDP growth *per capita* in the rapidly marketising China. However, *per capita* calculations ignored distortions caused by China's *hukou* residential registration regime. Millions of marginal farmers and unemployed youth, registered residents of underdeveloped regions, flocked to coastal cities looking for jobs, providing the labour which gave China comparative advantages in export manufacturing, built shiny new districts enabling coastal urbanisation and generated personal/family income and national wealth. But *hukou*-regulations, skewing distributive data, injected inaccuracies.[83]

Hu Jintao and Wen Jiabao, succeeding Jiang Zemin and Zhu Rongji in November 2002, built on their policy-inheritance. In 2003, Wen launched a 'Revitalise

---

[79]CIFIT Secretariat (2018).
[80]Report (2002).
[81]Report (2001a, p. i).
[82]Li et al. (2012b) and Drysdale (2012).
[83]Gibson (2012) and Plummer (2012).

the Old Northeast Industrial Bases' plan covering Liaoning, Jilin and Heilongjiang Provinces and five eastern prefectures of Inner Mongolia. In March 2004, Wen proclaimed before the NPC the 'Rise of Central China Plan', focusing resources on hinterland areas closest to the coastal belt, considered better able to absorb investment: 'Accelerating the development of Central China is an important aspect of our endeavours to ensure well-balanced development of regional economies'.[84] The plan targeted Anhui, Henan, Hubei, Hunan, Jiangxi and Shanxi Provinces. Together, the strategy and the plan would mitigate well analysed intra-PRC inequality. The NDRC, in drafting the 11th FYP (2006–2010), adopted Hu's theoretical approach to building China's future: scientific development. By applying advanced science and technology, this approach would help to achieve high-quality, highly efficient and sustainable development, securing the goal, i.e. a harmonious society. The NDRC set new priorities to focus leadership efforts and resources on developing the Yangtze River Delta, Beijing-Tianjin-Hebei, 'the old industrial base' in northeast China, and Chengdu-Chongqing. The latter belt would power the central region, enabling it to assist western development.[85]

Concern over inter-regional disharmony informed policy-discourse, shaping the FYP drafting process. Over 1978–2008, China's GDP increased nearly 17 times, and *per capita* GDP 12-fold. Outside a handful of coastal provinces, however, growth faltered. The average income in Central and Western China reached 77% of the national average in 1999, inching up to just 80% in 2008.[86] Iniquity may not have been the principal trigger behind rapidly rising 'mass group incidents'—large-scale protests, i.e. sit-ins, strikes, group petitions, rallies, demonstrations, marches, traffic blocking and building seizures, public melees, rioting and inter-ethnic strife, but central planners considered these a key contributor. Official figures showed 'mass incidents' rose from 8700 in 1993 to 32,000 in 1999, 50,000 in 2002, surpassing 58,000 in 2003, 90,000 in 2006 and an unconfirmed 127,000 in 2008, when Beijing stopped publishing strife data.[87]

Opacity over the incidents' scale, nature and triggers rendered causal connections conjectural. Economic triggers cut across ethnic divisions, and ethnic violence was rarely amenable to economic solutions. One study of certain categories of 'mass incidents' rocking China during 2003–2009, for instance, found that the highest frequency, 54, occurred in Guangdong, one of China's most prosperous provinces, and the lowest, 1, occurred in Jilin, Qinghai and Tibet, the last two among the poorest.[88] Societies experiencing rapid change, with dramatic asset growth for some, juxtaposed with slower accumulation or little for many, via a sharp transition from state-protected equality to state-authorised uneven protection, often betrayed angry despair.[89] China's single-party rule with loosened economic levers contrasting with

---

[84] Wen (2004).
[85] Ma (2006) and Information Office (2005).
[86] Huang et al. (2010) and Dunford and Bonschab (2013).
[87] Tanner (2005) and Tong and Lei (2010, pp. 23–33).
[88] Tong and Lei (2010, p. 25).
[89] Sekiyama (2016).

strong political control generated difficult to manage friction. The CPC's authoritarian political imperatives left only economic measures with which to address mass outrage.

3.6 In the *Great Recession*'s Wake

Beijing responded to the US subprime financial crisis wreaking global havoc with a RMB 4tn ($586bn) stimulus package designed to boost infrastructure building, social spending equivalent to 2–3% of GDP and credit availability, to encourage private consumption. As GDP growth averaged 9% during 2008–2011, the *Great Recession*'s peak, which either stalled or catalysed recession in most major economies, China's response was considered successful.[90] However, while China's GDP expanded by 14.2% in 2007, the stimulus enabled it to grow by only 9.6% in 2008 and 9.2% in 2009, indicating much pain.[91] The optimism driving China's stock markets in 2005–2007 dissipated. FDI flows fell by 15% to $121.68bn in 2008 and by 42% to $70.32bn in 2009, recovering to $124.93bn in 2010.[92] By early 2009, more than 20m migrant workers, 15.3% of the 130m who powered coastal construction and manufacturing, had lost work. Non-migrant job losses, too, rose; officials warned of risks of social unrest.[93]

CPC leaders learnt from the experience: China needed to 'be prepared for extreme cases' of exogenous shocks while striving for 'the best possible results' by dealing with 'abrupt shocks and impacts', and making 'long-term preparations for structural changes resulting from the crisis'; China must 'understand the changed implications' of its strategic opportunities to seek 'the widest intersection between Chinese and global interests' apparent in 'the domestic market driving a world economic recovery, acquisition of technologies of developed countries and investment in infrastructure'; and finally, China must 'concentrate on our own affairs', avoid entanglements in conflicts, focusing instead 'on really important matters to substantially improve our domestic conditions'.[94] These lessons guided China's recovery efforts, including in Western China. There were then few indications of any global ambitions driving Beijing.

China's Western Development Strategy transformed the region's economy, but did not heal gulfs separating the Han majority from minority nationalities. Differences deeply affected Lamaist-Buddhist Tibetans in Tibet, Qinghai, Gansu, Sichuan and Yunnan, and Uighur Muslims in Xinjiang. Mutually reinforcing ethno-cultural, political and economic tensions flowing from unresolved grievances exploded in Tibet in 2008 and Xinjiang in 2009. Security forces' 'strike hard' operations, with the PLA

---

[90] Yi and Jing (2017) and Cunha et al. (2012).
[91] Li et al. (2012a).
[92] Ibid.
[93] Yi and Jing (2017), Giles et al. (2012) and Anderlini and Dyer (2009).
[94] Liu (2014).

providing a 'forceful guarantee', restored control, but cost lives, limbs, assets and confidence. Beijing swiftly incorporated Gansu, Qinghai, Sichuan and Yunnan into its Tibet Development Plan. To drive Xinjiang's growth, the Muslim majority trade centre of Kashgar/Kashi was designated a Special Economic Zone (SEZ). China also engaged with Turkey, the Turkic-speaking Uighurs 'cultural homeland', attenuating any tacit Turkish endorsement of Uighur irredentism. Tibetans, with their iconic leader, the Dalai Lama, still symbolising 'splittism' from India, seemed less persuaded.[95]

Presiding over a PBSC meeting in late May 2010, Hu Jintao assigned greater import to Western Development Strategy implementation. The PBSC noted, 'the Western part of the nation is strategically important and there should be special policies to support the region'. Support should be 'increased with greater resolution and force'. In addition to infrastructure building, water conservancy and natural resources-based industrialisation, now education, technological innovation, health care, employment, social security and cultural development gained salience. The PBSC also ordered developing 'old revolutionary bases in poor, ethnic and border regions', as 'ethnic and religious harmony is important to the region's stability'.[96]

Five weeks later, marking the Strategy's 10th anniversary, Hu and Wen addressed heads of implementing ministries and agencies at a 'work meeting on developing Western China'. Although much had been done, Hu worried that 'the development gap between eastern and western regions remains broad, and the west is still a conundrum for the country's endeavour in building an overall moderately prosperous society'. Slamming poor execution, Wen urged 'serious and good implementation of all the policies and measures' adopted by Beijing. He asked that the region's resources be utilised to develop 'modern' and environmentally friendly industries, while advancing agricultural and tertiary sectors. He pledged 'vigorous efforts in social development' enabling western compatriots to 'share in the fruits of national economic progress'. Wen vowed to focus Beijing's poverty alleviation efforts on southern Xinjiang, eastern Qinghai–Tibet Plateau and Western Yunnan Province.[97]

Given the strategy's rising profile in elite calculus, urgency flowing from the Great Recession's impact and growing social disharmony, Beijing invited neighbouring states and potential overseas partners to help develop Western China and fashion terrestrial connectivity binding Central and Western China, Eurasia and Europe. Leadership-level exchanges began in October 2009 in Chengdu, at the 10th Western China International Economy and Trade Fair and the 2nd Western China International Cooperation Forum. Hosting the summit, Wen welcomed the Prime Ministers of East Timor, Cambodia, Cote d'Ivoire, Kenya, Sri Lanka and Vietnam.[98] Discussions boosted ties and addressed concerns precipitated by deepening Chinese economic

---

[95] Shan and Clarke (2011).
[96] Xinhua (2010).
[97] Report (2010b).
[98] Xanana Gusmao, Hun Sen, Guillaume Soro, Raila Odinga, Ratnasiri Wickramanayake, and Nguyen Tan Dung.

engagement. Wen's message: Beijing encouraged 'the Western Region to expand exchange and cooperation with these countries'.[99]

Wen briefed foreign leaders on the Strategy's goals: 'changing the backward outlook of Western China, building a moderately prosperous society in all respects, improving national land development pattern, promoting balanced regional development, expanding the scope of opening up both internally and externally, and fostering new economic growth areas'. In the first decade, Beijing provided over RMB 3.5tn in 'transfer payment and special subsidies', earmarking over RMB 730bn for the region's 'construction projects'. In 2000–2008, the west's GDP grew at an annual average of 11.7% from RMB 1.66tn to RMB 5.82tn. Revenues rose at 19.6%, and fixed asset investment, by 22.9%, exceeding national averages. During 1999–2008, urban and rural per capita disposable income grew by 105% and 74%, respectively, with 9.54m rural poor crossing the poverty threshold. Western China's highway mileage increased by over 888,000 km, and railway track length, by over 8,000 km. Regional airports now numbered 79, nearly half of China's total. Landmark projects like the Qinghai–Tibet railway, west to east gas transmission line and west to east electricity transmission grid, became operational.

Major water conservancy, energy and telecommunications projects progressed. Nearly 200,000 eastern Chinese businesses invested over RMB 2.2tn in the west which, in 2008, absorbed FDI worth $6.62bn while its foreign trade reached $106.8bn.[100] Despite the devastatingly coincidental Sichuan earthquake and the *Great Recession* in 2008, with Beijing devoting 43% of its stimulus to the region, it began recovering. In the first half of 2009, regional GDP grew by 11.8%, urban fixed asset investment by 42.1%, consumer goods sales by 19% and local government revenues by 12.8%, all above national figures. For the Strategy's 2nd decade, Wen made six promises. Two of these underscored Beijing's vision of Western China becoming a nexus of dense Sino-Eurasian connectivity channels:

- Build major transport hubs and passage ways that connect China with the rest of the world; speed up the construction of modern infrastructure
- Deepen reform and opening-up to expand exchanges and cooperation between the region and the rest of China, and the world at large, at a higher level.

Encouraging his guests, Wen outlined recent and planned progress: The China-ASEAN FTA would be completed in 2010; trade cooperation among SCO members was increasing; Beijing would 'actively push forward' Lancang-Mekong sub-regional cooperation, ensure success of the China-ASEAN Expo in Nanning, the Western China International Economy and Trade Fair in Chengdu, and the China–South Asia Business Forum in Kunming. Wen pledged to finance a China-ASEAN Fund on Investment Cooperation, provide credit support to SCO member-states and develop an East Asian foreign exchange reserve pool. Stressing Beijing's commitment to fashioning physical and normative connectivity linking Western China

---

[99]Report (2009b).
[100]Wen (2009).

and wider Eurasia, Wen urged the region's deeper engagement with neighbouring countries in:

- Energy and transportation
- Trade and investment
- Energy conservation and environmental protection and
- Regional and international affairs[101]

Foreign leaders, seeking national economic gains, responded to China's transregional aspirations. Thereafter, partner states sent high-level delegations to these annual gatherings, reviewing progress, refining plans and expanding networks of collective, connective, action.

China's 11th FYP laid 8,000 km of new rail tracks and 365,000 km of new motorways across the west, helping to boost annual growth rates to 13.6%.[102] The region needed to sustain the pace for five years, and Central Asia, especially Kazakhstan, promised a key role, as Hu Jintao stressed during another visit: 'Recognizing the huge potential in bilateral economic and trade cooperation, we have set a target of $40bn for bilateral trade by 2015'. Energy cooperation being 'very important to the development of Sino-Kazakh relations, we agreed to continue expanding and deepening energy cooperation. To reinforce cooperation on oil and gas, we will ensure the smooth construction of several oil and gas pipelines'.[103]

New motorways, expanded border trade facilities and reduced regulatory bureaucratic impediments eased economic integration. The process would mature by the end of China's 12th FYP (2011–2015). The NDRC, leading the FYP drafting effort, stressed substituting export dependence with domestic consumption, accelerating service sector growth and establishing capacity quotas for 'new energy' sources, i.e. nuclear, hydro, solar and wind. Abandoning past emphases on infrastructure-building and industrialisation in the strategy's previous iterations, planners now sought 'more balanced growth' between developed and poorer regions.[104]

The State Council approved the NDRC's draft 12th FYP for further promoting the economy of the Western Regions, i.e. Xinjiang, Tibet, Inner Mongolia, Guangxi, Ningxia, Gansu, Qinghai, Sichuan, Shaanxi, Guizhou, Yunnan and Chongqing, in February 2011. The plan emphasised building 'key economic regions', promoting agricultural production of 'main areas', sustainable development of key 'ecological zones', intensive development of resource-rich areas, hastening completion of 'border development zones' and 'leap-forward development' of areas with 'special difficulties'. Noting, 'infrastructure construction is the main guarantee to the western development', planners organised the region into 11 zones[105]:

- Chengdu-Chongqing
- Guanzhing–Tianshui

---

[101] Report (2009a).

[102] Dezan (2012).

[103] Hu and Nazarbayev (2011).

[104] Report (2010a).

[105] Liu (2011) and Report (2011).

3.6 In the *Great Recession*'s Wake

- Beihai Gulf
- Hohhot–Baotou–Yinchuan–Yulin
- Lanzhou–Xining–Geermu
- North Tianshan
- Central Yunnan
- Central Guizhou
- Yellow River valley in Ningxia
- Central and Southern Tibet
- Shaanxi–Gansu–Ningxia.

The Chengdu–Chongqing economic zone, comprising Chongqing Municipality, 15 cities and 31 districts and counties in Sichuan Province covering 200,000 square km, received top billing. With thriving auto and machinery plants, science and technology hubs and defence factories, this was already a leading industrial base. By 2015, the plan envisaged it becoming Western China's key economic centre and, by 2020, one of China's 'strongest comprehensive regions', to join the Bohai Bay, Yangtze River Delta and the Pearl River Delta as the 'fourth pole' of China's economy. The plan integrated these 'poles' via new Beijing–Kunming and Shanghai–Chengdu transport and communication corridors; 15,000 km of new highways would link Inner Mongolia, Xinjiang and Qinghai; urbanisation would top 45%.[106]

Western China, as a connectivity hub, would link Northeast Asia, Central Asia, Southeast Asia and South Asia. New railway lines joining Chongqing, Lanzhou, Urumqi, Guilin and Guangzhou would knit Xinjiang closer to eastern China, enabling Chengdu, Xian, Lanzhou, Urumqi and Kunming to 'serve as major container stations for goods criss-crossing their way across Asia'.[107] Phase II of the China–Kazakhstan crude oil pipeline, the Dushanzi–Urumqi pipeline, the Jiaopiao–Ruili–Kunming section of the China–Myanmar crude oil pipeline and sub-sections of the Kunming–Chongqing pipeline would either be completed or brought near completion. Provinces would industrialise by processing mineral and other resources for producing energy, petrochemicals and fertilisers. By 2015, Western China would drive Beijing's terrestrial economic outreach.

Emphasising China's drive to forge trans-continental economic/infrastructural connectivity along a new Silk Road, Wen Jiabao framed Beijing's regional developmental goals in Eurasian terms. Welcoming the Presidents of Kyrgyzstan and Maldives, Prime Ministers of Cambodia, Kazakhstan and Tajikistan,[108] over a thousand Eurasian officials and business executives, and multilateral organisations attending the 2012 China–Eurasia Development and Cooperation Forum in Urumqi, Xinjiang, Wen recounted Silk Road history. Reviewing the contemporary landscape, he enthused, 'As world multipolarity and economic globalization gather momentum, Eurasian countries have strived to enhance themselves through unity and achieved faster development. Their mutual political trust has deepened, trade and investment have rapidly expanded, regional and subregional cooperation has flourished'.

---

[106] Dezan (2012).
[107] Dezan (2012).
[108] Almazbek Atambayev, Mohammed Waheed, Hun Sen, Karim Massimov and Akil Akilov.

Over the past decade, China's trade with Central Asia, West Asia and South Asia had surged annually by 30.8%, growing from $25.4bn to over $370bn. Chinese firms, signing project contracts worth $470bn, had invested nearly $250bn in Eurasian economies.[109] Wen noted, the financial crisis had inflicted much pain, and 'profound and complex historic changes' presented Eurasia with 'both rare opportunities of development and cooperation and many severe challenges'. China and its partners were 'speeding up' construction of the 'grand Eurasia passage'. The China–Kazakhstan oil and gas pipelines were operational; a second China–Kazakh railway link was ready; the China–Kyrgyzstan–Uzbekistan motorway was nearly so. 'A connectivity network consisting of roads, railways, air flights, communications and oil and gas pipelines is taking shape'.[110] He proposed a four-step process:

- Enhance political trust to safeguard regional peace and stability
- Open markets wider to promote common development
- Advance cooperation in cross-border infrastructure to accelerate connectivity growth
- Deepen cultural and people to people exchanges to enhance friendship among societies.

Wen drew attention to Xinjiang's locus within the new Silk Road and its rapidly improving developmental status as an exemplar of the 'glory of the Silk Road' that could boost 'the vision energising the new Eurasia for the benefit of its peoples'. This was the baton Wen Jiabao and Hu Jintao handed to Xi Jinping and Li Keqiang two months later.

3.7 Xi Jinping's 'New Era' of Reform and Opening-up

In November 2012, CPC General Secretary Xi Jinping and a new PBSC with Li Keqiang as Premier of the State Council, assumed leadership. In his farewell address, Hu underscored his theoretical legacy, the Scientific Outlook on Development, proposed to the 16th CPCCC in 2003, and which was written into the CPC Constitution by the 17th National Congress in 2007. Hu noted the framework had been informed by China's rapid economic growth generating 'a series of problems including excessive consumption of resources, serious environmental pollution and a widening gap between the rich and poor'.[111] China must address these with a scientific approach bearing 'major immediate significance and far-reaching historical significance for upholding and developing socialism with Chinese characteristics', a recurrent refrain colouring China's political–economic discourse.

Hu reported that notwithstanding considerable progress made, 'much room for improvement in our work' and 'a lot of difficulties and problems on our road ahead'

---

[109] Wen (2012).
[110] Wen (2012).
[111] Hu (2012).

3.7 Xi Jinping's 'New Era' of Reform and Opening-up 95

remained. 'Scientific development' demanded 'freeing up the mind, seeking truth from facts, keeping up with the times and being realistic and pragmatic'. This would enable the CPC to 'boldly engage in practice and make changes and innovations, respond to the call of the times, follow the aspirations of the people, ensure that the party is always full of vigour and that China always has the driving force for development to open a bright future for developing socialism with Chinese characteristics through the creative practices of the party and the people'.[112] Hu's legacy was the inheritance on which Xi would erect his own policy-edifice.

Xi spent a year reenergising the CPC and state organs still recovering from the Bo Xilai crisis. Bo, CPC Secretary of Chongqing, a member of the Politburo, a rising star and, like Xi, a 'princeling', was stripped of all his titles, tried, convicted and sentenced to prison after his sacked police chief, Wang Lijun, sought refuge at the US Consulate in Chengdu in February 2012. Wang was taken into custody after a night spent at the Consulate. His testimony and further investigations led to the arrest of Bo and his wife, Gu Kailai, the latter in connection with the late 2011 murder of a British business associate, Neil Heywood. After confessing, Gu received a life term. Bo, convicted of abuse of power, bribery and corruption,[113] too, received a life sentence. This shocking fall of a CPC leader traumatised cadres both ideologically and functionally as Bo associates, seen as more 'red' than was acceptable, began toppling over.[114]

The NDRC, fountainhead of Chinese economic planning, had a busy year as Xi and Li took charge. Since 2010, NDRC economists had worked with WBG colleagues, examining the trajectory of China's post-1978 'reform and opening-up' progress. Diagnosing looming challenges, they recommended further reforms. This NDRC-WBG report, 'China 2030: Building a Modern, Harmonious, and Creative Society', launched in 2013, charted the pathway guiding Beijing's economic future.[115] The framework, enabling China to overcome myriad pitfalls precipitated by three decades of breakneck growth, directed Beijing's economic vectors towards its two centenary goals. Many major initiatives that Xi Jinping and Li Keqiang proclaimed in their first term, flowed from this seminal study.

Other NDRC economists spent 2013, the year Xi proposed his SREB/MSR visions, reviewing implementation of the Western Development Strategy, reporting in early 2014. Twelfth FYP goals, e.g. faster than national average GDP growth and residential income growth, were being met. However, railway construction, water resources management, and energy-saving and ecological protection targets were not. The NDRC refined the plan stressing infrastructure building, urbanisation and environmental protection. Beijing would 'push forward the construction of railway, highway and water projects', while also addressing transport and water supply issues. Railway bottlenecks prevented mineral-rich western provinces from shipping goods to 'resource-thirsty regions'. In 2013, construction began on 20 'major projects' with

---

[112] Hu (2012).
[113] Choi (2018).
[114] Wishik (2012), Osnos (2012), Potter (2012) and Liu (2012).
[115] Liu et al. (2013).

$53.87bn invested in new highways, airports, hydro-electric plants and wind-power stations.[116] In 2014, NDRC planned to open up Chongqing, Chengdu, Xian and Nanning, and 'step up efforts' to build SREB linkages binding 24 cities in eight countries.

By August 2015, when Xiamen in coastal Fujian began freight railway services to Lodz, Poland, 12 Chinese cities were connected by train to Europe. In March 2017, the EU–China Smart and Secure Trade Lane (SSTL) facility, enabling swift customs inspection of goods shipped by sea, was extended to China–Europe railway services. In April, Xiamen launched direct freight rail services to Moscow, thus becoming the first Chinese coastal hub to link up with Europe via both maritime and terrestrial routes. In January 2018, Chinese and Kazakh railways opened a freight service from Urumqi, Xinjiang, to Baku, Azerbaijan, where containers were loaded on to ships heading to Europe, slashing journey time. In 2017, China sent over 3000 trains to Europe;[117] in 2018, China–Europe freight trains made 6363 trips between 59 Chinese cities with 49 destinations in 15 European countries.[118]

Efforts begun years earlier now knit China and Europe via Central Asia.[119] This resulted from persistent, accretive, evolving, elite-driven planning and policy-execution while *Zhongnanhai* changed hands. Twelfth FYP programmes, especially those in Central and Western China, shaped these outcomes. Li Keqiang reported that in 2000–2016, Beijing invested $914bn (RMB 6.35tn) in '300 major projects mostly in infrastructure and energy', across Western China. RMB 743.8bn went into 30 major regional projects in 2016, the first year of the 13th FYP.[120] Li urged rapid growth of big data, cloud computing and 'sharing economy' as regional development drivers, inviting foreign investment and pledging to 'integrate big data and the Internet with manufacturing', upgrade traditional industry and protect IPR.[121] Results became apparent. Chongqing and Guizhou, growing at 10.7% and 10.5%, respectively, led 28 provinces, province-level municipalities and autonomous regions, in GDP growth during January–September 2016. Chongqing built automotive, 'strategic emerging manufacturing' and high-value high-tech industries; Guizhou promoted 'supply-side structural reform and speeded up the shift from old growth engines to new ones'.[122] Their success energised region-wide efforts.

The NDRC's 13th FYP Western Development Plan (2016–2020), submitted to the State Council in 2016, was approved on 24 January 2017. The final version stressed basic infrastructure, ecological protection, coordinated urbanisation, industrialisation, 'informatisation' and agricultural modernisation. Improving the people's lives, social harmony, border security and ethnic unity, too, received attention. Implementation needed to cohere with 13th FYP execution as well as BRI and the Yangtze River

---

[116] Chen (2014).
[117] Xinhua (2018d).
[118] Report (2019c) and Song (2018).
[119] Yang (2019) and Xinhua (2017b).
[120] Xinhua (2016a) and Li et al. (2016).
[121] Xinhua (2016b).
[122] Information Office (2016).

3.7 Xi Jinping's 'New Era' of Reform and Opening-up

economic belt development plans.[123] Against this backdrop, Xi Jinping invited the world to the May 2017 inaugural BRF. Despite progress in linking Western China to rest of the country, Eurasia and Europe, and plans to build it up over the next 18 months, the scale of the task became apparent in an August 2018 review.

Premier Li, head of the State Council's Leading Group promoting western development, while noting 'enormous achievements' in the region's socio-economic progress, noted large gaps demanding greater efforts by both central and regional authorities, and increased infusion of private capital. Major railway links, e.g. Sichuan–Tibet and Chongqing–Yunnan lines, and 'key water diversion projects' in Yunnan, Qinghai, Gansu and Guangxi Zhuang, needed faster progress. The development of the Internet for enhancing online retail and wholesale trade, for the extension of education and healthcare facilities in remote areas, and further development of employment opportunities, offered possibilities. The region, covering 6.8m km^2 and a population of 30%, accounted for a fifth of China's GDP in 2017. During 2013–2018, the west recorded an annual GDP growth of 8.8%, with yearly fixed asset investment rising by 13%, lifting 35m people out of poverty.[124] But with 55m rural poor in 2015, achieving the centennial goal of eliminating poverty by 2020 demanded 'significant policy-adjustments'.[125]

3.8 The Silk Road Economic Belt's Longitudinal Spurs

BRI projects could take two decades to build and another to mature. Unstable governments, poor governance, uncertain financial management and incompetent oversight posed potential risks. With maritime shipping still cheaper, if slower, than rail freight, and rail containers carrying Chinese exports to Europe often returning empty, challenged SREB's commercial merit and prospects.[126] Meaningful cost-benefit analyses would take longer but the experience of SREB's longitudinal spurs offered insights into the ground reality. Conceived in 1954 as PRC-USSR railway tie-ups, by 2018, SREB became a 'Eurasian Continental Land Bridge' connecting China to Europe. By the end of June 2018, trains making the China–Europe run in both directions had made over 9000 trips. However, as some Chinese cities and provinces sought to take advantage of government subsidies, many trains carried several empty containers to make up the numbers.[127] In the seas, the MSR linked Coastal China to Eastern Africa and South-Western Europe via ports, airports and industrial hubs along the shores of the SCS and the Indian Ocean. Trade between China and its BRI partners

---

[123] Information Office (2017b).
[124] Xinhua (2018c).
[125] Report (2018h).
[126] Hillman (2016).
[127] Leng (2019).

reached $664bn in the first seven months of 2018, up 11.3% year-on-year, exceeding a quarter of China's total foreign trade.[128]

In July–September 2018, BRI trade grew 20.1% year-on-year to $332bn, with Chinese exports reaching $182bn and imports, $150.5bn.[129] Aggregate trade between China and other BRI partners in 2013–2018 reached $6.47tn. In these five years, more than 80 'economic and trade cooperation zones' were built in various BRI partner states and over 244,000 jobs had been created for non-Chinese locals.[130] Estimates for 2019 suggested China's exports to BRI states could increase by $56bn and its imports from them rise by $61bn.[131] In the first half of 2019, China's actual trade with BRI partners hit $617.5bn, an increase of 9.7% year-on-year, compared to overall trade growth of 3.9% during the same period.[132] Long before Xi's twin proposals, however, partner states had initiated north-south 'economic corridors' designed to integrate Eurasia and its maritime periphery, corridors that now connected the SREB and MSR alignments. One unheralded SREB/MSR spur, the China–Singapore Connectivity Initiative (CCI), showed how some projects steadily progressed while others stalled. Launched in 2017, CCI was designed to link Chongqing to Singapore, reduce shipping distances by 1000 km and slash cargo voyages by 12 days. An unstated aim was to bypass the Malacca Strait choke point.

In 2018, CCI-borne cargo, using 660 sea-rail trains, 100 freight trains and 500 specialised cross-border cargo vehicles, travelled among 58 ports in 35 countries, but a planned high-speed Malaysia–Singapore section suffered an unexpected delay.[133] CCI and several other connective spurs, e.g. the China–Kazakhstan–Azerbaijan and the CKTI railway lines, germinated in the early twenty-first century, but the China–Mainland Southeast Asia, China–Pakistan, the Balkan Silk Road and Bangladesh–China–India–Myanmar alignments accreted over decades. The origins, evolution and progress recorded by these 'corridors' showed that each uniquely combined diverse elements and interests. Even within each spur, segments grew at varying paces. Chinese initiative and finance, while essential, proved insufficient for success. With the reticence of even one partner precluding completion, the 'corridors' presaged BRI's mixed prospects.

3.8.1 The Lancang–Mekong Economic Corridor

On 15 March 2003, with world attention riveted on the USA's imminent invasion of Iraq, all passengers and crew aboard a New York-to-Frankfurt flight were off-loaded on arrival and 'taken to hospital isolation'. The pandemic spread of the

---

[128] Liang and Bianji (2018).
[129] Report (2018b).
[130] Li (2019b).
[131] Chen (2019).
[132] Report (2019b).
[133] Palma (2018) and Hongyu (2018).

severe acute respiratory syndrome (SARS) virus had caused atypical human pneumonia in Canada, China, Hong Kong, Indonesia, the Philippines, Singapore, Thailand and Vietnam. Hours after that German action, Gro Bruntland, Director General of the World Health Organisation (WHO), proclaimed SARS a 'worldwide health threat'.[134] The pandemic led ASEAN to summon a special summit in Bangkok in April.

The host, Thailand's Prime Minister, in addition to counter-SARS action, proposed an 'Economic Cooperation Strategy' to counterparts from co-riparian Cambodia, Laos and Myanmar to reduce economic gaps between Thailand and its neighbours, harmonise relations and strengthen ASEAN. The four leaders met again in Bagan, Myanmar, in November, adopting the Bagan Declaration on cooperation in five sectors, endorsed a decade-long Economic Cooperation Strategy Plan of Action listing 46 common projects and 224 bilateral ones. Underscoring shared riverine interests, they named the initiative the 'Ayeyawady-Chao Phraya-Mekong Economic Cooperation Strategy' (ACMECS). Vietnam joined in May 2004.

ACMECS leveraged bilateral and group cooperation to join other regional programmes to secure 'common benefits, shared prosperity, enhanced solidarity, peace, stability and good neighbourliness'.[135] Working groups focused on

- Trade/investment facilitation
- Public health and social welfare
- Human resources development
- Industrial/energy cooperation
- Tourism cooperation
- Transport connectivity
- Agricultural development and
- Environmental protection.

China's relations with ACMECS members, except Vietnam, were close. Despite Hanoi–Beijing tensions,[136] Chinese trade and investment featured prominently across the sub-region. The biggest Chinese projects, including a gas pipeline and an oil pipeline running north-east from Myanmar's Bay of Bengal coast to Kunming in Yunnan, were in Myanmar. Feasibility talks began in 2004. In December 2005, PetroChina contracted to buy Myanmar's natural gas for 30 years. The NDRC approved the twin pipelines project in April 2007. In November 2008, China and Myanmar agreed to build a $1.5bn oil pipeline and a $1.04bn gas pipeline. PetroChina's parent company, China National Petroleum Corporation (CNPC), signed a contract on Christmas Day, 2008, with a Daewoo-led consortium, to buy gas from Myanmar's offshore Shwe gas field.[137] Officials signed state-level contracts on

---

[134]Media Centre (2003).

[135]Department of Foreign Trade (2016).

[136]Ali (2017a).

[137]Report (2008).

the two pipelines in March and June 2009. Work on the Maday Island deep sea oil terminal began on 31 October 2009.[138]

The Kyaukpyu–Kunming gas pipeline, with an annual capacity of 12bn cubic metres, completed in June 2013, began moving both Myanmar and imported gas to China in October.[139] After a 300,000 ton supertanker discharged its crude cargo at the Maday Island terminal for the first time, the oil pipeline began operating on 29 January 2015. The pipelines cut oil and gas sailing distance by 700 miles, and time, by 30%.[140] Promises to pay royalties worth $53bn over 30 years to Naypyitaw, and $25m to regional authorities for social development projects, addressed local grievances.[141] Reduced time, costs and avoidance of the Malacca Strait mitigated China's 'Malacca Dilemma'-rooted energy insecurity.[142] Other ACMECS states shared less acute but still significant interests with China.

At the November 2014 ASEAN-China summit in Naypyitaw, Li Keqiang proposed building on the Thai initiative on sustainable development across the Lancang–Mekong sub-region by forging a China-ACMECS grouping. ASEAN leaders 'welcomed the countries of the Mekong region and China to explore possibilities for setting up relevant dialogue and cooperation mechanisms'.[143] The Lancang–Mekong Cooperation (LMC) programme followed. The first LMC Foreign Ministers' meeting in Yunnan in November 2015 affirmed the group's framework documents. Hosting his five LMC counterparts at their first summit in March 2016 in Sanya, Hainan, Li announced a special LMC fund, with $1.6bn in preferential loans and $10bn in credit facilities.[144] Leaders insisted LMC operate on 'the principles of consensus, equality, mutual consultation and coordination, voluntarism, common contribution and shared benefits, and respect for the UN Charter and international laws'.[145]

They endorsed a 26-point developmental cooperation charter to 'Encourage synergy between China's BRI and LMC activities and projects, as well as relevant development programs of the Mekong countries, including the Master Plan on ASEAN Connectivity (MPAC)', and 'Step up both hardware and software connectivity among the LMC countries. Improve the Lancang-Mekong rivers, roads and railways network, push forward key infrastructure projects to build a comprehensive connectivity network of highway, railway, waterway, ports and air linkages in the Lancang-Mekong region; expedite the construction of network of power grids, telecommunication and the Internet; implement trade facilitation measures, promote trade and investment and facilitate business travel'.[146] The LMC advanced BRI's shared goals, with periodic official, ministerial and summit meetings driving progress.

---

[138] Xinhua (2009).
[139] Aung (2013) and Report (2014).
[140] Meyer (2015).
[141] Meyer (2015).
[142] Hook (2013).
[143] Sein (2014).
[144] Liu (2016).
[145] FMPRC (2016b).
[146] Ibid., Agreed Measures, paragraphs 6–7.

Member-states established LMC National Secretariats/Coordination Units in respective capitals before the third Foreign Ministers' meeting in Dali, China, in December 2017, when they approved a Chinese-drafted FYP of Action and its first set of projects. Beijing extended fresh concessional loans worth $1.1bn and credit worth $5bn for 45 'early harvest' projects.[147] Officials reviewed the FYP's project proposals for leadership approval at their second summit in Phnom Penh in January 2018. The plan indicated coalescing convergence.[148] LMC institutionalised consultation and joint management of political-security issues, economic and sustainable developmental activities and sociocultural and people to people exchanges. Chinese funding, technical support and training and investment incentivised collaboration; voluntary, consensual decision making offered leverage to LMC's weaker members. The first Chinese-funded LMC projects were in Cambodia and Laos. In late 2017, using Chinese designs and hardware, Thailand began building a high-speed railway linking Bangkok to Nong Khai on the Lao border. The Thai–Lao Friendship Bridge over the Mekong beckoned to Vientiane 25 km upriver, as the first phase of a regional connectivity project.[149]

The LMC's 'foundation-laying stage' (2018–2019) would be followed by an 'expansion stage' (2020–2022), with Beijing initially funding 132 projects. Several, e.g. the Kunming–Bangkok road, China–Laos railway, and Vietnam's Long Jiang Industrial Park, were listed in BRI portfolios; others could follow. This first Chinese-built sub-regional cooperative institution offered templates for others. Progress from initial proposal in November 2014 to formal launch of the 2018–2022 FYP in January 2018 was rapid,[150] but trans-frontier collaboration across the 'Greater Mekong Subregion' (GMS) went back decades. Historically inhabited by four loosely related ethnic-Dai principalities, the Lancang–Mekong basin below Tibetan highlands was incorporated into China (Yunnan), Thailand, Myanmar and Laos. As frontiersland troubled by ethnic conflicts fuelled by drug-financed warlordism, the GMS was a volatile cockpit of interstate intrigue and security operations. In 1957, ESCAP proposed a Mekong Committee to drive regional development,[151] but Cold War era political challenges thwarted action.

The GMS received scant attention beyond state consolidation efforts. Then, in 1987, Myanmar's informal and formal border trade with Yunnan, catalysed by Yangon's liberalisation policies resonating with China's fast-growing economy, rose 72% over 1986. New official border transit points, opened in 1988, boosted trade by another 65%.[152] Sino-Lao relations, helped by China funding the 1960s–1970s era 'Mao Zedong Road' in Phong Saly region, chilled when Laos sided with Vietnam in the 1978–1979 Sino-Vietnamese crisis. Sino-Lao relations were normalised in 1988 but trade volumes remained modest. Sino-Vietnamese commerce, suspended until

---

[147] Information Office (2017a).

[148] Xinhua (2017c, 2018a) and Report (2018c).

[149] Lowe (2018), Deng (2018), FMPRC (2016a) and Song (2015).

[150] LMC National Secretariat (2018).

[151] Severino (2000).

[152] Berman (1998, p. 3).

1990, rapidly grew since. Sino-Thai trade, exploding from $700m in 1992 and 1993 to over $2bn in 1994, showed the potential. However, poor infrastructure, primitive roads and railways, old cross-Mekong ferries, diverse hardware designs and regulatory norms, and varied levels of interest in the six capitals, constrained 'natural' trade growth between Yunnan and its five GMS/LMC neighbours.

Post-Cold War dynamics catalysed change. Beginning in 1992, the six neighbours, reviving the Mekong Committee's GMS Economic Cooperation Programme, met repeatedly to coordinate infrastructure development and investment plans. They asked the ADB to be the programme's Secretariat, precipitating an 'economic quadrangle', integrating a market of 50m populating 600,000 km^2 in the Mekong basin.[153] In May 1993, the Governor of Yunnan proclaimed a 'Lancang Economic Belt' spanning GMS neighbours, with mining operations in the basin's mountainous upper sections, hydro-power and timber-processing projects in the middle sections, and the lower sections open to international river traffic. This coincided with a joint field survey by specialists from China, Laos, Myanmar and Thailand and Premier Li Peng's presence when the Governor made his 'Kunming proclamation'. Yunnan's provincial administrators took this leadership-level endorsement as the start of a drive to build new transport networks from Kunming across the Mekong basin southward to the sea.[154]

China promoted GMS/LMC trade links as mutually beneficial 'South-South' engagement because Yunnan's transport links to coastal China could only handle half of the province's coal and phosphates output. Access to foreign markets via LMC ports would help while meeting regional developmental needs, but connectivity required concerted effort.[155] Pressure for such cooperation grew when the USSR and Soviet-bloc states suddenly dissolved, forcing Vietnam, Cambodia and Laos to seek new trade partners. In December 1993, a Sino-Lao border treaty was signed; a Lao–Myanmar accord was concluded in October 1995, enabling formal boosting of infrastructure development and trans-frontier trade, including riverine commerce.

The fourth GMS conference in September 1994 listed priority projects as ASEAN economic ministers and Japan's Ministry of International Trade and Industry launched plans to help develop Cambodia, Laos and Myanmar. In April 1995, the moribund Mekong Committee was revived into the Mekong River Commission. The ADB, UNDP, ESCAP, the Tourism Authority of Thailand and the Mekong River Commission together pressed for progress. In May 1996, the ADB offered a technical assistance grant of $800,000 for training staff in environmental protection measures. At the August 1996 GMS conference in Kunming, the ADB extended $300m in loans for Greater Mekong Basin projects. Japan and New Zealand, too, helped. With China and ASEAN driving leadership and the ADB offering financial help and technical supervision, GMS was ready when LMC was proposed.[156]

---

[153] Report (1996).
[154] Central News Agency (1993) and Dao (1996).
[155] Berman (1998, p. 13, fn. 10–11).
[156] GMS Secretariat (2015) and Severino (2000).

3.8 The Silk Road Economic Belt's Longitudinal Spurs

The major BRI-ASEAN/ACMECS project designed to connect China to Southeast Asia was the 'Pan-Asia Railway Network' with three key routes, all beginning in Kunming: Eastern route linking Vietnam, Cambodia and Thailand—$600m; central route linking Laos, Thailand, Malaysia and Singapore—$33bn; and a western route linking Myanmar and Thailand—$2bn. Lateral spurs would knit the region into an interconnected whole.[157] Work progressed slowly and along the central route, was suspended in 2018, after Mahathir Mohamad assumed Malaysia's premiership. LMC projects advanced against a darkening regional backdrop.

3.8.2 The Bangladesh–China–India–Myanmar Economic Corridor

The outcome of a decades-long initiative launched by Yunnan's Development and Reform Commission (YDRC), research centres and universities, the Bangladesh–China–India–Myanmar Economic Corridor (BCIM-EC)'s slow progress underscored pitfalls confronting China's regional connectivity enterprise. Years of track-2, track-1.5 and, finally, track-1 engagement, aimed at linking Kunming to Kolkata (K2K) via road, rail and riverine transport networks binding Myanmar and Bangladesh to India and China failed to deliver. The K2K vision grew around a 2800 km road alignment from Kunming via Dali, Tengchong and Ruili (China), Mandalay and Ka Lay (Myanmar) Imphal (India), Tamabil (Bangladesh), Silchar (India), Dhaka and Jessore (Bangladesh), to Kolkata.[158] Xi Jinping secured Narendra Modi's endorsement of the BCIM blueprint during Xi's September 2014 state visit to India, where he pledged to invest $20bn in India's economy and infrastructure over five years. That visit followed the first meeting of the BCIM Joint Study Group (JSG) in Kunming and the two leaders pledged to 'continue their respective efforts to implement understandings reached at the meeting'.[159]

At the JSG's second session in November 2014 in Cox's Bazar, Bangladesh, delegates assigned each country to 'synthesise' specific 'chapters' of the BCIM framework document; that is where it stood. The vision appeared reasonable. BCIM represented 40% of humanity occupying 9% of Earth's landmass. The population of just Yunnan, Northeast India, Myanmar and Bangladesh totalled 440m. However, intra-BCIM trade in 2005 stood at 7.99% of world trade and, although it grew, in 2012, it was just 5% of total BCIM trade. This compared to ASEAN's intra-regional trade representing 35% of the group's total trade. Upgrading 312 km of the 1940s era Ledo-Kunming road linking North-Eastern India to Yunnan via Northern Myanmar

---

[157] Vineles (2019).
[158] Ghosh (2017).
[159] Media Centre (2014).

could boost trade by reducing Sino-Indian transportation costs by 30%.[160] Notwithstanding acknowledged potential, summit-level pledges, and track-1.5 proclamations, BCIM remained 'comatose' until 2017, when discussions were elevated to an inter-governmental process.[161]

Pre-existing Yunnan–Myanmar road and pipelines alignments to be shared between the BCIM and LMC ECs should have aided BCIM's central ambition: fashioning surface transport links. However, Sino-Indian divergences hobbled progress. At the third JSG session in Kolkata in April 2017, Bangladesh and Myanmar offered fresh proposals for moving forward, but the leader of the Indian Delegation insisted: 'We should be mindful of different domestic circumstances and development aspirations in our respective countries. While we focus on expanding trade volumes, equal attention should also be paid to its sustainability. Greater access to each other's markets is desirable to achieve more viable and sustainable trade cooperation in our region'. His Chinese counterpart, Vice-Chairman of the NDRC, urged concrete action: 'There is a need to put in place an Inter-Governmental Cooperation Mechanism for the success of the proposed initiative'.[162]

His team had drafter the proposed outline of an 'Inter-Governmental Cooperation Mechanism', but facing Indian resistance, did not table it. Delhi's anxiety over BCIM potentially increasing Chinese exports, further widening trade imbalances, and the need to strictly control India's insurgency-prone border districts, imposed caution. The mismatch between Chinese, Bangladeshi and Myanmar enthusiasm and Indian reluctance, and BCIM's consensual underpinning, stalled progress. This explained Kunming's provincial-level initiatives.

Efforts to boost support for BCIM via visits to China by leaders of Eastern Indian states of Bihar, Chhattisgarh, Odisha and West Bengal made slow progress. Raman Singh, Chhattisgarh's Chief Minister (CM) met Chinese leaders and potential investors in Shenzhen, Guangdong province, and Zhengzhou, Henan province, in April 2016. His aides signed deals with Chinese corporations offering to invest $945m in projects to be built in the largely rural and violence-afflicted state.[163] However, the biggest regional prize, visits by West Bengal's CM Mamata Banerjee, eluded Beijing. By January 2018, Chinese provinces had extended ten separate invitations to her, but she took up none.

The Chinese Consul-General in Kolkata encouragingly noted, 'the blueprint drawn at the 19th National Congress of the CPC last October will have great significance on China's ties with Bengal and other eastern states as this region lies within the areas covered by the BRI and the BCIM'.[164] Delegations representing 30 Chinese corporations, led by Jiangsu Province's Vice-Governor, Chen Zenning, attended

---

[160] Sahoo and Bhunia (2014).
[161] Report (2015, 2016), Li and Modi (2015), Yan (2017) and Iyer (2017c).
[162] Iyer (2017b).
[163] Drolial (2016) and PRC Con-Gen (2016).
[164] PRC Con-Gen (2018).

3.8 The Silk Road Economic Belt's Longitudinal Spurs

the January 2018 Bengal Global Business Summit (BGBS) in Kolkata.[165] Chinese investors showed interest in automobiles, electronics, engineering, garments and textiles across Eastern India, but compared to other trade delegations, they secured few agreements, with none listed in the official press release. Chinese emphasis on BRI and BCIM for boosting Sino-Indian trade and investment drew no attention at the summit, reflecting profound unease afflicting Indian leaders, analysts and businessmen over both. China also participated in the February 2019 Bengal Global Business Summit (BGBS), but was neither made a 'partner', nor received a mention in any of the summit's official reports.[166]

Banerjee eventually agreed to visit Beijing and Shanghai in June 2018, but a failed demand to meet a PBSC member during the visit triggered a last-minute cancellation.[167] Prime Minister Modi's refusal to attend the 2017 BRF, and Delhi's official explanation, underscored concerns over 'territorial and sovereignty' issues. Those concerns were rooted in the CPEC's alignment across Pakistan's Gilgit-Baltistan region, which Delhi considered parts of disputed Jammu and Kashmir. However, Indian enthusiasm for any Chinese initiative, e.g. BRI and BCIM ('Kunming Initiative'), was limited. China had proposed K2K connectivity and the BCIM-EC at a GMS meeting in Kunming. Chinese officials hoped that in addition to road links passing through Mandalay in Myanmar and Sylhet in Bangladesh, BCIM could lay a parallel 2800 km K2K high-speed railway line benefiting all BCIM states.[168] With Myanmar straddling the LMC and BCIM alignments, both Myanmar and India could advance or hinder BCIM's implementation. This was the key obstacle to the process[169] which was begun by the YDRC shortly after the Soviet collapse transformed the planet's political–economic landscape.

Chinese planners had revived a 1959 vintage UN Economic and Social Commission for Asia and the Far East (ESCAFE, later ESCAP) proposal for cooperative development by building road transport linkages. YDRC, the Yunnan Academy of Social Sciences (YASS), the Sichuan Academy of Social Sciences and regional universities hosted several workshops in Kunming with academics, think tank analysts, journalists and businessmen from Eastern India, Myanmar and Bangladesh. Organisers promoted synergistic development using trade and investment, linking South-Western China, Eastern India, Myanmar and Bangladesh. Track-II exchanges socialised participants from diverse backgrounds, historically insulated from mutual exposure, crystallising a collective sub-regional vision of shared development potentially transforming a region marginalised by post-Cold War globalisation. After several sessions, YDRC consensually named the gatherings the Conference on Regional Cooperation and Development.

---

[165] Despite this level of presence, the BGBS did not consider China as a 'partner country', a group comprising the Czech Republic, France, Germany, Italy, Japan, RoK, UAE, Poland and UK. BGBS (2018); BGBS report https://bengalglobalsummit.com/ Accessed 26 Jan 2018.
[166] Report (2019a), Rakshit and Dutt (2019) and Singh (2019).
[167] Chaudhury and Patranobis (2018).
[168] PTI (2015).
[169] Mishra (2011).

In August 1999, Chinese and Myanmar officials attended the first BCIM gathering with academics, analysts, businessmen and journalists. Participants—94 from China, 22 from India, 13 from Bangladesh and four from Myanmar, launched the BCIM Economic Connectivity Forum for Regional Cooperation, or 'the Kunming Initiative', as a track-1.5 process.[170] Next, YDRC and the Yunnan government established the 'China Kunming International Logistics & Finance Association' (ILFA) to facilitate connectivity, trade and investment among states in Southeast Asia, South Asia and the Indian Ocean Region (IOR). ILFA would aid infrastructure projects, especially roads, land and sea ports, industrial parks and free-trade zones.[171]

BCIM aimed to develop economic growth and connectivity via transport infrastructure, energy resources, agriculture and trade and investment.[172] Three Delhi-based official think tanks, as counterparts to the YDRC and its academic partners, hosted the second BCIM Conference on Regional Cooperation in December 2000. Participants cheered at the screening of a documentary on the 769 km Ledo-Kunming Road which linked Arunachal Pradesh (then Assam's North-East Frontier Agency) with Yunnan in the 1940s and which was named after the Commander of US–Chinese joint forces in their war against Imperial Japan, General Joseph Stilwell. The Stilwell Road had fallen into disrepair in Myanmar and India, but offered an alignment for enabling heavy commercial traffic in the future.[173]

An absence of road and railway connectivity, especially usable alignments linking BCIM countries, added to the challenges. China and Myanmar, and Myanmar and India, had developed border roads connecting neighbouring regions, but apart from stretches of Yunnan–Myanmar highways, these were unsuited to serving BCIM-envisaged heavy commercial traffic. Bangladesh and Myanmar lacked any surface connectivity. Recognising the dramatic economic growth in China and ASEAN members, and the potential for accessing new markets there, in 2003, Bangladesh and Myanmar proposed road links between Bangladesh's south-eastern Cox's Bazar district and Myanmar's Western Rakhine State. In 2004, during Myanmar's Prime Minister's visit to Dhaka, officials signed a MoU endorsing this proposal.

In July 2007, Myanmar's Construction Minister and his Bangladeshi counterpart signed an agreement affirming Dhaka's offer to fund construction of the crucial Phase-1 between Guandhum in Cox's Bazar and Bawlibazar in Rakhine, a 25 km stretch with 2 km in Bangladesh and 23 km in Myanmar, of the planned 153 km Bangladesh–Myanmar Friendship Road. Myanmar would build a border trade economic zone near Bawlibazar, formalising informal border trade. The two ministers reportedly discussed linking the planned road to Myanmar's highway to Kunming as part of BCIM connectivity, but this was left out of official records.[174] In July 2009, Bangladesh signed the UN's 2003 Inter-governmental Agreement on the Asian Highway Network, and the Agreement entered into force for Dhaka in November.

---

[170] Report (2000b).
[171] Mazid (2017).
[172] Islam et al. (2015).
[173] Report (2001c) and Ding (2016).
[174] Kamaluddin (2007) and Report (2007a).

3.8 The Silk Road Economic Belt's Longitudinal Spurs

Three Asian Highway routes traversed 1771 km of Bangladeshi territory. AH1: from Dawki (India) to Petrapole (India) via Tamabil–Sylhet–Dhaka–Bhanga–Jessore–Benapole (Bangladesh), ran for 492 km in Bangladesh. AH2: from Dawki (India) to Fulbari (India) via Tamabil–Sylhet–Dhaka South-Dhaka North–Bogra–Rangpur–Bangalbandha (Bangladesh), ran for 294 km in Bangladesh. These two could serve as sections of the BCIM's Myanmar–Bangladesh–India alignment. AH41: from Teknaf–Cox's Bazar via Feni-Dhaka South-Dhaka North-Kushtia–Jessore–Khulna to Mongla, ran for 762 km excluding 162 km common with AH2.[175] This section, with existing spurs, could serve as a Southern Myanmar–Bangladesh–India K2K alignment with an extension to the Chittagong port in South-Eastern Bangladesh.

At the ninth BCIM Forum in Kunming in January 2011, participants 'discussed the existing and potential routes of roads, rail, air and water to enhance connectivity within the region, reviewed the progress achieved so far and discussed to (sic) call for joint working groups to consider ways and means to further improve the infrastructure regarding connectivity'. They agreed to 'enhance the thrust for improved regional connectivity and to focus on establishing the Kunming–Mandalay–Dhaka–Kolkata Economic Corridor'.[176] Agreeing that progress required decision-makers' involvement for coordinating development of transport, telecommunications and energy networks, conferees took up a suggestion made by Indian Prime Minister Manmohan Singh during Chinese President Hu Jintao's 2006 visit to Delhi:[177] to hold a K2K car rally demonstrating BCIM's feasibility, and enthusing public and official-level support.

To hasten progress, in July 2011, the China Institute of International Studies (CIIS) and the YASS hosted a China-SAARC (South Asian Association for Regional Cooperation) conference themed 'Towards a Better Understanding through Enhanced People-to-People Exchanges', in Kunming. Around 40 scholars and officials from China and SAARC member-states attended.[178] Sponsors sought South Asian governments' support for BCIM and China-South Asian collaboration generally. Much of 2012 was spent in preparations, including a route survey, for the planned car rally. In February 2012, five Chinese, four Indian, two Bangladeshi and two Myanmar pioneers began the route survey from Kunming, reaching Kolkata after nine days. They identified physical, regulatory and security problems confronting rally drivers. BCIM countries, addressing these, scheduled the rally in February–March 2013.

Flagged off in Kolkata by Chief Minister Banerjee on 22 February, 75 enthusiasts from the BCIM countries riding 20 SUVs, shared equally between Indian- and Chinese-built models, crossed the Petrapole–Benapole border checkpoints into Bangladesh. After a halt at Jessore, they continued north-eastward, crossing the Jumna River by ferry, and reached Dhaka, where a reception awaited them. From

---

[175] Siddique (2016).
[176] YDRC (2013).
[177] Mahalingam (2013).
[178] Zhou (2011).

Dhaka, the teams rode north-east to Sylhet, turning eastward to cross the Tamabil–Dawki border checkposts into India for a break at Silchar in Assam. The next leg took them to Imphal, capital of Manipur State, where they turned south, crossed into Myanmar, and halted at Ka Lay in Sagaing Division.[179] The rally then proceeded southeast to Mandalay, where it turned north and then north-east towards the Chinese border, crossing into Yunnan at Ruili. The road ran north-east to Tengchong and Dali where the riders drove south-east on the final leg to Kunming, arriving there on 5 March. The 12-day, 3000 km rally showed that commercial road transportation along the K2K alignment was feasible.[180]

The 2013 BCIM-EC Forum in Dhaka was timed to coincide with the motorists' passage. Participants reflected the enthusiasm the rally generated, but Bangladesh's decision to upgrade the forum to track-1 level in 2005 and 2010 had left Delhi 'increasingly isolated and on the defensive'. Anxiety over China's foundational role in BCIM and its 'strategic objectives' turned BCIM in Indian eyes into a tool of Beijing's 'geo-strategic and economic interests in the region', giving China 'unfettered access to the Bay of Bengal'.[181] Additionally, while Yunnan represented the Chinese 'pillar' driving BCIM's development, North-Eastern Indian states, given their 'security status', could not pursue reciprocal diplomatic/economic initiatives. Delhi's deep scepticism dimmed BCIM's prospects.

Then, in May 2013, during Premier Li Keqiang's visit to India, 'The two sides appreciated the progress made in promoting cooperation under the BCIM Regional Forum. Encouraged by the successful BCIM Car Rally of February 2013 between Kolkata and Kunming, the two sides agreed to consult the other parties with a view to establishing a Joint Study Group (JSG) on strengthening connectivity in the BCIM region for closer economic, trade, and people-to-people linkages and to initiating the development of a BCIM Economic Corridor'.[182] In October 2013, Beijing received Bangladesh's Foreign Minister, Dipu Moni, and the Indian Prime Minister, Manmohan Singh, in quick succession. As hosts, Chinese Foreign Minister Wang Yi and Premier Li Keqiang, respectively, stressed the strategic import of building up the BCIM-EC, with Moni and Singh endorsing the development of a 'southern Silk Road'.[183]

The JSG, at its first meeting in Kunming in December 2013, agreed that 'the proposed corridor could run from Kunming (China) in the east to Kolkata (India) in the West, broadly spanning the region, including Mandalay (Myanmar), Dhaka

---

[179] For more information, you can find the 2013 'BCIM K2K Car Rally Route Map' issued by the State of Manipur, India, here: http://manipur.gov.in/wp-content/uploads/2013/03/souvenir-bcim-car-rally-2013.pdf (last accessed October 16th, 2019).

[180] Government of Bangladesh, Foreign Affairs Office of the People's Government of Yunnan Province, Confederation of Indian Industry, Government of Myanmar (2013), Government of Manipur (2013) and John (2014).

[181] Uberoi (2014).

[182] MEA (2013a).

[183] Report (2013a, b).

3.8 The Silk Road Economic Belt's Longitudinal Spurs

and Chittagong (Bangladesh) and other major cities and ports as key nodes'. Delegates believed, 'with the linkages of transport, energy and telecommunications networks, the Corridor will form a thriving economic belt that will promote common development of areas along the Corridor'.[184]

In June 2014, the BCIM Exchange Forum, ILFA and the South Asia Federation of Exchanges together reviewed progress. Participants agreed to support both SREB and MSR and work towards establishing an Asian Economic Corridor Transport Facility to help realise the BCIM objectives of collaborative development via trade, investment and connectivity. They also lent support to the UN-ESCAP Governmental Land Ports Agreement, which proposed building international land ports at Dhaka–Kamalapur (Bangladesh), Kunming–Tengjun (China), Kolkata–Durgapur (India), Mandalay (Myanmar) and others. Participants agreed to encourage the use of RMB-denominated products for accelerating trade and other BCIM activities.[185]

At the twelfth BCM Forum, meeting in Yangon in February 2015, officials presented specific proposals to infuse momentum.[186] No concrete action followed. Beijing, pointing to India's historic role in the ancient Silk Road commerce, offered to coordinate its BRI projects with Delhi's own 'Mausam' programme,[187] but received no response. During Prime Minister Narendra Modi's May 2015 visit to China, Li, while acknowledging Sino-Indian differences, insisted 'our common interests go far beyond our differences'. For 'the welfare of 2.5bn people', Li urged Modi to accelerate 'building of the BCIM-EC and promoting regional economic development'.[188] The two leaders 'agreed to continue their respective efforts to implement understandings reached' by the JSG. They expressed satisfaction with Sino-Indian collaboration in developing key sections of India's railway network,[189] but progress stalled.

After high-level meetings failed to implement BCIM projects, in 2015, Dhaka proposed a new alignment for the Bangladesh–Myanmar Friendship Road—linking its Bandarban Hill District to Myanmar's Chin region, then on to the Myanmar–India Kaladan Multimodal Transit Transport project.[190] This would reduce new work, but without support from BCIM partners, this road, too, went nowhere. Fighting between Rohingya militants and Myanmar forces triggering an exodus of refugees to Bangladesh in 2017–2018 halted road link discussions. And even though Modi and Li presided over the first session of a new forum of Chinese and Indian provincial leaders, BCIM faded. Top-level engagement failed to breathe life into the vision.[191]

Perhaps to mitigate that disappointment, Beijing focused on Myanmar's National League for Democracy (NLD) after it won general elections and took office following decades of military rule. Even before the NLD assumed power, in December 2014,

---

[184] MEA (2013b) and Yan (2013).
[185] Mazid (2017).
[186] Xinhua (2015b).
[187] Zhang (2015).
[188] Xinhua (2015a).
[189] Modi and Li (2015).
[190] Rohingya News Agency (2015).
[191] Yao (2018), Iyer (2017a) and Yhome (2017).

signalling diplomatic pragmatism, Beijing hosted its head, Aung San Suu Kyi, as the official leader of the opposition. In early 2015, as the NLD moved towards replacing military-backed politicians hitherto enjoying Beijing's backing, fighting between ethnic Chinese autonomists and Myanmar forces flared along sections of the 2185 km Sino-Myanmar borders.[192] Naypyitaw's forces occasionally crossed into Chinese territory. Suu Kyi also closed several major Chinese-funded projects seen as detrimental to local interests. The NLD, beneficiary of US, Indian and Japanese support, was considered a nationalist party opposed to alleged Chinese tacit co-option of Myanmar's military rulers.[193] Against that backdrop, Beijing hosted Suu Kyi in mid-2015 before general elections in November, which the NLD won.

Xi Jinping, assuring his guest that China viewed bilateral relations 'from a strategic and long-term perspective', expressed the hope that Naypyitaw, too, would 'maintain a consistent stance' and remain 'committed to advancing friendly ties no matter how its domestic situation changes'.[194] Agreements on resuming operations at the Letpadaung Mining Project, reopening the Myanmar–China oil pipeline, Chinese-funded construction of the Kyaukpyu deep sea port and establishing a border economic cooperation zone, followed.[195] Suu Kyi kept frozen the Chinese-financed Myitsone Dam project in northern Myanmar, suspended in September 2011 by transitional President Thein Sen and maintained friendly ties to the West, but reliance on China grew. Taking office as State Counsellor in March 2016, Suu Kyi returned to Beijing on another official visit in August. Agile pragmatism restored bilateral warmth.[196]

Naypyitaw's troubles over the Rohingya crisis, triggered by Muslim attacks on the repressive Arakan state police exploding into an 'ethnic cleansing' response, reportedly killing and raping many stateless Rohingyas, forced over 700,000 survivors to seek refuge in Bangladesh.[197] Widely condemned in the west, Myanmar deepened its China ties. As international organisations and news media focused on the Rohingya's plight, especially in vast 'temporary' refugee camps sprawling across South-Eastern Bangladesh, the UNSC debated their fate. In mid-March 2017, the USA and its allies tabled a draft mildly critical of Myanmar's actions, including denial of 'humanitarian access to all effected areas'. China, with Russian support, blocked a vote.[198] Two months later, Suu Kyi attended the inaugural BRF in Beijing. In October, a Chinese consortium acquired 70% equity in the Kyaukphyu port with its oil and gas pipeline terminals linking the Bay of Bengal to Kunming.[199] With UN critics silenced, China's Foreign Minister Wang Yi proposed a 'three-step solution' to the Rohingya issue in November.

---

[192] Naing (2016).
[193] Naing (2016).
[194] Pang and Wang (2015).
[195] Nian (2018).
[196] Naing (2016).
[197] Darusman et al. (2018), Report (2017b, 2018g) and High Commissioner (2017).
[198] Nichols (2017).
[199] Liu (2017).

This relieved the pressure on Myanmar. Suu Kyi returned to Beijing in December 2017. In her meetings with Xi and his aides, they agreed to speed up construction of the Sino-Myanmar economic corridor, part of the BRI/BCIM blueprint.[200] Her entourage included Myanmar's energy and construction ministers and the CM of Mandalay, a key waypoint along the BRI's Kyaukpyu-Kunming alignment. The team first stopped off at Kunming for dinner meetings with provincial leaders. Deepening economic and security diplomacy between the neighbours enabled the partial pursuit of Beijing's BRI vision along moribund alignments so that work could resume if and when recalcitrant parties reconsidered their options.[201]

3.8.3 The Balkan Silk Road

China's construction of the 'Balkan Silk Road', an offshoot of BRI's European alignment, built on economic, commercial and investment ties to Central and Eastern European Countries (CEEC), was formalised via Beijing's 16 + 1 framework.[202] Established in 2012, the grouping caused consternation in the West, especially among the original EU member-states. In July 2018, Li Keqiang co-hosted his CEEC counterparts at a 16 + 1 summit in Bulgaria. On the sidelines, 700 CEEC businessmen and executives from 250 plus Chinese firms discussed current and future projects. Bulgaria's Deputy Foreign Minister Georg Georgiev noted, 'Our main goal is to increase Chinese business presence in Bulgaria and in the whole region of Central and Eastern Europe'. Li replied, 'China, now and in the future, supports European integration and welcomes a united, stable, open and prosperous Europe and a strong Euro'.[203]

Notwithstanding benign sentiments, the 16 + 1 process deepened EU unease over Chinese presence sufficiently for the EU to discuss the issue at the highest levels.[204] The Balkan Silk Road, despite early excitement, did not advance very fast. Also, Beijing gained little influence over CEEC voting practices at the UN and other international fora on issues affecting China's interests. In terms of resource flows, Chinese investments contributed less than one per cent of the CEEC's FDI stock; and more than 90% of Beijing's EU investment went to Western Europe. China-CEEC trade grew from around $50bn in 2012 to $90bn in 2017, falling short of stated aspirations of crossing $100bn by 2015.[205] CEEC exports to China often comprised products bearing components from the rest of the EU while some Chinese exports to CEEC destinations were reexported to Western Europe. EU anxiety appeared to

---

[200] Nitta (2017).

[201] Mu (2018) and Report (2018a).

[202] China's 16 partners: Albania, Bosnia & Herzegovina, Bulgaria, Croatia, Czech Republic, Estonia, Hungary, Latvia, Lithuania, Macedonia, Montenegro, Poland, Romania, Serbia, Slovakia, Slovenia.

[203] Tsolovia et al. (2018).

[204] Mazumdar (2018) and Spisak (2017).

[205] Turcsanyi (2018).

be founded more on presumptions and subliminal concerns than on an empirically evident complex reality.

The 16 + 1 inaugural summit held in Warsaw in April 2012 identified ten fields of cooperation, to be managed via regular summit meetings:

- Policy-coordination
- Economy and trade
- Culture and education
- Agriculture
- Transportation
- Tourism
- Science & technology
- Health
- Think tanks and local exchanges
- Youth exchanges.

Prime Ministers from the 17 partners endorsed 'Twelve Measures for Promoting Friendly Cooperation' between China and CEEC states. Beijing initiated the establishment of a '16 + 1 Secretariat for Cooperation between China and CEEC' as an official structure with links to CEEC governments for policy-coordination and driving development. In September 2012, the Secretariat hosted the first China-CEEC National Coordinators' Meeting in Beijing. The second meeting, in Bucharest, followed in October 2013, and the second China-CEEC summit in November, when leaders issued the 'Bucharest Guidelines for Cooperation between China and CEEC'. Beijing hosted the third China-CEEC National Coordinators' meeting in May 2014, followed by the fourth meeting in Belgrade in November. The leaders' summit in December issued the 'Belgrade Guidelines for Cooperation between China and CEEC'. This sequence drove collaborative projects with Chinese investment across the CEEC region.

Following up on summit 'Guidelines', China established Joint Working Groups (JWG) and high-level conferences (HLC) with specific sets of CEEC countries to initiate, implement, supervise and monitor projects. The China–Hungary–Serbia JWG on Infrastructure Cooperation met in Beijing in June 2014 while a parallel HLC on 'Transport, Logistics and Trade Routes: Connecting Asia with Europe' was held in Riga. This led, in December, to officials from China, Hungary, Macedonia and Serbia reaching agreement on cross-border customs clearing cooperation while Hungary and Serbia signed a MoU on building a high-speed railway link between their capitals with Chinese aid. In addition to terrestrial connectivity, China, Hungary, Macedonia and Serbia took a 'China–Europe Land Sea Express Line' initiative that led, in May 2015, to a four-party 'Cooperation Action Plan for 2015–2016'.

BSR building received a boost in April 2015 when FMPRC appointed a 'Special Representative for China-CEEC Cooperation'. In December 2015, the parties began constructing the Hungary–Serbia high-speed railway. China and CEEC countries also signed MoUs on jointly building BRI projects. These accreted into an economic/commercial/regulatory infrastructure on which the 17 countries focused attention. However, BSR comprised the CEEC, as well as Cyprus, Greece and Turkey.

3.8 The Silk Road Economic Belt's Longitudinal Spurs

By the end of 2017, Chinese investment across the wider BSR region in port infrastructure, steel mills, highway construction, bridge building and thermal power plants exceeded $10bn.[206]

A key BSR terminus connecting the SREB/MSR alignments resided in the Greek port of Piraeus near Athens. And there, China's COSCO shipping and port management conglomerate began investing, acquiring equity and expanding facilities, in 2009. Although Beijing entered the region two decades after the Berlin Wall came down, and long after Russia, the EU and EBRD extended developmental aid there, BSR's progress stirred unease.[207] BSR consolidation picked up speed in 2016, when the China-CEEC Secretariat initiated two new mechanisms: the Secretariat began a quarterly meeting with CEEC embassies in Beijing, lending structure to coordinating and monitoring progress, with the first meeting held in February. The Secretariat also invited senior CEEC officials to visit Beijing and Chinese provinces playing major roles in developing the SREB and MSR.

A conference of Presidents of the Supreme Courts of China and CEEC met in Suzhou in May 2016; senior CEEC officials visited Fujian and Ningxia in August. Reviews and correctives ensured that notwithstanding EU/IMF concerns over BSR member debts, projects proceeded apace. Regular exchanges at official, diplomatic and leadership levels built cooperative mechanisms and experience. Bureaucrats and politicians collaborated on key aspects of the foundation the 17 countries laid for the rapidly evolving Balkan Silk Road.[208] The near simultaneity of the EU formally designating China a 'systemic rival' and Italy joining BRI in early 2019 betrayed an ambivalence troubling Europe's approach to China.[209]

Steady progress culminated in Xi Jinping's assurances to French, German and EU leaders that notwithstanding the EU's formal designation of China as a 'systemic rival', Beijing would further open itself up, and support Sino-European collaboration and European consolidation. Italy's accession to BRI was topped by Greece joining and expanding the China-CEEC grouping to 17 + 1 at the April 2019 Dubrovnic summit in Croatia.[210] Some tension persisted. CEEC leaders questioned why Beijing had signed on to a strongly reciprocal framework of cooperation with EU but would not quite do so with CEEC. The response, 'this was a different platform', reflected China's challenges in balancing myriad, occasionally divergent, interests while moving forward across Eurasia in building BRI's SREB and MSR networks. Beijing's vigour united its critics, including EU/CEEC envoys in Beijing, but not always predictably.[211]

---

[206] Bastian (2017).

[207] Eng (2017).

[208] FMPRC (2017).

[209] Mogherini (2019) and Zhong (2019).

[210] Li (2019a).

[211] Chaguan (2019), Elmer (2019a, 2019b) and Zhen et al. (2019).

References

Alam M (2014) China's changing strategic engagements in central Asia. J Cent Asian Stud XXI:15–16
Ali S (2005) US-China cold war collaboration, 1971–1989. Routledge, New York, pp 42–165
Ali S (2017a) US-Chinese strategic triangles. Springer, Heidelberg, pp 215–227
Ali S (2017b) US-Chinese strategic triangles: examining Indo-Pacific insecurity. Springer, Heidelberg, pp 128–156
Anderlini J, Dyer G (2009) Downturn causes 20m job losses in China. FT, 3 Feb 2009
Apel R, Gallagher P (2001) China-Russia-India: a new step for the 'survivors club'. Exec Intell Rev 28(28):51
Aung S (2013) Controversial pipeline now fully operational. Myanmar Times, 27 Oct 2013
Banyan (2018) The perils of China's 'debt-trap diplomacy'. Economist, 6 Sept 2018
Bastian J (2017) China's BSR: examining Beijing's push into Southeast Europe. Reconnecting Asia, 20 Nov 2017
BBC (1994) Summary of world broadcasts (far east): No. 1978. London, 22 Apr 1994
Berman M (1998) Opening of Lancang (Mekong) River in Yunnan: problems and prospects for Xishuangbanna. University of Massachusetts, Amherst, May 1998, p 3
BGBS (2018) Press release. Kolkata, 17 Jan 2018
Bureau of Economic & Business Affairs (2017) China: 2017 investment climate statements. DoS, Washington
Campbell C (2017) Ports, pipelines, and geopolitics: China's new silk road is a challenge for Washington, Time, 23 Oct 2017
Central News Agency (1993) Premier Li Peng from the highest strategic level, approved all of Yunnan's proposals to establish the Lancang basin as a vital national district for development and opening up. Qiao Bao, 29 May 1993
Chaguan (2019) Hope remains for Western solidarity. Look at embassies in Beijing. Economist, 17 Apr 2019
Chaudhury S, Patranobis S (2018) Mamata Banerjee cancels China visit over rejection of meeting request. Hindustan Times, 23 June 2018
Chen Y (2014) China sets tasks for 2014 Western development. GT, 10 Feb 2014
Chen Y (2019) BRI driving world growth, connectivity, analysts say. CD, 19 Feb 2019
Choi C (2018) Former Bo Xilai family aide jailed over murder of UK businessman Neil Heywood 'released early'. SCMP, 20 Jan 2018
CIFIT Secretariat (2018) 2018 CIFIT to highlight BRI. CD, 17 Aug 2018
CMC (2004) China's National Defense in 2004. State Council, Beijing, Dec 2004
Cooley J (2002) Unholy wars: Afghanistan, America and international terrorism. Pluto, London, pp 55–56
Cooper C (2009) The PLAN's 'new historic mission': expanding capabilities for a re-emerging maritime power. RAND/USCC, Washington, 11 June 2009
Cunha A, Bichara J, Lelis M (2012) China and Brazil after the great recession. In: XVI Encontro De Economia Da Regiao Sul, Porto Allegre, pp 19–21
Dao Y (1996) In Report. China pushes new route to Thailand. Straits Times, 11 Nov 1996
Darling D (2015) Is the SCO emerging as eastern counterweight to NATO? Real Clear Defense, 30 Aug 2015
Darusman M, Coomaraswamy R, Sidoti C (2018) Report of the independent international fact-finding mission on Myanmar. UN Human Rights Council, Geneva, 24 Aug 2018
Deng X (2018) LMC champions inclusive regional development. Xinhua, Phnom Penh, 10 Jan 2018
Department of Foreign Trade (2016) ACMECS-WGTIF. Bangkok, Aug 2016
Dezan S (2012) China Approves 12th FYP for Western Regions. China Briefing, 27 Feb 2012
Ding G (2016) Stillwell's old road offers chance to revitalize trade. GT, 16 June 2016
Ding L, Neilson W (eds) (2003) West China development: issues and challenges. Centre for Asia-Pacific Initiatives, University of Victoria, Sedgewick, 6–8 Mar 2003, pp 1–13

References

Drolial R (2016) Chhattisgarh signs MoUs for Rs6600cr Chinese investments. ToI, 8 Apr 2016
Drysdale P (2012) Chinese regional income inequality. EAF, 13 Aug 2012
Dunford M, Bonschab T (2013) Chinese regional development and policy. Reg Mag 289(1):10–13
Editorial (2006) On the PLA's historic mission in the new stage of the New Century. PLA Daily, 9 Jan 2006
Elmer K (2019a) China says it respects EU laws and standards as 16+1 becomes 17+1 with new member Greece. SCMP, 12 Apr 2019
Elmer K (2019b) China's commitments to EU's big-hitters embolden countries in Europe's '16+1' group, says source. SCMP, 12 Apr 2019
Eng I (2017) China in the Balkans, 'firmly in play in the coming years'. Osservatorio Balcani-Caucaso, 28 Nov 2017
EU External Action (2016) ASEM: Working together for peace and prosperity. EU Commission, Brussels, Ulaanbaatar, 15 July 2016, p 1
Fang Y (2009) China, Kazakhstan pledge to enhance co-op in trade, investment. Xinhua, Beijing
Feigenbaum E (2007) The SCO and the future of Central Asia. DoS/Nixon Centre, Washington, 6 Sept 2007
Fillingham Z (2009) SCO: Asian NATO or OPEC? Geopolitical Monitor. 19 Oct 2009
FMPRC (2016a) Five features of LMC. Beijing, 17 Mar 2016
FMPRC (2016b) Sanya declaration of the first LMC leaders' meeting: for a community of shared future of peace and prosperity among Lancang-Mekong Countries. Sanya, 23 Mar 2016
FMPRC (2017) Five-year outcome list of cooperation between China and CEEC. State Council, Beijing, 28 Nov 2017
Gao F (2010) The SCO and China's new diplomacy. Clingendael, Hague
Gardner N, Kries S (2014) Moscow to Beijing by train: 60 years on. European Rail News, 18 Jan 2014
General Political Department (2006) Di Si Jiang: Wei Weihu Guojia Liyi Tigong Youli de Zhanlue Zhicheng (Lesson 4: Provide a powerful strategic support for safeguarding national interests). CMC, Beijing, 21 Aug 2006
Ghosh D (2017) China frames draft of Inter Govt Mechanism for BCIM. Millennium Post, 26 Apr 2017
Gibson J (2012) Rising regional inequality in China: fact or artefact? EAF, 12 Aug 2012
Giles J, Park A, Cai F, Du Y (2012) Weathering a storm: survey-based perspectives on employment in China in the Aftermath of the global financial crisis. WBG, Washington, Mar 2012, pp 5–9
Gill B (2001) Shanghai five: an attempt to counter US influence in Asia? Brookings, Washington
Gill B, Oresman M (2003) China's new journey to the west. CSIS, Washington, p 29
Goldstein L (2017) A China-Russia alliance? NI, 25 Apr 2017
Government of Bangladesh, Foreign Affairs Office of the People's Government of Yunnan Province, Confederation of Indian Industry, Government of Myanmar (2013) BCIM car rally 2013: K2K. Kolkata, Feb 2013
Government of Manipur (2013) BCIM car rally 2013: Manipur—gateway to Southeast Asia. Imphal, Feb 2013
Hali S (2016) SCO set to widen fold of 'Shanghai Spirit'. CRI, Beijing, 23 June 2016
Haltmaier J (2013) Challenges for the future of Chinese economic growth. Federal Reserve Board, Washington, pp 1–32
High Commissioner (2017) Flash report: interviews with Rohingyas fleeing from Myanmar since 9 Oct 2016. UN Human Rights. Geneva, 3 Feb 2017
Hillman J (2016) OBOR on the ground: evaluating China's BRI at the project level. CSIS, Washington
Hongyu B (2018) More partners on board CCI Southern transport corridor. PD, 25 May 2018
Hook L (2013) China starts importing natural gas from Myanmar. FT, 29 July 2013
Hu J (2004a) Historic missions for the PLA in the new stage of the new century. CMC, Beijing, 24 Dec 2004

Hu J (2004b) Renqing Xinshiji Wojun Lishi Shiming (clearly see our armed forces' historic missions in the new period of the new century). CMC, Beijing, 24 Dec 2004

Hu J (2004c) Scientific development theory is a new significant strategic thinking of our party. Xinhua, Beijing, 4 Apr 2004

Hu J (2012) Scientific outlook on development becomes CPC theoretical guidance. Xinhua, Beijing, 8 Nov 2012

Hu J, Nazarbayev N (2007) Joint communique between the PRC and the Republic of Kazakhstan. FMPRC, Astana, 18 Aug 2007

Hu J, Nazarbayev N (2009) Joint statement between the PRC and the Republic of Kazakhstan. FMPRC, Beijing, 17 Apr 2009

Hu J, Nazarbayev N (2011) Joint statement on the occasion of state visit to Kazakhstan by president Hu Jintao of China. Xinhua, Astana, 14 June 2011

Huang K (2009) Watching PLA fulfill its new mission from the gulf of Aden: PLA's mission will extend to wherever national interests expand. PLA Daily, 4 Jan 2009

Huang N, Ma J, Sullivan K (2010) Economic development policies for Central and Western China. China Business Review, 1 Nov 2010

Ilie E (2010) New Eurasia land bridge provides rail connection between China and Europe. Railway Pro, 15 July 2010

Information Office (1996) Symposium on economic development along new Euro-Asia continental bridge opens in Beijing. UN Secretariat, New York, 9 May 1996

Information Office (2005) The new FYP. State Council, Beijing, 9 Nov 2005

Information Office (2016) Western regions lead in GDP growth in first three quarters. State Council, Beijing, 28 Oct 2016

Information Office (2017a) Joint press communique' of the third LMC foreign ministers' meeting. FMPRC, Dali, 15 Dec 2017

Information Office (2017b) State council approves western development plan for 2016–2020. State Council, Beijing, 24 Jan 2017

Islam N, Hossain M, Matin S (2015) BCIM economic corridor: next window for economic development in Asia. Daffodil Int Univ J Bus Econ 9(1):132–136

Iyer R (2017a) BCIM-EC: facilitating sub-regional development. IPCS, Delhi, May, p 2017

Iyer R (2017b) Reviving the comatose BCIM corridor. Diplomat, 3 May 2017

Iyer R (2017c) Reviving the comatose BCIM corridor: India and China's tug of war is preventing any progress. Diplomat, 3 May 2017

Jiang Z (2001) Speech at third ASEM foreign ministers' meeting. FMPRC, Beijing, 25 May 2001

John E (2014) Racing to a new prosperity. CD, 4 June 2014

Kamaluddin S (2007) Bangladesh-Myanmar road link. New Age, 2 Aug 2007

Kan S, Best R, Bolkom C, Chapman R, Cronin R, Dumbaugh K, Goldman S, Manyin M, Morrison W, O'Rourke R, Ackerman D (2001) China-US aircraft collision incident of Apr 2001. CRS, Washington, 10 Oct 2001, pp 1–2

Karrar H (2006) The new silk road diplomacy: a regional analysis of China's Central Asian foreign policy, 1991–2005. McGill University, Montreal, Apr 2006

Kazakh MoFA (2007) China-Kazakhstan relations grow stronger. CD, 15 Oct 2007

Kujis L (ed) (2008) Mid-term evaluation of China's 11th FYP. WBG, Washington, Beijing, pp ii–iv

Kurmanov B (2018) SCO looking frosty from Astana. EAF, 21 July 2018

Kamphausen R, Lai D, Tanner T (eds) (2014) Assessing the PLA in the Hu Jintao Era. AWC, Carlisle, pp 3–4

Leng S (2019) China's belt and road cargo to Europe under scrutiny as operator admits to moving empty containers. SCMP, 20 Aug 2019

Li P (1994) China's basic policy towards Central Asia. Beijing Rev 37(18):18–19

Li Z (2005) Peace, Development, Cooperation: The Flag of Chinese Diplomacy in a New Era. FMPRC, Beijing, 23 Aug 2005

Li K (2019a) Speech at the eighth summit of heads of Government of China and CEEC. State Council, Dubrovnik, 12 Apr 2019

Li X (2019b) Key takeaways on BRI development. Xinhua, Beijing
Li K, Modi N (2015) Joint Statement between the PRC and the Republic of India. FMPRC, Beijing, 20 May 2015
Li L, Willett T, Nan Z (2012a) The effects of the global financial crisis on China's financial market and macroeconomy. Econ Res Int (Lond) 2012:2
Li S, Sato H, Sicular T (eds) (2012b) Rising inequality in China: key issues and findings. Cambridge University Press, Cambridge, pp 1–5
Li S, Wang H, He J (2016) Analysis of implementation of China's 12th FYP and prospects of its next five years. J Sci Econ For Trade Stud 9(1):40–59
Liang J, Bianji (ed) (2018) Belt and Road creates enjoyable shopping experience. PD, 21 Aug 2018
Liu X (2011) The 12th FYP: China's scientific and peaceful development. Chinese Embassy/Chatham House, London, 17 Mar 2011
Liu Y (2012) China faces political uncertainty after Bo Xilai affair. EAF, 1 July 2012
Liu H (2014) Overcoming the great recession: lessons from China. Harvard Kennedy School, Cambridge, July 2014, p 22
Liu Z (2016) China pledges billions to Mekong River countries in bid to boost influence and repair reputation amid tensions in SCS. SCMP, 24 Mar 2016
Liu Z (2017) Aung San Suu Kyi to visit China as international criticism over response to Rohingya crisis grows. SCMP, 27 Nov 2017
Liu X, Wu H, Shen J, Lee W (2011) Current affairs focus: China moves toward 'big diplomacy'. PD, 8 Feb 2011
Liu S, Rohland K, Nehru V (eds) (2013) China 2030: building a modern, harmonious, and creative society. WBG, Beijing, Washington, pp 1–442
LMC National Secretariat (2018) FYP of action. FMPRC, Beijing, 11 Jan 2018
Lowe M (2018) LMC to develop closer regional ties. CGTN, Phnom Penh, 10 Jan 2018
Ma K (2006) The 11th FYP: targets, paths and policy orientation. NDRC, Beijing
Mahalingam S (2013) Kolkata-Kunming rally begins. Hindu, 23 Feb 2013
Majumdar D (2018) Is America's greatest fear coming true? is a Russia-China Alliance Forming? NI, 4 Apr 2018
Maksutov R (2008) The SCO: a Central Asian perspective. SIPRI, Stockholm, p 2008
Maraczi F (2017) Hungary—a key state on the silk road. Belt and Road Centre, Budapest, 7 Nov 2017
Mazid M (2017) Role of BCIM-EC initiatives. Financial Express, 29 Aug 2017
Mazumdar S (2018) EU fears divisions as China woos Eastern European nations. DW, 5 July 2018
McGrath J (2017) Russia, China ink $11bn in partnership deals. International Investor, 5 July 2017
MEA (2013a) Joint statement on the state visit of Chinese premier Li Keqiang. GoI, Delhi, 20 May 2013
MEA (2013b) Minutes of the first meeting of the JSG. Indian Embassy, Beijing, Kunming
Media Centre (2003) Emergency travel advisory. WHO, Geneva
Media Centre (2014) Joint statement between the Republic of India and the PRC on building a closer development partnership. MEA, Delhi, p 20
Meyer E (2015) With oil and gas pipelines, China takes a shortcut through Myanmar. Forbes, 9 Feb 2015
Mishra B (2011) Effectiveness of track II in promoting BCIM: the K2K example. South Asia Masala, ANU, Canberra
Mo J (2018) Visiting leaders hail success of FOCAC summit. CD, 7 Sept 2018
Modi N, Li K (2015) Joint statement between India and China during Prime Minister's visit to China. MEA, Beijing
Mogherini F (2019) EU-China: a strategic outlook. High Representative for Foreign Affairs and Security Policy, EC, Brussels, p 1
Morrison W (2017) China's economic rise: history, trends, challenges, and implications for the US. CRS, Washington, 15 Sept 2017
Mu X (2018) China, Myanmar pledge to strengthen exchanges, cooperation. Xinhua, Naypyitaw

Naing K (2016) What's next for Myanmar-China relations? Myanmar Times, 31 Aug 2016
Nian P (2018) China and Myanmar's budding relationship. EAF, 24 Aug 2018
Nichols M (2017) China, Russia block UN council concern about Myanmar violence. Reuters, New York
Nitta Y (2017) Myanmar and China to cooperate on economic corridor. Nikkei Asian Review, 2 Dec 2017
Osnos E (2012) China's crisis. New Yorker, 30 Apr 2012
Palma S (2018) Singapore, Malaysia delay high speed rail link. FT, 5 Sept 2018
Pang Q, Wang D (2015) Xi urges stable Myanmar ties in changing political winds. GT, 12 June 2015
Parietti M (2016) 4 economic challenges China faces in 2016. Investopedia, 13 Jan 2016
Peyrouse S (2008) Chinese economic presence in Kazakhstan. China Perspect 2008(3)
Plummer B (2012) How economists have misunderstood inequality: an interview with James Galbraith. WP, 3 May 2012
Potter P (2012) Bo Xilai crisis embodies China's weaknesses. Toronto Star, 2 May 2012
Pottter R (2012) The importance of the Straits of Malacca. E-International Relations, 7 Sept 2012
PRC Con-Gen (2016) CG Ma Zhanwu witnesses successful China visit by CM of Chhattisgrah, India. Kolkata, 10 Apr 2016
PRC Con-Gen (2018) Consul General Ma Zhanwu briefs media on Chinese participation in BGBS 2018. Kolkata, 10 Jan 2018
PTI (2015) China seeks rail link between Kunming and Kolkata. ToI, 17 June 2015
Radyuhin V (2016) Setting up SCO as a counter to NATO. Hindu, 29 Sept 2016
Rakshit A, Dutt I (2019) 5th BGBS generates proposals of Rs 2.84tn. Business Standard, 8 Feb 2019
Report (1994) Li Peng affirms entrepreneurs' role in central Asia visit. Xinhua, Beijing
Report (1996) The crosby fund: crosby asset management. Business Times, 17 May 1996
Report (2000a) Full text of Dushanbe Statement of Shanghai Five. Xinhua, Dushanbe, 5 July 2000
Report (2000b) The Kunming Initiative for a growth quadrangle between China, India, Myanmar and Bangladesh, 14–17 Aug 1999. Sage J, Delhi, 1 Aug 2000
Report (2000c) Western development key to China's development strategy: Jiang. PD, 16 Sept 2000
Report (2001a) China's Western development campaign. CRS, Washington
Report (2001b) Go West Young Han. Economist, 23 Dec 2000
Report (2001c) Outline of the 2nd BCIM forum, 4–7 Dec 2000. China Rep 37(2)
Report (2002) Sustainable development strategy for Western China. Xinhua, Beijing
Report (2007a) Burma-Bangladesh friendship road to affect nearby villages. Kaladan Press, 11 Aug 2007
Report (2007b) China-Kazakhstan relations grow stronger. CD, 15 Oct 2007
Report (2007c) Moscow–Beijing and Rossiya. Trains world-express at http://www.trains-worldexpresses.com/500/515.htm. Accessed 9 Feb 2018
Report (2008) Daewoo seals Myanmar-China gas export deal: Xinhua. Reuters, Beijing
Report (2009a) The 10th Western China international economy and trade fair and the 2nd Western China forum on international cooperation open. FMPRC, Beijing
Report (2009b) Wen Jiabao meets with foreign leaders attending the 10th Western China international economy & trade fair. FMPRC, Beijing
Report (2010a) China's priorities for the next five years. China Business Review, Washington, 1 July 2010
Report (2010b) Chinese leaders call for more efforts to develop west. Xinhua, Beijing
Report (2011) The Western development Plan during the 12th FYP period was formally approved. Xinhua, Beijing
Report (2013a) China, Bangladesh to progress BCIM-EC. Xinhua, Beijing
Report (2013b) India-China ties a 'strategic-vision': Indian PM. Xinhua, Beijing
Report (2014) China-Myanmar joint pipeline starts delivering gas. CCTV, Beijing, 8 June 2014
Report (2015) Xi Jinping meets with PM Narendra Modi of India. FMPRC, Beijing

Report (2016) Xi calls for joint efforts to enrich China-India partnership. Xinhua, Goa
Report (2017a) CKTI nations seal deal to boost rail traffic. Iran Daily, 28 Nov 2017
Report (2017b) Massacre by the river: Burmese army crimes against humanity in Tula Toli. Human Rights Watch, New York, 19 Dec 2017
Report (2017c) More FTZs to open in China. Xinhua, Beijing
Report (2018a) Aung San Suu Kyi meets senior Chinese official on bilateral ties. Xinhua, Naypyitaw, 16 June 2018
Report (2018b) BRI quarterly: Q4 2018. EIU, London
Report (2018c) FYP of action on LMC (2018–2022). CD, 11 Jan 2018
Report (2018d) Hungary's Victor Orban accuses EU of 'abuse of power'. DW, Strasbourg, 11 Sept 2018
Report (2018e) Seven key tasks mapped out for 2019 at economic meeting. CD, 24 Dec 2018, p 4
Report (2018f) The evolution of the SCO. IISS Comments, 24(19):1 (IISS, London)
Report (2018g) The state of the world's human rights 2017/2018: Myanmar. Amnesty International, London 2018:269–273
Report (2018h) The World Bank in China. WBG, Beijing, 26 Sept 2018 (Overview)
Report (2019a) BGBS 2019 begins today. United News of India, Kolkata
Report (2019b) BRI trade reached $617.5 billion in first half of 2019. Maritime Executive, 15 July 2019
Report (2019c) China-Europe freight train services surge in 2018. Xinhua, Beijing
Rohingya News Agency (2015) Bangladesh's new proposal to build cross-border friendship road. Dhaka, 11 Apr 2015
Rosenthal E (2001) US plane in China after it collides with Chinese jet. NYT, 2 Apr 2001
Sahoo P, Bhunia A (2014) BCIM Corridor a game changer for South Asian Trade. EAF, 18 July 2014
Secretariat GMS (2015) GMS: regional investment framework implementation plan (2014–2018). ADB, Manila, pp 1–16
Sein T (2014) Chairman's statement of the 17th ASEAN-China summit. MoFA, Naypyitaw, 13 Nov 2014, p 5
Sekiyama T (2016) economic drivers of social instability in China. Tokyo Foundation, 24 Nov 2016
Severino R (2000) Developing the GMS: the ASEAN context. ASEAN Secretariat, Bangkok, 10 Feb 2000
Shan W, Clarke R (2011) China's Western development strategy in Tibet and Xinjiang. East Asian Institute, Singapore, 1 July 2011, pp. i–ii
Shea D (2014) China's energy engagement with Central Asia and implications for the US. House Foreign Affairs Subcommittee on Europe, Eurasia and emerging threats testimony. USCC, Washington, p 1
Shigeru O (2001) Central Asia's rail network and the Eurasian land bridge. Jpn Railw Transp Rev 28:42
Siddique M (ed) (2016) Regional road connectivity: Bangladesh perspective. Ministry of Road Transport & Bridges, Dhaka, pp 15–16
Singh V (2019) BGBS 2019 woos investors. Media India Group, 9 Feb 2019
Song M (2015) China, Mekong countries launch LMC framework. Xinhua, Kunming, 13 Nov 2015
Song L (2018) China is uniting, not fracturing, Europe. EAF, 11 Aug 2018
Spisak A (2017) EU uneasy over China's efforts to woo central and eastern European states. FT, 8 May 2017
State Council (undated-a) The development of Western China. http://www.china.org.cn/english/features/38260.htm. Accessed 5 Jan 2018
State Council (undated-b) What is 'three represents' CPC theory? http://www.china.org.cn/english/zhuanti/3represents/68735.htm. Accessed 5 Jan 2018
Storey I (2006) China's 'Malacca Dilemma'. China Brief 6(8)
Stronski P (2018) China and Russia's uneasy partnership in Central Asia. EAF, 29 Mar 2018
Tang D (2018) China completes new silk road to Europe. Times, 10 Oct 2018

Tanner M (2005) Chinese government responses to rising social unrest: USCC testimony. RAND, Washington, 14 Apr 2005
Tong Y, Lei S (2010) Large-scale mass incidents in China. East Asian Policy 2(2):25
Tsolovia T, Barkin N, Emmett R (2018) China's ambitions in Eastern Europe to face scrutiny at summit. Reuters, Sofia, Berlin, Brussels, 4 July 2018
Turcsanyi R (2018) China is raising its flag in CEEC. EAF, 31 Aug 2018
Uberoi P (2014) The BCIM-EC: a leap into the unknown? ICS, Delhi, Dec 2014, p 2
USN/NSA (2001) EP-3E collision: cryptologic damage assessment and incident review final report. DoD, Washington, p v
Vineles P (2019) ASEAN and China struggle to buckle the BRI. EAF, 26 Jan 2019
Vingoe S (2015) The contest for economic supremacy in Central Asia. NATO Association, Brussels
Washburne R, Churchill O (2018) China hasn't changed belt and road's 'predatory overseas investment model', US official says. SCMP, 13 Sept 2018
Webb J (2017) The new Silk Road: China launches Beijing–London freight train route. Forbes, 3 Jan 2017
Wen J (2004) Work report to the 10th session of the NPC. State Council, Beijing
Wen J (2009) Towards greater development and opening-up of Western China. State Council, Chengdu
Wen J (2012) Speech at the opening session of the 2nd China-Eurasia Expo and the China-Eurasia economic development and cooperation forum. FMPRC, Urumqi, 2 Sept 2012
Wishik A (2012) The Bo Xilai crisis: a curse or a blessing for China? NBR, Seattle, 18 Apr 2012
Wishnick E (2017) The unpredictable triangle: US-China-Russia relations in the Trump Era. China-US Focus, 29 May 2017
Xi J (2013) Promote people-to-people friendship and create a better future: speech delivered at Nazarbayev University. PRC Embassy, Astana
Xing G, Sun Z (2007) SCO Studies. Changchun Publishing, Changchun, pp 92–93
Xinhua (2009) China starts building Myanmar pipeline. Downstream Today, 3 Nov 2009
Xinhua (2010) China's western development to get more support. CD, 28 May 2010
Xinhua (2015a) Chinese premier says talks with India's Modi 'meet expectations', 24 deals signed. PD, 16 May 2015
Xinhua (2015b) Regional cooperation forum kicks off in Myanmar. CD, 10 Feb 2015
Xinhua (2016a) New five-year plan brings hope to China's west. State Council, Beijing
Xinhua (2016b) Premier encouraged foreign investment in central, western China. State Council, Guiyang
Xinhua (2017a) First train from China to Iran stimulates Silk Road revival. CD, 16 Feb 2017
Xinhua (2017b) Fujian: a new star on ancient maritime silk road. CD, 8 Mar 2017
Xinhua (2017c) Joint press communique of the 3rd LMC Foreign Ministers' meeting. FMPRC, Dali, 15 Dec 2017
Xinhua (2017d) New China-Europe freight train links Central China and Munich. CD, 23 Aug 2017
Xinhua (2018a) 6th LMC senior officials' meeting paves way for summit. CD, 9 Jan 2018
Xinhua (2018b) China-Russia oil pipeline begins operating, expected to double imports. PD, 2 Jan 2018
Xinhua (2018c) Li: accelerate western development. PD, 24 Aug 2018
Xinhua (2018d) New China-Europe freight train route launched, opens quicker access. GT, 21 Jan 2018
Yan M (2013) China, Bangladesh, India, Myanmar to boost cooperation. Xinhua/PD, 20 Dec 2013
Yan (2017) Talks on BCIM-EC to be upgraded to intergovernmental level. Xinhua, Kolkata, 26 Apr 2017
Yang Y (2019) China-Europe freight train services surge in 2018. Xinhua, Beijing
Yao Y (2018) Progress on BCIM may change India's attitude. GT, 16 Oct 2018
YDRC (2013) Joint statement of the 9th BCIM Forum on Regional Economic Cooperation. Kunming, 19 Jan 2011
Yhome K (2017) The BCIM-EC: Prospects and challenges. ORF, Delhi, 10 Feb 2017

References

Yi W, Jing W (2017) Withstanding the great recession like China. Federal Reserve Bank, St. Louis, 21 Oct 2017, pp 2–3

Ying (ed) (2017) 'Belt and Road' incorporated into CPC Constitution. Xinhua, Beijing

Zhang Y (2006) Vice Foreign Minister of China elaborated the meaning of 'harmonious world'. China News, Beijing, 2 Apr 2006

Zhang Y (2015) China willing to strengthen communication with India over BRI: envoy. Xinhua/PD, 1 Apr 2015

Zhang Y (2018) BRI progress and opportunities for Taiwan Businessmen. ICS, Kuala Lumpur, 18 Jan 2018

Zhen L, Elmer K, Lau S (2019) Hopes high that EU-China joint statement on opening up is more than just words. SCMP, 10 Apr 2019

Zheng X (2017) New pipeline broadbases gas supplies. CD, 15 Apr 2017

Zheng X (2019) CNOOC to step up deep-water exploration activities. PD, 13 Mar 2019

Zhong S (2019) Xi's visit ushers in new chapter of China Italy ties. PD, 26 Mar 2019

Zhou C (2011) CIIS held China-SAARC conference. CIIS, Beijing

Zhu R (1999) Speech at ASEM science and technology ministers' meeting. FMPRC, Beijing, 14 Oct 1999

Zhu R (2001) Outline of the 10th FYP for National Economic and Social Development (2001–2005): implementing a strategy for developing the Western Regions to promote coordinated progress of different areas. State Council, Beijing, 5 Mar 2001

ZX (ed) (2018) Facts & figures: from blueprint to tangible benefits, 5-year development of B&R. Xinhua, Beijing

Chapter 4
East Meets West: BRI's Eclectic Origins

4.1 The New Silk Road's United Nations-Driven Germination

The OBOR/BRI, SREB and MSR conceptual constructs proclaimed by Xi Jinping in 2013, as noted, elicited mixed responses. China's BRI partners appeared delighted with prospects for securing economic gains, but critics discerned Beijing's covert geopolitical designs dressed up in geoeconomic garb. An examination of BRI's core alignments, notably SREB, reveals both Chinese and external, especially Western, contributions to the accretive palimpsest of ideas, frameworks, drivers and executable policies that cumulatively, over decades, grew into BRI and SREB. The previous chapter recounted how Sino-Soviet collaboration in 1954 germinated the beginnings of the Eurasian railroad network which, decades later, provided SREB's ideational bases. SREB's road networks, in contrast, began as a multilateral concept envisaged by the United Nations Economic and Social Commission for Asia and the Far East (ESCAFE).

ESCAFE, the UN organ leading post-war reconstruction across the Asia-Pacific region, was established in Shanghai in 1947. However, as the Chinese Civil War moved towards a CPC victory over the *Kuomintang*-led Republic of China (RoC), in January 1949, ESCAFE relocated to Bangkok. In 1974, recognising the socio-economic aspects of development and the region's geographical locus, the name was changed to UN Economic and Social Commission for Asia and the Pacific (ESCAP). Given the theatre's expanse from Turkey to Kiribati, and from Russia to New Zealand, the UNGA broadened ESCAP's functional remit in 1977. ESCAP's primary driver has been lifting the region's 680 m poor into its dynamic economic mainstream.[1] It sought to achieve this by extending technical assistance and capacity-building aid to its 53 member-states and nine Associate-Members in

- Macroeconomic policy, poverty-reduction and development financing
- Trade, investment and innovation

---

[1] ESCAP (Undated-c).

- Transport
- Environment and development
- Information and communications technology and disaster risk reduction
- Social development
- Statistics
- Sub-regional activities for development and
- Energy.[2]

That comprehensive approach to post-decolonisation development and poverty alleviation led ESCAP in 1959 to envision and launch a regional transport cooperation initiative aimed at enhancing the efficacy of 'the road infrastructure in Asia, supporting the development of Euro-Asia transport linkages and improving connectivity for landlocked countries'.[3] The second of these three objectives became the foundational driver behind what eventually emerged as SREB. In 1960–1970, ESCAP and member-states jointly identified and analysed potential routes for the future Asian Highway Network (AHN). However, Cold War-era Eurasian polarisation and intermittent warfare in Asia precluded progress. The end of the Cold War transformed not just the geoeconomic landscape but also elite thinking across the region.

In 1992, ESCAP launched the Asian Land Transport Infrastructure Development (ALTID) initiative comprising the AHN, Trans-Asian Railway (TAR) and proposed facilitation of land-transport projects. ALTID envisaged region-wide integrated rail-and-road networks weaving across the Eurasian landmass. The Asian Highway proposal made faster progress, leading President Nazarbayev of Kazakhstan, a country central to the endeavour, to proclaim, 'We are reviving a New Silk Road by setting up a Western Europe–Western China transportation corridor'.[4] Kazakhstan was not alone among Central Asian republics cognisant of the opportunities now available to build on the region's connected history. Uzbekistan's President Islam Karimov acknowledged to China's Premier Li Peng, 'Friendly ties between our states (Uzbekistan and China) date back to the time of the Great Silk Route'.[5]

Between 1994 and 2002, a series of ESCAP-sponsored studies examined the feasibility, possible alignments and the challenges and costs of building sub-regional sections of the AHN. Member-state officials and ESCAP experts met repeatedly to thrash out details, address concerns and build consensus. They launched the International Agreement on the Asian Highway Network on 18 November 2003; it entered into force on 4 July 2005. By the end of 2017, 30 parties had signed on to the 'Asian Highway Agreement', leading ESCAP to report: 'The Asian Highway and Trans-Asian Railway networks play a pivotal role in fostering the coordinated development of regional road and rail networks … formalised through the Intergovernmental Agreement on the Asian Highway Network and the Intergovernmental Agreement

---

[2]ESCAP (Undated-a).
[3]ESCAP (Undated-b).
[4]Nazarbayev (2012) in Bradsher (2013).
[5]Karimov (1994) in Li (1994) and Report (1994).

on the Trans-Asian Railway Network which entered into force in July 2005 and June 2009, respectively'.[6] The agreements incorporated 143,000 km of express-motorways within the borders of 32 countries[7] promising consequential advances towards Trans-Asian/Eurasian economic integration:

- A framework for coordinated development of national, sub-regional and regional road networks
- Interest in greater regional/sub-regional connectivity enabling development of sub-regional networks
- Common highway design and technical standards which were adopted by member-states and sub-regional organisations
- Improved domestic and interstate terrestrial connectivity supporting economic growth and interstate trade
- Enhanced member-state negotiating position vis-à-vis multilateral financial institutions in securing development finance and capacity for maintaining design standards
- Greater development institutional interest in financing regional terrestrial connectivity.[8]

ESCAP was not the only UN organ driving Eurasian connectivity projects. The World Tourism Organisation (WTO—sharing the acronym with the World Trade Organisation) charted a parallel path, building regional consensus on reviving Eurasia's Silk Road linkages. At its first meeting in London in 1946, the International Congress of National Tourism Bodies set up the International Union of Official Travel Organisations (IUOTO), which met in The Hague in 1947 and set up temporary headquarters in London. Securing the status of a UN organ in 1948, it established regional commissions for Europe, Africa, Middle East, Central Asia and the Americas in 1948–1952. In 1951, it moved to Geneva, staying there until 1975 and having, in 1970, become the WTO. In 1976, after several international tourism facilitation conferences, the WTO, now an executive branch of the UNDP, moved its General Secretariat to Madrid. WTO General Assemblies refined the WTO's role. This led to the concept of 'Silk Road Tourism' being endorsed by member-states at the 1993 General Assembly in Bali, Indonesia.

Post-Cold War 'Silk Road' initiatives advanced with the interests driving Central Asia's rejuvenated polities occupying the heart of Eurasia's continental landmass. The Turkic-speaking *Stans* revived an imagined community as seen in a widely circulated brochure titled 'Turkestan Nash Obshchii Dom' (Turkestan is our common home). This view informed the January 1993 summit hosted by Uzbek President Islam Karimov, bringing together leaders of the five *Stans* in Tashkent. The leaders issued a 'Protocol of Five Central Asian States on a Common Market'. Karimov

---

[6]Li (2017).

[7]For more information, you can find the UNESCAP 'Asian Highway Route Map' issued on 1 November 2016, here: http://www.unescap.org/sites/default/files/AH%20map_1Nov2016.pdf (last accessed 16 October 2019).

[8]Li (2017) and ESCAP (Undated-b).

and his Kazakh counterpart, Nazarbayev, stated that at the summit, 'all preconditions were laid down for creating a new community of Central Asian states'. They asserted, no one should doubt that the Central Asian states, 'traditional neighbours with their common culture, traditions and language, would pursue their own path of development'.[9]

At their January 1994 summit in Tashkent, Uzbekistan and Kazakhstan proclaimed a 'Central Asian Common Economic Space'. Initially, comprising the two neighbours, whose bilateral trade had grown 2.5 times over the past 12 months, the formulation soon attracted Kyrgyzstan. Bishkek hosted another summit in April 1994 following which the leaders of Kazakhstan, Kyrgyzstan and Uzbekistan met in Almaty, Kazakhstan, in July. They established a Central Asian Bank for Cooperation and Development, pooling financial resources and institutionalising collaborative development. The April 1995 regional summit approved a five-year economic integration plan, establishing interstate committees to drive implementation, e.g. Executive Committee of the Interstate Council, the Council of Ministers of Defence, the 'Centralazbat' peacekeeping force and the Assembly of Central Asian Culture.[10] With Central Asian leaders driving integration, the *Stans* robustly reinforced international initiatives designed to create new connectivity networks knitting together the Eurasian supercontinent.

This energetically growing interest in the Silk Road legacy and template for promoting cultural exchanges, trade and tourism, led the WTO to revive the 12,000 km alignments as a tourism pathway linking three continents. In 1994, at a WTO-UNESCO meeting in Samarkand in Uzbekistan, 19 participating states adopted the 'Samarkand Declaration on Silk Road Tourism', resonating with post-Cold War 'Silk Road' connective aspirations. This WTO-UNESCO Silk Road Programme, linking 25 Asian, European and Middle-Eastern countries along land and sea routes, offered capacity-building, local empowerment, business-networking and inbound investment while creating a more enriching travel experience for tourists. As partner-governments and institutions built expanding networks of tourism facilities along ancient Silk Road alignments, the Eurasian 'New Silk Road' concept emerged as a vector of the dynamic trans-continental socio-cultural-economic integrative discourse.

Existing, modernised and newly built motorways linking up the 25 countries proved crucial in securing official-level and popular engagement with the 'New Silk Road' paradigm. ESCAP's proposed TAR Network[11] reinforced that trend. The 48th ESCAP Commission Session held in Beijing in April 1992 endorsed ESCAP's ALTID project, which included three Asia–Europe rail land-bridges. The network would allow freight trains to move containers among South Asia, Southern China,

---

[9]Karimov and Nazarbayev (1993) in Azizov (2017).

[10]Azizov (2017).

[11]For more information, you can find the UNESCAP 'Trans-Asian Railway Network Map' issued on 1 November 2016, here: http://www.unescap.org/sites/default/files/TAR%20map_1Nov2016.pdf (last accessed 16 October 2019).

4.1 The New Silk Road's United Nations-Driven Germination

Thailand and Europe. The routes would run via Myanmar, Bangladesh, India, Pakistan, Iran and Turkey. ESCAP also wished to examine prospects for rail containers travelling between Central Asian states and the Iranian ports of Bandar Abbas and Bandar Imam Khomeini and between Nepal, Bhutan and North-east Indian states and South Asian ports on the Bay of Bengal and the Arabian Sea.

Beginning in 1995, with financial help from France, Germany, Japan, the Netherlands and South Korea, ESCAP examined possible railway routes linking Bangladesh, China, India, Iran, Myanmar, Nepal, Pakistan, Sri Lanka, Thailand and Turkey. These composite alignments became the TAR's southern route.[12] The route's spurs would connect each country's hinterland with the nearest seaports, but its primary purpose was to link Southeast Asia and Southwestern China to Europe, with the allied objective that 'must always be to provide a conduit through which it is possible for trains to pass without interruption between Asia and Europe'.[13]

Energised by ESCAP's initiatives, ASEAN began exploring connectivity proposals linking its member-states with China. At the December 1995 ASEAN summit in Bangkok, leaders envisaged a Singapore–Kunming Rail Link (SKRL), designating Malaysia project coordinator. SKRL would be part of ASEAN's Mekong Basin Development Cooperation (AMBDC) programme. At the December 1997 ASEAN summit, leaders decided a consortium comprising all member-states should implement the project and invited the USA, Japan and EU-members to participate in SKRL's construction. Malaysian consultants, having conducted a feasibility study during March 1997–August 1999, proposed a Singapore–Kuala Lumpur–Bangkok section proceeding to Kunming along one or more of six possible alignments, shown in Table 4.1.[14]

Discussions among the various governments, agencies of the ASEAN Secretariat, supportive multilateral institutions, financiers and contractors took nearly two decades to push the project towards phase-1 of construction work. This exposed some of the challenges confronting the SKRL and similar projects. On 21 December 2017, Thai Prime Minister Prayuth Chanocha presided over the ground-breaking ceremony of the construction of the Bangkok–Nong Khai high-speed railway line.

Table 4.1 Singapore–Kunming railway link feasibility study route alignments

Alignment	Total length (km)	New tracks needed (km)	Total cost (bn)
1. Cambodia–Laos–Vietnam	5382	431	$1.8
2. Myanmar–Thailand–China	4559	1127	$6.0
3. Laos–Vietnam	4259	531	$1.1
4. Laos–China	4164	1300	$5.7
5. Laos–Vietnam–Thailand	4481	616	$1.1
6. Laos–Vietnam–Thailand	4225	589	$1.1

---

[12] Hodgkinson (1999, p. 1).
[13] Hodgkinson (1999, p. 3).
[14] ASEAN (2002).

Work on the Bangkok–Kuala Lumpur–Singapore alignment proceeded even more slowly. Nonetheless, Chinese Premier Li Keqiang, congratulating all parties involved in phase-1 construction, wrote, 'the China–Thailand Railway is a flagship project of BRI and will improve the level of regional infrastructure-construction'.[15]

Mainstream TAR projects advanced equally deliberately. Financial feasibility founded on commercial prospects was considered crucial, but ESCAP planners believed that the vision was key to driving the progress sought by member-states at ESCAP's 54th session, which stipulated:

- Capital-to-capital connectivity
- Connection between main agricultural/industrial centres and 'growth-triangles/zones'
- Integrated land and water transport networks linking major sea and river ports
- Integrated road and rail connectivity between inland container terminals and depots.

The main east–west trunk routes from Thailand and Southwestern China to Europe via Myanmar, Bangladesh, India, Pakistan, Iran and Turkey comprised 'routes of international significance', while those connecting India's hinterland to its coastal ports were 'routes of sub-regional significance'.[16] At the 16–22 April 1998 session in Bangkok, ESCAP-members applauded the construction of the Bandar Abbas–Bafgh and Mashad–Sarakhs railway lines in Iran and the Sarakhs–Tedjen line in Turkmenistan, fashioning 'a New Silk Railway linking China and Central Asia with Europe' via Iran and Turkey.[17] Members showed interest in parallel efforts to boost terrestrial transport links within the GMS and SAARC subregions, underscoring synergistic endeavours underpinning a trans-Asian connectivity-building drive.

Within the TAR's southern corridor, ESCAP identified three 'routes of international significance':[18]

- TAR S1: Starts at Kunming, running southwest from Xiaguan near Dali to the Myanmar border at Ruili, then to Lashio and Mandalay, then runs westward into India, crosses into Bangladesh, India (again), Pakistan, Iran and on to the Turkish–Bulgarian border at Kapikule, 11,700 km from Kunming. From Kapikule, lines run to Bulgaria, Romania, Hungary and Austria, providing access to Western Europe, specifically Frankfurt in Germany, 13,500 km from Kunming. Of this, 1800 km of new tracks would have to be laid across mountains, 95 km comprised ferry-links, and the remainder existed. The 12 countries encompassing the route would require five inter-gauge trans-shipment points.
- TAR S2: Starts at Nam Tok in Western Thailand, 210 km by rail from Bangkok, running westward into Myanmar, joining existing railway tracks northwards to Mandalay, where it would connect with route TAR S1. ESCAP recommended secondary S2 spurs linking other Myanmar cities, e.g. Yangon, Moulmein, Martaban

---

[15] Pan (2017).
[16] Committee on Transport (1998).
[17] Committee on Transport (1998, p. 5, para 27).
[18] Hodgkinson (1999, pp. 4–8).

and Bago, and indeed, two distinct alignments. However, the one connecting Thailand to India via Myanmar was the international route. TAR S2 was 75% (810 km) already in place, comprised 4 km in ferry-links and needed 25% (263 km) of new tracks to be laid for completion.

- TAR S3: Starts at Sarakhs on the Turkmenistan–Iran border, runs 164 km southwest on an existing trunk-line to Fariman from where a new, 790 km line, would take it to Bafq (which is located on TAR S1) from where the route runs 635 km south along the recently laid tracks to Bandar Abbas on the Gulf. This 1589 km route required much new work but an alternative, 2452 km link to Bandar Abbas via Tehran already existed.

ESCAP's TAR studies also proposed several sub-regional alignments linking Southern China to Southeast Asian routes, connecting north-eastern Indian states to Kolkata across Bangladesh, six additional alignments notionally connecting the former to networks criss-crossing India and other routes running across Myanmar, Thailand, Sri Lanka, Pakistan, Iran and Turkey.[19] While ESCAP members signed on to the vision, progress was patchy. In parallel, other UN organs pursued their own visions. The UN Educational, Scientific and Cultural Organisation (UNESCO) drove efforts to link cities located along the ancient Silk Road with a modern network of socio-cultural-scientific ties.[20]

Notably, UNDP supported Chinese-led efforts at building a regional institutional constituency for implementing the New Silk Road Initiative crystallised a collaborative perspective which prepared the ground for future development. In 2006, UNDP, WTO and UNCTAD sponsored the first Silk Road Mayors' Forum in Tashkent, Uzbekistan, a caravan hub along the ancient Silk Road, and the first Silk Road Investment Forum in Xian, in China's Shaanxi Province. High-level presence of Chinese, Kazakh, Kyrgyz, Tajik and Uzbek officials at both, and the RMB 700,000 in contributions collected from four leading Chinese corporations for the Silk Road roadshow which followed, encouraged the organisers.[21]

By the time Xi Jinping proclaimed his BRI vision, many of the feasibility studies and most consensus-building needs had already been addressed. UN organs played a foundational role in establishing conceptual frameworks and in completing many essential preliminary tasks in bringing together governments, specialists and multilateral financiers for fashioning trans-Eurasian connectivity. On 24 November 2015, ESCAP published the consolidated, revised and updated version of the Trans-Asian Railway Network Agreement. It reflected decades of work now largely mirrored in major BRI proposals.[22] The UN, along with other contributors, played a significant role in envisioning and laying the foundations of concerted multiple-level efforts to forge Eurasian connectivity designed to economically integrate the supercontinent.

---

[19] Hodgkinson (1999, pp. 8–37).
[20] For more information, you can find the UNESCO 'Silk Road Connectivity Network Map' here: https://en.unesco.org/silkroad/network-silk-road-cities-map-app/en (last accessed 16 October 2019).
[21] Malik (2008).
[22] ESCAP (2015).

The UN leadership discerned a resonance among the ideational bases underpinning the UN's 2030 Agenda for Sustainable Development (SDG), trends towards regional economic integration for mutually reinforcing economic growth, and Beijing's BRI vision.[23] There was little surprise when, eight months before Xi Jinping invited the world to the inaugural BRF, the UNDP Administrator, Helen Clark, signed a MoU pledging support for BRI implementation with the NDRC Chairman, Xu Shaoshi, on the sidelines of Premier Li Keqiang's visit to the UNGA in New York in September 2016. Clark noted: 'The BRI represents a powerful platform for economic growth and regional cooperation, involving more than 4bn people, many of whom live in developing countries. It can serve as an important catalyst and accelerator for the sustainable development goals'.[24] Just weeks later, presiding over a World Bank-IMF conference on BRI, the then-President of the World Bank, Jim Yong Kim, underscored the consensus among Bretton Woods multilateral institutions and their sister organisations that 'Belt and Road is a great opportunity to enable free movement of goods and capital, connect people and support inclusive growth. The world needs multilateral approaches like the BRI'.[25]

4.2 US Visionaries Imagine a New Silk Road

Historical Western imagery of Eurasia and imagination of Asia contributed towards creating a collective European identity in opposition to the non-European-inhabited continental landmass stretching away eastward. Early in the twentieth century, British geopolitician, Halford Mackinder, posited, 'European civilisation is, in a very real sense, the outcome of the secular struggle against Asiatic invasion'.[26] His construct of the 'world island', its twin-tiered 'peripheries' and the need to dominate this territory or, at least, prevent its hostile occupation, shaped imperial strategic thinking. This motivation acquired added salience following the Soviet collapse and the advent of US *primacy*, which began confronting challenges in the early twenty-first century.[27] While official and semi-official analyses viewed the possible rise of 'near-peer rivals' in Eurasia, namely Russia and China, non-establishmentarian Western thinkers envisioned a non-competitive, indeed collaborative and positive-sum, Eurasian future.

---

[23] Horvath (2016).
[24] Saling (2016).
[25] Kim (2017a).
[26] Mackinder (1904).
[27] Iseri (2009) and Togt et al. (2015).

Controversial US politician Lyndon LaRouche, co-founder with his wife Helga LaRouche, of the Washington-based Schiller Institute, promoted such a vision.[28] LaRouche died in early 2019 a fringe figure, as obituaries published by mainstream US newspapers underscored,[29] but he influenced segments of trans-Atlantic opinion and views elsewhere. In October 1988, LaRouche briefed the media in West Berlin on 'US Policy Towards the Reunification of Germany', prophesying the collapse of COMECON economies and urging food support to Poland so that a 'majority of Germans on both sides' desired reunification. In December, he assigned Schiller Institute specialists to examine prospects for establishing a Paris–Berlin–Vienna 'productive triangle'. In January 1990, Schiller published LaRouche's book on a proposed 320,000 km^2 European economic area comprising a population of 92 m concentrated in ten large industrial zones, from which he envisaged infrastructural corridors, linked via high-speed railways, radiating in all directions, 'providing a basis for upgrading living standards' across Eurasia.[30] These uncelebrated views fertilised pan-European thinking.

LaRouche was imprisoned in the USA since January 1989 on unrelated mail fraud, conspiracy and tax evasion charges, sentenced to 15 years but was released after five years. At Schiller's March 1991 'Infrastructure for a Free Europe' conference in Berlin, his paper, urging the construction of 'a sphere of cooperation for mutual benefit among sovereign states' of Europe and Asia, was read out to over 100 participants from 17 countries. In October 1991, at the first All-European Conference on Transport in Prague, Schiller staff distributed literature describing energy- and technology-intensive economic corridors radiating from Europe's 'productive triangle'. In November, 400 delegates from almost three-dozen countries, including former Soviet republics, gathered at Schiller's Berlin conference on 'The Productive Triangle: Cornerstone of an All-Eurasian Program of Eurasian Development'.

In 1992, Schiller economists detailed the 'spiral arms' or economic/infrastructure corridors radiating from the 'productive triangle', claiming resonances in Beijing's 'Eurasian Land-Bridge' initiatives as China–Kazakhstan rail connectivity became operational in June 1992. This made it possible to travel 11,000 km from China's Yellow Seaport of Lianyungang across Eurasia to Rotterdam.[31] LaRouche's political action committee and the Schiller Institute mounted a privately funded campaign aimed at persuading the American political mainstream to abandon its 'policy of antagonism towards China and Russia' and embrace Eurasian economic integration instead.[32] This proved to be a difficult and ultimately failed undertaking.

---

[28] Friedrich von Schiller's plays and romantic poetry contributed to shaping post-revolutionary European thought. His Ode to Joy, which praised freedom, unity and the brotherhood of mankind in a period that divided and froze society in arbitrary stratification, was set to music by Beethoven in his ninth symphony and is celebrated as the European Union's collective anthem. Rabitz (2009).

[29] Severo (2019) and AP (2019).

[30] LaRouche (1990).

[31] Schiller Institute (Undated).

[32] LaRouche Political Action Committee (2018) and Leesburg (Undated).

In late 1993, the EU championed European Commission President Jacques Delors' eponymous 'Delors Plan' to extend Western Europe's high-speed railway network into former Soviet-*bloc* Central- and Eastern European countries, starting with a Berlin–Warsaw section and raising the prospects for the eventual fashioning of a 'continental bridge' linking Europe to ports in Asian Russia and later China. Unlike Schiller Institute's theoretical proposition, the EU's pragmatic plan excluded the war-torn Balkans. In December 1994, Lyndon LaRouche, resuming his activist stance after being released from prison, presided over Schiller's conference on 'Global Economic Recovery and the Cultural Renaissance' in Eltville, Germany, where he led a seminar on Eurasian development corridors attended by influential delegates from China, Russia, Ukraine and Eastern Europe.

This led, on 7–9 May 1996, to the UNDP, the UN Department for Development Support and Management Services, the WBG, EU Commission, ADB and other multilateral and international organisations helping the Chinese government to host a three-day 'Symposium on Economic Development along New Euro-Asia Continental Bridge' in Beijing. More than 400 delegates representing governments, academia and businesses discussed the status of socio-economic development across Eurasia, 'thereby laying the groundwork for a new continental bridge to cover a vast area of the Eurasian continent'.[33] In addition to ministerial-level officials from China and other states, and non-governmental contributors to the Eurasian economic-infrastructural integration discourse, Helga Zepp-LaRouche addressed the meeting, focusing on 'Building the Silk Road Land-Bridge: The Basis for the Mutual Security Interests of Asia and Europe'.

Coincidentally, Iran and Turkmenistan announced the opening of their Mashhad–Ashgabad railway line, enabling direct rail transport from the Persian Gulf to Central Asia and further east and west, brightening the prospects for knitting much of the southern half of Eurasia into a cohesive transport network. In January 1997, Lyndon LaRouche addressed a Washington conference, urging the Clinton Administration to sponsor a 'New Bretton Woods system', reorganising the world economy to prevent disruptive boom-bust cycles and recognise the global merit of the 'Eurasian Land-Bridge' programme. Reinforcing her husband's thematic refrain, Helga LaRouche published a commentary titled, 'Eurasian Land -Bridge: A new era for mankind', which the Schiller Foundation widely circulated across the Atlantic.[34] In October–November 1997, she presented a paper on the 'Principles of Foreign Policy in the Coming Era of the New Eurasian Land-Bridge' at a conference themed 'Asia–Europe Economic and Trade Relations in the twenty-first Century and the 2nd Eurasian Bridge' hosted by Beijing with multilateral financial assistance.

By then, railway connectivity between coastal China, Central Asia and Russia was a reality, Europe beckoned.[35] A year-and-a-half later, in July 1999, Schiller's Indian representative, Ramtanu Maitra, convened an academic seminar in Delhi with Russian, Chinese and Indian scholars discussing triangular collaboration across Eurasia.

---

[33] Information Office (1996).
[34] LaRouche (1997).
[35] Xu (1997).

R. B. Rybakov, Chairman of the Russian Academy of Sciences Institute of Oriental Studies, Ma Jiali, a professor at the Chinese Institute for Contemporary International Relations (CICIR) and Devendra Kaushik, Head of the School of International Studies at Jawaharlal Nehru University, Delhi, discussed challenges and prospects. With Rybakov presiding, the scholars established a 'Triangular Association' with the goal of promoting Indo-Russian–Chinese cooperation in forging a shared vision of Eurasia's post-Cold War future of peace, progress and prosperity.

The effort failed for a combination of difficulties: fallout from the 'Asian Economic Crisis', the September 2001 al-Qaeda attacks on New York and Washington and the US' subsequent 'Global War on Terrorism', wars in Afghanistan and Iraq, and then, the *Great Recession*.[36] Seeds had nonetheless been sown in the febrile post-Cold War intellectual hotbeds, with one outcome being formation of the Russia–India–China (RIC) informal consultation forum. The three countries, myriad mutual differences notwithstanding, regularly met at foreign minister-level and occasionally at summits. Although largely informal, the grouping both reflected and reinforced trends towards Eurasian integration while eroding *systemic unipolarity*.[37]

Ideas analysed at Schiller's many conferences and events began gelling into policy-frameworks early in the twenty-first century. Although political, legal and financial challenges were compounded by the US' mainstream media almost totally ignoring its work, Schiller launched a series of publications in early twenty-first century.[38] These highlighted the import of the LaRouches' vision of a networked and united Eurasia that, being gradually realised via BRI and associated endeavours, would transform the lives of billions and transmogrify geopolitics into a non-competitive, positive-sum dynamic. Although the campaign resonated in China, where Helga LaRouche was an honoured guest at high-level conferences focused on the SREB, mainstream commentators in the USA and Europe greeted Schiller's perseverance with deafening silence.

The LaRouches were not the only US visionaries imagining an inter-connected post-Cold War Eurasian future. Schiller's protracted endeavours eventually percolated into the US academia. S. Frederick Star, a Johns Hopkins University scholar specialising in Central Asian studies, presided over an April 2006 conference themed 'Partnership, Trade and Development in Greater Central Asia' in Kabul, Afghanistan, amidst the US-led counter-*Taliban* military campaign raging across much of the country. A number of mostly Western specialists presented papers which were later edited into a volume titled 'The New Silk Roads: Transport and Trade in Greater Central Asia' and published in 2007.[39]

By this time, the US Congress was already exploring legislative options supportive of US engagement with Central Asian states and regional stability, with a view to strengthening Washington's strategic influence and footprint there. So, the general

---

[36] Maitra (2013).
[37] MEA (2019) and Chen and Feng (2016).
[38] Beets et al. (2018), Report (2017a, 2018a), LaRouche (2017, 2018) and Askary and Ross (2017).
[39] Starr (2007).

concept of Eurasian transport-and-trade integration, with many separate but interconnected points of origin, including private US political–economic activism and academic scholarship, came together early in the new century. Execution, dogged by practical, political, financial, regulatory and engineering challenges, was slow. Beijing's assumption of leadership, with multilateral support, propelled the drive to forge BRI/SREB.

4.3 Brussels Builds a Trans-Eurasian 'Land-bridge'

La Rouche and his foundation sponsored several high-profile events in European cities to drive the vision of Eurasian integration. This resonated with the European Union's (EU) interest in knitting together post-Soviet Europe into a united, liberal-democratic and capitalist, Europe. Given the former Soviet Union's spread across the Eurasian landmass, EU interest extended to Central Asia and further eastward. The North Atlantic Treaty Organisation (NATO) and the Organisation for Security and Cooperation in Europe (OSCE), as indeed the UN and its various organs, too, launched efforts to ease former Soviet-bloc polities into the Western fold. Success varied. However, given its non-confrontational approach, combining ideational attraction and financial and technological prowess, Brussels made a greater impact on the ex-Soviet space.

Shortly after the USSR's dissolution, the EU initiated an Agreement on Partnership and Cooperation with former Soviet republics. European republics were the first to negotiate, sign and ratify the agreement. Talks began with Kazakhstan and Kyrgyzstan in May 1996 and with Uzbekistan, in June. In July, the European parliament and all 15 EU members ratified the accord. The EU also signed temporary trade agreements with several regional states. Unlike former Soviet republics in Central- and Eastern Europe, Central Asian polities showed little interest in pursuing EU membership and Brussels reciprocated that indifference. The EU's concerns resided in stabilising the region to ensure its own periphery remained undisturbed by the reverberations of Central Asia's troubled transition and any fallout from incipient complex competition among the USA, Russia, China and India roiling the region.

The *Stan*s' engagement with these and other—mostly Western—organisations and institutions, conducted bilaterally and expressed in political–economic support, helped redefine the former's revived sovereignty, offering hopes of economic development and poverty-alleviation, and a more optimistic and peaceful future.[40] In the context of that general aspiration, the EU co-hosted a 1998 conference on 'Restoration of the Historical Silk Road' in Baku, Azerbaijan. Given the official nature of the gathering and high-level of attendance and presentations, the 'Silk Road' motif gained currency in Eurasian developmental discourse.

---

[40] Rakhimov (2010).

4.3 Brussels Builds a Trans-Eurasian 'Land-bridge'

That enabling environment proved crucial to integrating the *Stans* and realising the region's potential for becoming a hub for Eurasian transportation and commercial integration. The EU's stabilisation efforts began in economic-commercial transactions. EU-wide coherence came in the form of its 2007 'Strategy for a New Partnership', encouraging the four Central Asian states not yet WTO members, to move towards accession. The *Stans*' limited market size with 60 m people of modest means outside Kazakhstan offered few incentives for private investment beyond hydrocarbons. German–Kazakh cooperation in construction, tourism, agriculture, transport infrastructure, agribusiness and textiles was an exceptional bright spot. Like other interlocutors, Germany, the EU's economic powerhouse, saw Kazakhstan as its salient Central Asian partner. This view was widely shared among Britain, France and Italy.[41]

Arguably, the most directly relevant EU programme was Brussels' Transport Corridor Europe–Caucasus–Asia (TRACECA). Launched in 1993, TRACECA would fashion a corridor linking Europe to China via the Black Sea, the Caucasus, the Caspian Sea and Central Asia. Brussels aimed to wean post-Soviet polities along the planned corridor, i.e. Armenia, Azerbaijan, Bulgaria, Georgia, Kazakhstan, Kyrgyzstan, Moldova, Romania, Tajikistan, Turkmenistan, Ukraine and Uzbekistan, from reliance on Moscow by strengthening their politico-economic independence via exposure and access to European and other non-Russian markets and investment sources. After five years of negotiations, the parties, as well as Turkey and Iran, signed a 'Basic Multilateral Agreement on International Transport for Development of the Europe–the Caucasus–Asia Corridor' (MLA) in 1998. The EU funded the Intergovernmental Commission (IGC), housing the TRACECA Permanent Secretariat, in Baku, Azerbaijan. EU grant of Euro 180 m financed 80-plus projects focusing on transport-infrastructure, legal harmonisation, transport safety and security, trade facilitation and logistics. TRACECA would

- Assist development of economic relations, trade and transport communications linking Europe, the Black Sea region and Asia
- Ensure access to the world market of road, rail transport and commercial navigation
- Ensure traffic security, cargo safety and environmental protection
- Harmonise transport policy and legal structure relating to transportation and
- Create conditions of equal competition for transport operations.[42]

These EU-initiated and -funded programmes both bridged the physical gaps hitherto preventing Europe-to-China terrestrial movement and aided the evolution of joint regulatory frameworks and state-level perspectives favouring the construction of transport networks knitting the Eurasian supercontinent into an integrated whole. Having established the framework for laying down the 'Eurasian Land-bridge', the EU midwifed regulatory norms guiding and formalising Europe–Central Asia–China transportation, terrestrial, aerial and maritime links:

---

[41] Peyrouse (2009).

[42] Directorate General for International Cooperation and Development (2017).

- Logistical Processes and Motorways of the Seas (2011–2014): developing strategies for establishing intermodal integrated transport and logistics chains 'underpinned by Motorways of the Seas' and manifest in the delivery of a 'Master Plan'
- Civil Aviation Safety and Security (2012–2015): covering civil aviation regulation and oversight, market access, safety, security, air traffic management and environmental protection
- Transport Dialogue and Networks Interoperability II (2012–2015): improving coordination among members of the European Neighbourhood Policy states, involving international financial institutions and private investment in transport projects
- Maritime Safety and Security II (2012–2013): securing ratification and implementation of international maritime safety and security conventions and enhancing the quality of maritime administration across the Black Sea and Caspian Sea states
- Road Safety II (2013–2016): assisting TRACECA states to implement the TRACECA Regional Road Safety Action Plan with a view to significantly reducing road accidents, deaths and injuries.[43]

These endeavours resonated with the EU's initiatives to further integrate transport and communications networks linking its Western European member-states with the newer Central- and Eastern European members. The European Parliament and the European Council adopted plans in December 2013, as amended in December 2016, to fashion 'core network corridors'. These terrestrial, maritime/marine and aerial corridors, comprising four 'core and comprehensive' trans-European transport networks covering freight, passengers, roads and waterways, were designed to be completed by 2030.[44] The eastern extremities of the networks joined up, or were collocated with, several SREB alignments. Convergent Chinese and European interests, advanced energetically by two of the world's three largest economic actors, sped up the realisation of the SREB blueprint.

The success of these initiatives could at least be partly measured from the fact that China's trade with Central- and East European states multiplied from $6.8 bn in 2002 to $58 bn in 2013, and China's aggregate trade with all former Soviet republics exploded from $16 bn to $153.5 bn over the same period. Nonetheless, analysts cautioned Europeans to 'fully understand' and acknowledge 'the geopolitics of transport and trade' so as to ensure that no power gained the ability to 'monopolise or control' the emergent trans-Eurasian transport corridors.[45] While European observers urged broad support for China's SREB and MSR blueprints, to ameliorate the potential risks others perceived, they recommended equally strong focus on forging 'the Southern Corridor', linking South Asia, particularly India and Pakistan, to Central Asia and on to Europe via Turkmenistan, which demanded a shift in policy.[46]

---

[43] Ibid., p. 2.
[44] Juncker (2016).
[45] Starr and Cornell (2015).
[46] Cornell and Engvall (2017).

4.3 Brussels Builds a Trans-Eurasian 'Land-bridge' 137

Such a shift, were it to occur, could indicate a hardening in Brussel's geoeconomic perspective vis-à-vis China generally and BRI specifically. However, the EU's High Representative for Foreign Affairs and Security Policy, Federica Mogherini, underscored Brussels' interest in Central Asia by visiting Uzbekistan twice in five months: to attend an international conference on 'Central Asia: Shared Past and Common Future, Cooperation for Sustainable Development and Mutual Prosperity' in Samarkand in November 2017, and to participate in the Tashkent Conference on Afghanistan in March 2018. EU ambitions were set out in two policy documents: EU Strategy on Connecting Europe and Asia, and the EU Central Asia Strategy.[47]

They showed that concerns over China's 'rise' notwithstanding, Brussels remained engaged in efforts to knit Mackinder's 'World Island' into a benign, positive-sum, collective. However, the EU's emphasis on 'the European Way: sustainable, comprehensive and rules-based connectivity' it sought to forge across Eurasia underscored a not-too-subtle questioning of BRI's attributes.[48] The near-simultaneity of Mogherini's office launching a strategy document describing China as 'a systemic rival' and an 'economic competitor', Italy signing on to China's BRI vision, EU leaders boosting multilateralism with Xi Jinping, and Brussels fitfully following the general direction of Washington's toughening approach to technological trade and investment with Beijing illuminated manifestations of the fluidity of *systemic transition*.[49] The EU's identification, although without naming, of China and Huawei, as sources of 5G threats, illuminated the trajectory of Brussels' threat-perceptions running parallel to Washington's.[50]

4.4 Multinationals Revive the Silk Road

Despite Chinese requests to collaborate and jointly implement the BRI blueprint to which 126 states and 29 international organisations had acceded,[51] the US government, apart from issuing bitter criticism, abjured any involvement. Other US nationals found the vision more appealing. Indeed, scholars and thinkers were not the only US actors driving the 'New Silk Road' paradigm. Western, especially US, corporations deeply invested in manufacturing or assembling consumer goods and other products in China, primarily for Western markets until Chinese demand picked up, played a pioneering role in forging what became the SREB's spine. They had good reason to. In April 2007, the CPC and China's State Council issued a joint 'guideline' for the development of Central Chinese Provinces with a view to spreading the benefit of

---

[47] High Representative for Foreign Affairs and Security Policy (2018) and General Secretariat (2017).
[48] High Representative for Foreign Affairs and Security Policy (2018, pp. 2–3).
[49] Rosario (2019), Chatzky (2019), Report (2019a) and High Representative for Foreign Affairs and Security Policy (2019).
[50] Network Information Security Group (2019).
[51] Cui (2019).

industrialisation, urbanisation and economic well-being beyond China's coastal belt. The NDRC, 'by taking reference of favourable policies applied in the revitalisation' of China's north-eastern 'rust-belt', other now-moribund regions of China's former industrial heartland, and Western China, offered attractive benefits for the flow of FDI into Central China.[52]

By early-2010, the policy had drawn a large number of Chinese and foreign manufacturers to setting up plants and ancillary facilities in vast new industrial parks in cities like Chongqing, a conurbation with provincial status, Chengdu, capital of Sichuan Province and Zhengzhou, the capital of Henan Province. Big names included Intel, the world's largest microchip-maker, Hewlett-Packard (HP), one of the world's largest personal-computer manufacturers and Foxconn, sired by Taiwan's Hon Hai Corporation and a key supplier to Apple, IBM, Google, Microsoft, Intel, Cisco, GE, Amazon, HP, Dell, Motorola, Nokia, Sony, Panasonic, Toshiba, Fujitsu, Samsung, Acer, Huawei, Lenovo and others. Foxconn represented the increasingly globally integrated manufacturing supply chains binding assemblers and producers from several continents to Chinese production bases. The move by Foxconn and many of these other firms to new central Chinese manufacturing hubs shifted the centre of gravity of global IT-hardware production from coastal China to central China.[53] This had several effects.

HP's experience was illustrative. Until it relocated to Chongqing, its notebook PCs were built in Shanghai's manufacturing hub. Rising wage-costs of labour, of technical, supervisory and managerial staff, charges for electricity and other utilities on the coast, and attractive benefits offered by Chongqing's Xiyong Micro-electronic Industrial Park, persuaded HP to move. HP and its principal subcontractor, Foxconn, jointly invested $3 bn in a new factory in Chongqing. Completed in 2012, the plant had the designed annual capacity to produce 20 m laptops. Other IT luminaries, Quanta and Inventec, too, relocated some production facilities from the coast to the same Park. Established in 2005, the Xiyong Micro-electronic Industrial Park had attracted investment worth $1 bn in its first three years, but between mid-2009 and late-2010, investment-flows reached $3 bn. Initially, as most parts and materials were brought from the coast, per unit production costs at inland factories were marginally higher. As manufacturers began sourcing components and materials locally, costs came down, while economic activity increased and regional supply chains both expanded and deepened.

Comparable incentives attracted Dell to Chengdu, while Foxconn similarly relocated a major production base to Zhengzhou. Zhengzhou provided land to Foxconn for its new 'campus', whose first phase covered 133 hectares. Foxconn also planned to move its major production line from Shenzhen to Langfang in the northern Hebei Province. Dell, based at Xiamen in the Fujian Province, similarly planned to relocate westward. After investing $600 m in a new plant in Chengdu, Intel moved its major chip-manufacturing efforts there in 2010. Troubled by the rising costs of hiring talented researchers in coastal China, the largest US pharmaceuticals manufacturer,

---

[52] NDRC (2007).
[53] Report (2010), Pomfret (2010), Meng (2010) and Pijl (2015).

Pfizer, established a new research-and-development (R&D) centre in Wuhan, in the central Hubei Province. Wuhan's East Lake Hi-Tech Development Zone offered tax and housing benefits, while Pfizer negotiated research partnerships with Wuhan University.[54]

Resonantly growing industrialisation of central China in large-scale manufacturing hubs around Chongqing, Chengdu, Zhengzhou, Wuhan and other regional cities, rapid expansion of export trade and the accretion of trans-Eurasian railroad networks across Central- and Western China created the perfect commercial-logistical backdrop against which Western firms based in central China began exploring faster methods of shipping their wares to Europe. Since airfreight remained too expensive and sending goods first by truck or train to eastern Chinese ports before shipping them to European markets by sea often took 40 days, the prospect of containerised freight-trains running from Western China across Eurasia to Europe looked attractive.[55]

The early China-to-Europe freight-trains, hired by Western manufacturers and their suppliers with plants in central China, moved containers carrying electronics, cars, car parts and medical equipment. Since 2011, the trains, starting in Chongqing, passed Xi'an, Lanzhou, Urumqi and the Alataw Pass where they crossed into Kazakhstan, then on to Russia, Belarus and Poland, finally entering Germany to stop at Duisburg. The 11,179 km one-way trips took 16 days. In July 2013, a second route to Germany, from Zhengzhou via Kazakhstan, Russia, Belarus and Poland, to Hamburg, 10,214 km away, opened. With two gauge changes, at Alataw Pass and on the Belarus–Poland border, this route took 18 days.[56] By mid-2016, over 1700 trains had travelled from Chinese hubs, across the Eurasian expanse, linking dozens of cities in Central Asia, Europe and China.

And it all began in 2007 with Terry Gou, Foxconn's founder, asking Ronald Kleijwegt, head of Foxconn's logistics, to explore the prospects of rail-freighting their export products from China to former Soviet republics and European markets. Kleijwegt identified possible routes in 2007, and by 2008, Foxconn was sending trainloads of electronic goods and components from its Shenzhen factories via Mongolia and Russia to Eastern Europe. When Mongolia doubled transit fees, reducing the route's attraction, Foxconn re-routed shipments from Shenzhen using sections of the Trans-Siberian Railway. Other manufacturers began using this route with Harbin, Zhengzhou and Suzhou as hubs. But the margins on this route were smaller. Kleijwegt moved from Foxconn to HP as director of global logistics in March 2009, coincidentally just as HP relocated its main plants to Chongqing. At the time, Chongqing lacked logistics facilities for rail-freighting exports across Eurasia. Hit by the *Great Recession*, shippers were then 'slow steaming' freighters to save fuel costs, extending China-to-Europe shipping time from 25 to 35 days and disrupting supply-chain schedules. This led HP, using a Foxconn test-run, to pilot freight-trains

---

[54]Zhou (2009).
[55]Roberts et al. (2012).
[56]Zhou (2013).

from Chongqing to Europe via Kazakhstan, Russia, Belarus and Poland. The New Silk Road's 'southern route' was born.[57]

But before rail-freight along the route could become 'routine', HP executives had to explain their goals and plans, and how these would benefit transit-revenue coffers, to senior officials and ministers in countries lying along its alignment. Trans-border customs inspections and inter-gauge trans-shipments consumed time until the launch in July 2011 of a Russia–Kazakh–Belarus customs union, which granted freight-railway the same status as passenger trains, allowing the former to run almost directly across Central Asia. This paralleled the establishment of a joint venture, the EurAsia Land-bridge Rail Service, uniting track operator DB Schenker, the Chongqing Holding Group and the state railway corporations of China, Kazakhstan and Russia. As this joint venture streamlined the New Silk Road's hardware, more Western and other China-based multinational producers and exporters evinced interest.[58] Switzerland's InterRail Group later assumed management responsibilities for key sections.

HP, running 'pilot freight-trains' into 2012, showed the promise of the 'Silk Road' railway routes. To boost their commercial viability and political–economic heft before seeking their 'routine status', in 2013, HP invited fellow-exporters to hop aboard. Dell, Asustech, Foxconn and other exporters, and international carriers-and-forwarders like DHL, Essers, Geodis, Haltrans and Wagonborg did, ensuring steadily increasing train-frequencies. As additional Chinese and European cities signed on, frequency and volumes rose, slashing the cost of shipping a container by train from $7000 to less than $4000. A Kazakh–Russian joint venture built new locomotives using GE designs and components from Pennsylvania. Container-freight unit costs were still 20–25% higher than shipping by sea when Xi summoned the first BRF. However, prices were falling, and the 15–17-day door-to-door trips proved attractive for shippers despatching consignments chasing time-schedules or seasonal deadlines.[59]

The Eurasia Land-Bridge railway service reached a milestone on 18 January 2017 when a 34-twenty-foot-equivalent-unit (TEU) freight-train, travelling over 12,000 km from China's Yiwu in 18 days, arrived at DB Cargo's London Eurohub in East London. Taking half the time needed by merchant-shipping and costing half of what airfreight demanded, the train demonstrated commercial viability of China–UK freight-rail-services, with London becoming the 15th European destination directly linked to Chinese cities. A year after the inaugural Yiwu-London service, a 36-container train started from Urumqi, Xinjiang's capital, to a Kazakh port where the containers were transferred on to a ship bound for Baku, the capital of Azerbaijan.

The freight-rail network expanded, linking a growing collection of cities, ports and production-hubs, deepening trade-ties, while spurring innovation enhancing the network's technical features. Most innovative advances were initiated and operationalised by Western corporations. Several such firms were engaged in tackling

---

[57] Shepard (2016a) and Report (2016a).
[58] Macguire and Anderson (2013).
[59] Bradsher (2013), Shepard (2016a) and Smith (2017).

4.4 Multinationals Revive the Silk Road

the one factor which neither the corporations fashioning the New Silk Road nor governments whose jurisdiction it traversed could alter: Eurasian weather. The Silk Road's northern route spanned Siberia, while the southern route bridged Kazakhstan, and both crisscrossed Eurasia's varied topography and multiple, often extreme, climatic patterns. High-value, temperature-sensitive items, e.g. micro-electronics and components, pharmaceutical active-ingredients and products and perishable foodstuffs required constantly monitored and controlled temperatures. Standard heated or refrigerated containers had to be manually reset and refuelled repeatedly across the 12,000 km routes. As the railways' key selling point was speed, this challenged sustaining and expanding the service.

To address it, Ronald Kleijwegt of Foxconn and HP fame met up with Jan Koolen, head of the Dutch shipping container manufacturer, Unit 45, in 2012. Koolen was fabricating heated diesel-electric containers used by European road hauliers, but their 250 l tanks would be useless for the fortnight-long, 12,000 km Eurasian rail-journeys. Unit 45's technical team worked to develop 45-foot, self-sufficient containers armed with 800 l diesel tanks that could maintain steady temperatures within their storage spaces for three weeks as the trains travelled from one climate zone to another and yet another, before returning all the way. The new containers carried remotely controlled thermostats capable of changing temperatures on receiving instructions from operators' laptops stationed at control stations thousands of miles away.

These were also GPS-equipped so that operators knew their exact location throughout a trip and armed with light-triggered alarms that could alert operators to any untoward incident, providing real-time diagnostics.[60] Even with added costs of the new containers and their diesel-fuel, given the value of the merchandise, margins remained attractive, making it feasible to freight, for instance, Dutch veal to China, and Chinese-made electronics and pharmaceuticals to Europe. These and other innovations made the New Silk Road a year-round enterprise, enhancing its commercial value and financial viability. The containers were a key aspect of the technological revolution necessitated by the New Silk Road. Other advances eased terrestrial inter-gauge trans-shipment and automated cargo-handling procedures and mechanisms.

The New Silk Road crossed states operating railways with varied gauges, requiring operators to move containers from one rail-bed to another without damaging delicate and fragile goods or consuming excessive time. One response was a Sino-Kazakh trans-shipment joint venture along and across their border at Khorgos, initially a 6 km^2 site on a barren desert on which the two neighbours fashioned a sophisticated dry port, with supporting infrastructure, congruent commercial, residential and cultural hubs and recreational facilities. Kazakh workers were drawn to free housing, subsidised schooling and other facilities at a newly built border village called Nurkent. A free-trade zone called the International Centre for Boundary Cooperation offered visa-free cross-border travel and commerce.[61]

---

[60] Jeremy (2017), Shepard (2016b) and Sand (2015).
[61] Suzuki (2019).

Formally called the Khorgos Eastern Gates Special Economic Zone, this was envisaged as the world's largest dry port. Under the management of DP World, which adapted systems from its Jebel Ali Free-Trade Zone in Dubai as the template on which to build the Khorgos Dry Port, this Eurasian land-port processed a cargo train in 47 min. Khorgos also emerged as a manufacturing, trans-shipping, warehousing, importing and exporting hub. After recording 113% growth in cargo-handling in 2017, Khorgos processed 133,922 TEU in 2018, a 44% increase year-on-year.[62] Some of the credit went to California's Cargotec Corporation, which designed, furnished and installed its custom-built N4 Terminal Operating System, one of the world's most advanced.[63] Underscoring rapid growth in New Silk Road commerce, in mid-2016, the dry port's managers projected that Khorgos would handle 500,000 TEU containers annually by 2020.[64] SREB's critical Khorgos fulcrum holding the Eurasian Land-bridge network together relied heavily on the quality, efficacy and cost-effectiveness of US-Finnish electro-mechanical equipment and support systems.

As the SREB network matured, its spurs inched southwards, connecting the Eurasian hubs to maritime ports along the MSR. Reducing both time and costs, this coalescing rail-ship connectivity showed the potential for 'Silk Road' transportation to become truly multi-modal.[65] Still, in 2016, railways carried just under 1% of the $570 bn China–Europe trade by volume and just over 2% by value. Ships carried 94% of trade by weight and 64% by value. Airfreight carried the rest.[66] By the end of 2018, SREB railway services connected 51 cities in 15 European countries to 62 Chinese cities, along with several transit-hubs in countries along the route. A total of 14,691 trains had carried 1.1 m TEUs since March 2011. In 2018 alone, the trains freighted goods worth $33 bn, a 106% jump year-on-year. Of the containers, 94% starting from China were fully loaded, as were 71% of those arriving in China.[67]

As is apparent, although over a decade, China–Europe freight traffic grew from zero to a substantial category, its economic impact on over all Europe–China commerce remained modest. However, the perceptual and normative linkages with which it knit Eurasia with deepening intensity might produce disproportionately impactful political–economic outcomes over the long term. The rapidity of this development, creating and reinforcing the transportation, regulatory-software and policy-convergences along proliferating new routes, laid the palimpsest that became the SREB's spinal column. But even Western corporations not directly engaged in laying out the SREB hardware-software mix were involved in boosting BRI's prospects. The experience of Siemens, a globe-girdling German engineering conglomerate and one of the world's largest, typified that contribution.

Siemens entered China in 1872 as part of a Western imperial-commercial drive to penetrate that semi-mysterious market. That year, Siemens delivered China's first

---

[62] Suzuki (2019).

[63] Report (2016b) and Shepard (2017a).

[64] Report (2017b).

[65] Report (2018b, c).

[66] Hillman (2018).

[67] Belt and Road Portal (2019) and Report (2019b).

pointer-telegraphic system to be used by Western diplomatic outposts and trading firms. In 1899, Siemens installed a power plant in Beijing, one of China's earliest, electrifying several city districts and enabling the launch of Beijing's first electric streetcar network. Trade boomed and early in the twentieth century, Siemens China became the corporation's biggest subsidiary outside Europe. After decades of post-revolution inactivity, drawn by Beijing's 'reform and opening up' outreach, Siemens re-entered China in October 1985, signing a 'Memorandum of Comprehensive Cooperation between Siemens and China's mechanical engineering, electrical and electronics industries'. It built production centres, design units, research-and-development laboratories, and trained Chinese technical personnel. Siemens transferred and co-developed technology, contributing to China's FDI-driven renascence as a manufacturing and trading power.

By 1995, Siemens had 30 joint ventures in China. One of the firm's most visible contributions to Beijing's efforts to power the economy and integrate China's remote western reaches into its political–economic mainstream resided in its development of high-voltage direct current (HVDC) technology for power transmission across distant provinces. Siemens initiated HVDC projects in China in 1989 and in 2018, was building, as part of Beijing's Western Development/BRI vision, a 3000 km transmission line powered with the world's first 1100 kv transformer manufactured in China. When completed, the line would transmit surplus electricity from Xinjiang to a power-hungry Anhui Province. Of the 32,000 Siemens personnel stationed at its 40-plus factories, 21 innovation centres and 60 regional offices spread across China, 4500 were engaged in R&D work. By mid-2018, Siemens had commercial interest in 89 companies in China, directly and indirectly employing more than 100,000 people.[68]

Siemens thus reflected and reinforced China's integration into the global technological manufacturing-and-supply chains. In sharp contrast to Washington's fulminations against Beijing's 'Made in China 2025' R&D framework, Joe Kaeser, Siemens's President and CEO, welcomed it as a pathway to 'modernising China's industrial base' and vowed to contribute to its fulfilment. At the June 2018 Beijing Belt and Road International Summit which Siemens hosted, linking poverty alleviation and improvement in lives to manufacturing, both globally and especially in China, Kaeser enthused, 'I do commend the Chinese government for making this a top priority. If any country in the world proves how important manufacturing is, it is China'. He noted Siemens had built its first digital manufacturing centre outside Germany in Chengdu, a key hub of Beijing's Western Development drive and a fount of BRI/SREB trans-Eurasian connectivity.

In Siemens's perspective, BRI was a key shaper of the emerging world order: 'the current geopolitical constellations and the BRI are changing China's role in the world—not just strategically but also economically'. Kaeser saw China's vision as an attractive one: 'the BRI is an invitation to the rest of the world to take part in the biggest infrastructure project of all time'. He described BRI as 'a landmark movement that represents a Euro 1tn investment in infrastructure' covering 90 countries, with

---

[68]Kaeser (2018).

'the potential to improve the lives of 70% of the world population'. Kaeser assessed, BRI created 'opportunities in practically every sector'.[69] Judging by the response of his Chinese and non-Chinese audience, his was not a solitary voice.

Rail networks, the SREB's crucial spinal column linking Chinese ports and manufacturing hubs to Eurasian and European ports via Central Asia and CEEC countries, could only function as well as European corporations' partnerships and joint ventures with their Chinese counterparts allowed them to. Hupac, a Swiss logistics corporation, for instance, worked with China Railway Container Transport and City Railway grouping on the Yiwu–Madrid and other alignments linking China to France and Germany; Geodis, a French firm, actively participated in the Chengdu–Rotterdam link and Wuhan Asia Europe (WAE), a logistics firm based in Wuhan, Hubei Province, collaborated with French partners Geodis in running the Wuhan-Lyon services. Germany's DHL, too, energetically utilised and encouraged the expansion of SREB links for moving many of its China–Europe consignments. The SREB functioned and grew profitably because Chinese and European partners saw significant benefits flowing in both directions from their collaboration.[70]

Oil-and-gas pipelines, too, attracted Western corporate engagement long before 2013. Honeywell, the US-based technology conglomerate, teamed up with CNPC and Huawei in 2009 to support the Uzbekistan–China gas-pipeline project. In 2012, Honeywell again joined Huawei to similarly support the 'Line C' section of a 7000-plus km pipeline complex supplying Central Asian gas to China. In 2013, China became Honeywell's largest non-US market. With 23 units employing over 32,000 local staff along BRI alignments, Honeywell was positioned to provide hydrocarbon transmission and processing technological support as the SREB's energy component took shape. In 2017, Honeywell joined Wilson Engineering, a chemicals firm supplying China, offering methanol-to-olefin technologies, engineering and construction services to BRI-partner-states.[71] General Electric and Caterpillar, too, showed interest. Western corporate engagement and contribution thus hastened BRI implementation.

4.5 The World Bank Pushes China's Global Integration

Major UN organs, as noted, played a crucial role in establishing both conceptual paradigms and policy-frameworks for Eurasian states to seek faster economic growth using regional transport, communications and commercial integration as vectors. The *Great Recession's* global disruptions, and China's FDI-financed, foreign-owned, export-manufacturing's shift inland, generated commercial incentives for exploring terrestrial shipping of Chinese-made and -assembled export goods to Europe.

---

[69] Kaeser (2018).

[70] Urban (2018), Media Office (2017), Shepard (2017b), Media Centre (2016) and Information Office (2016).

[71] Tedjarati (2017), Tsuruoka (2017) and Wang (2016).

4.5 The World Bank Pushes China's Global Integration

Beijing's domestic socio-economic dynamics resonated with these changes, driving NDRC development strategies and plans to progressively adopt, adapt and refine trans-continental integrationist tendencies to Chinese interests. A crucial encouragement came from the World Bank, specifically, its then-President, Robert Zoellick.

In 2005, concerned over the disruptive potential of China's dramatic economic breakout, then-Deputy Secretary of State, Zoellick urged Beijing to become a 'responsible stakeholder' upholding the US-led post-Cold War order.[72] Chinese officials apparently internalised and ascribed this attribute, to China's external policies since then.[73] Beijing's 2011 White Paper on 'Peaceful Development' noted, 'For China, the most populous developing country, to run itself well is the most important fulfilment of its international responsibility'. As for diplomacy, 'As a responsible member of the international community, China abides by international law and the generally recognised principles governing international relations and eagerly fulfils its international responsibility. China has actively participated in reforming international systems, formulating international rules and addressing global issues'. China 'should play a constructive role by fulfilling their due international responsibility in accordance with their own capability and on the basis of aligning their own interests with the common interests of mankind'. China 'will assume more international responsibility as its comprehensive strength increases'.[74]

Having internalised Zoellick's recommendation on assuming the attributes of a 'responsible stakeholder' in the international security system, Beijing paid attention to his advice when Zoellick assumed WBG presidency. In 2010, while marking the 30th anniversary of the WBG-PRC partnership, Zoellick proposed to Chinese leaders 'to work jointly on identifying and analysing China's medium-term development challenges looking forward to 2030. Together, China and the World Bank would conduct research drawing on lessons from international experience as well as China's own successful development record and prepare a strategic framework for reforms that could assist China's policy making as well as guide future China–World Bank relations'. As Zoellick, his WBG successor Jim Yong Kim, and their NDRC counterpart noted, 'China's state leaders welcomed and supported the proposal'.[75] The fruit of that WBG-NDRC collaboration was ready shortly after Kim succeeded Zoellick. Made public in 2013, the joint-report reverberated in and powerfully informed NDRC's subsequent work. The World Bank's endorsement of and support for Beijing's BRI implementation proved crucial to 'multilateralising' BRI.[76]

---

[72] Zoellick (2005).

[73] Zoellick (2007), Rudd (2008), Ding (2009), Editorial (2011), Li (2009), Xinhua (2013), Chen (2016, 2017), Garlick (2016) and Zhao (2018a).

[74] Information Office (2011).

[75] Zoellick et al. in Liu et al. (2013, p. xiii).

[76] For more information, you can find the World Bank Group's 'Belt and Road Initiative Map' issued on 29th March 2018, here: http://www.worldbank.org/en/topic/regional-integration/brief/belt-and-road-initiative (last accessed 16 October 2019).

World Bank and NDRC economists told Chinese policymakers that facing decelerating growth, slower export expansion and lower FDI inflows, Beijing needed to support Chinese firms 'going global', investing overseas, acquiring technology and market-share and 'move production up the value chain'.[77] Abandoning its marginal role in global financial markets, 'China will have to become a more dominant player'. Instead of pegging the RMB to the dollar, China's financial 'stability should come from a more international and independent role for China's currency'. Crucially, 'it is in the interest of both China and the rest of the world that' rather than reacting to events, 'China adopt a more proactive approach and take responsibility for proposing solutions to global governance problems and for the provision of global public goods'. WBG and NDRC economists urged Beijing to robustly act across the spectrum of economic, commercial and financial policy. 'Such a successful transformation can create win–win opportunities for China and the world'.[78]

Noting how rising protectionism threatened multilateral trade-regimes, they encouraged China to 'emphasise both multilateral and regional arrangements. It will benefit from abiding by and protecting existing multilateral agreements' and 'pushing for further opening of global markets using multilateral channels'.[79] World Bank economists urged the RMB's internationalisation: 'greater use of the RMB as an international currency would provide more economic stability than a managed exchange rate … With unrestricted capital movements, a floating exchange rate will be necessary to enable the government to use monetary policy to control inflation. This strategy entails risk, however … Thus, a relatively conservative approach, stretching over many years, is recommended in transitioning to a more open and efficient financial and exchange rate system'.[80] They suggested that instead of leaving 'the determination of these global policies (on providing public goods) to a multilateral consensus', Beijing 'actively help to shape global agreements'.

Actively shaping global rules implied 'shouldering some of the costs of preserving global public goods'. But Beijing had 'much to gain from helping to shape international agreements on public goods'.[81] However, growing economic salience notwithstanding, China faced major challenges. WBG economists asserted that Beijing could not achieve its 'goals by looking inward. Instead, China needs to embrace further steps towards global integration to improve the competitiveness of its economy and sustain increases in living standards', by precluding 'middle-income trap' stagnation. WBG- and NDRC economists concluded, 'Using its capital surplus to invest in foreign markets, increasing exports of more sophisticated goods, encouraging domestic competition in services sectors and deepening the financial sector through the participation of foreign financial institutions' would enable China to sustain economic growth.[82] The report provided examples from Japanese and European

---

[77] Liu et al. (2013, pp. 361–362).
[78] Liu et al. (2013, pp. 361–362).
[79] Liu et al. (2013, p. 365).
[80] Liu et al. (2013, p. 366).
[81] Liu et al. (2013, p. 366).
[82] Liu et al. (2013, p. 378).

4.5 The World Bank Pushes China's Global Integration

experience in managing FDI, currency internationalisation and voluntarily restraining exports for Beijing to consider. These WBG propositions informed many Chinese policies in Xi's 'new era'.

The WBG country report on China, launched shortly before Xi assumed CPC leadership, sketched out China's achievements, its evolving economic landscape, and challenges ahead; it proved equally influential. The report noted that GDP growth 'averaging about 10% a year' over three decades had 'lifted more than 600 m people out of poverty'. All MDGs had 'been achieved or are within reach. With a population of 1.3bn, China recently became the 2nd largest economy'. Still, with per capita GNI of $4930, China ranked 114th globally, with over 170 m people still mired in poverty.[83] Beijing must manage uncertainties afflicting the world economy, create a more sustainable, consumption-based, trajectory for its own economy 'and reversing inequalities of income and opportunity', address pollution and resource-depletion. Previewing the 'China 2030' blueprint, the country report recommended 'supporting greener growth', 'promoting more inclusive development' and 'advancing mutually beneficial relations with the world'.[84] WBG economists lauded China's 'remarkable' post-1979 socio-economic achievements but noted fresh challenges this dramatic transformation precipitated:

- The characteristics of poverty had changed, complicating further poverty-reduction
- China betrayed high inequality of income, consumption, assets and opportunity
- Using income and poverty parameters, China now comprised three distinct regions—the coastal belt, home to 45% of the population, enjoying clear benefits of reforms and growth; and the central and western regions, home to 55% of the population, which enjoyed only half the former's per capita GNI.[85]

Acknowledging that Beijing's 12th FYP (2011–2015) addressed these difficulties, WBG experts noted that WBG support had enabled China to extend its South-South 'knowledge exchange' with other developing countries, especially in Africa. The WBG, China's Ministry of Finance and the NDRC closely collaborated, annually soliciting development project-plans from provinces and municipalities and jointly vetting these for International Bank of Reconstruction and Development (IBRD) funding. Most major projects, except those of national significance managed by Beijing, e.g. railways, were jointly approved by the State Council and IBRD.[86]

WBG's key components active in China, the IBRD, International Finance Corporation (IFC), and Multilateral Investment Guarantee Agency (MIGA), in convergence with the State Council's Western Development Strategy and plans designed to boost the central- and other under-developed regions,[87] focused investments there. Consequently, WBG advice and support boosted Beijing's pursuit of its regional

---

[83] Funk et al. (2012, p. i).
[84] Funk et al. (2012, p. i).
[85] Funk et al. (2012, p. 2).
[86] Funk et al. (2012, p. 14).
[87] Funk et al. (2012, p. 25).

development priorities. This, and other developments explained in the report, set the contextual backdrop against which FDI-financed export-manufacturing moved inland from coastal hubs, enabling exploration of the potential for trans-Eurasian overland shipping of myriad wares from Chinese factories to their European markets.

The World Bank encouraged China's growing profile as an economic and financial power with rising responsibilities in global economic governance and helped in 'enhancing the development impact of China's international economic engagement and cooperation'.[88] WBG advice and finance helped China to move up the global supply-chain from low-cost, labour-intensive manufactures to 'higher value products based on innovation'. Innovation and a WBG-boosted focus on environmental-sustainability[89] became refrains in China's developmental discourse. Parallel emphases on these and related issues in successive WBG reports and the State Council's economic-planning documents reflected collaborative synergy among analysts, economists and planners from WBG organs, China's Ministry of Finance and the NDRC.

WBG also worked with the UNDP, UNICEF, WHO, and the ADB, on helping China develop its socio-economic attributes.[90] The WBG posted regular updates on China's socio-economic transition generally following the schedule drawn up by NDRC and Ministry of Finance staff in consultation with their World Bank colleagues. In mid-2015, WBG analysts reported:[91]

- The transition to a 'more moderate and more balanced' growth trajectory continued
- Despite efforts to lower financing costs, aggregate credit impulses had moderated
- With quasi-fiscal expenditures tightening, fiscal policy remained 'accommodative'
- Economic deceleration affected urban market wages more sharply than jobs
- With domestic and external conditions changing, capital outflows had sharply risen
- As the transition to sustainable growth advanced, growth would keep moderating
- Balancing reforms and short-term demand-management remained a priority
- Transitioning to a 'new normal' required better credit-allocation and native innovation.

Whatever the critics' concerns over China's dramatic economic 'rise', the World Bank, a key source of advice, expressed broad satisfaction with the outcomes of WBG-NDRC joint efforts. WBG economists saw Beijing's BRI vision as a positive-sum reflection of China's success and the success of WBG's collaboration with Chinese planners in charting the trajectory of China's economic integration with Eurasia and the wider global economy generally. This was the view Mahmoud Mohieldin, WBG Senior Vice-President, expressed in September 2017.

At a UN Industrial Development Organisation (UNIDO)-sponsored conference in Vienna, he noted: 'The BRI has the potential to be global in its reach but local in its impact. It will reach 65 countries—potentially affecting 4.4bn people and leveraging

---

[88] Funk et al. (2012, pp. 30–31).
[89] Funk et al. (2012, pp. 18–20, 27).
[90] Funk et al. (2012, p. 32).
[91] Report (2015a).

4.5 The World Bank Pushes China's Global Integration

40% of the world's GDP. Huge investments will be channelled towards infrastructure projects across Asia, Africa and Europe. Six new land corridors will be rehabilitated, and maritime connectivity will be improved. And it has the potential to help countries reach their national objectives embedded in the SDGs, in areas such as jobs, poverty, infrastructure and sustainable cities'.[92]

Three weeks later, during the WBG-IMF Annual Meetings, the World Bank itself hosted an international conference on BRI, elucidating the vision underpinning the trillion-dollar, multi-decade, intercontinental undertaking, its challenges, and prospects. Then-WBG President Kim said, 'the world needs multilateral approaches like the Belt and Road'. Pointing to BRI's potential for spurring growth, he urged policy reforms among BRI-partners 'for countries to take full advantage of the opportunities presented' by BRI.[93] The locale, timing and substance of Kim's remarks and the sustained efforts by WBG economists collaborating with their Chinese counterparts over decades indicated a measure of authoritative ownership which underscored the WBG's belief in BRI's fundamental purpose and confidence in its prospects.

With many BRI projects nearing completion and the framework itself approaching its fifth anniversary, WBG analyses became more nuanced and mixed. To enhance the positives, mitigate the negatives and expand BRI's beneficial footprint, WBG economists generated a series of reports underscoring both potential gains and risks confronting BRI-partner-states. The reports offered advice on refining and risk-proofing the BRI blueprint. The basic premise WBG experts enunciated was that as a conceptual framework and a practical, physical set of transport, communications, informational, commercial, regulatory and inter-cultural exchange-based networks, BRI was a praiseworthy undertaking and a potentially transformative initiative. It nonetheless faced significant practical and economic-policy challenges. However, given its dramatic promise for a large part of the developing world, it merited strong support.[94]

Their key conclusions: 'an eventual reduction of trade costs' effected by improved connectivity would likely 'increase not only trade but also the vertical specialisation linkages in the region from the exporting and importing side'.[95] Some BRI economies, especially ASEAN-members with strong trade links to China, i.e. Vietnam, Malaysia, Philippines, Thailand and Indonesia, could expect import competition from Chinese manufacturers if BRI further integrated them into the Chinese economy. Still, benefits to consumers from a wider range of products and firms and economies securing efficiency gains from specialisation 'in different varieties or stages of production' would extend 'mutual gains from further integration'.[96]

BRI states must counteract risks flowing from 'thick borders' obstructing efficient trade-flows, weak governance challenging large projects and macroeconomic fallout

---

[92] Mohieldin (2017).

[93] Kim (2017b).

[94] Bastos (2018), Ruta and Boffa (2018), Ruta (2018), Boffa (2018, pp. 36–38) and Freund and Ruta (2018).

[95] Boffa (2018, p. 38).

[96] Bastos (2018, p. 38).

from project-finance. Still, BRI's scale and scope, the unexploited potential of many partner-economies, and the trade-off between saving time and saving money via improved connectivity, together, promised mutual benefits.[97] If BRI successfully integrated partner-states into the global value chain, as seemed likely given China's locus at the 'centre of the network', amplified effects of trade cost reduction from BRI connectivity would lead 'to sizeable efficiency gains'. WBG analysts found that, in fact, an expansion of value-added was already evident.[98]

4.6 Japan and the US Proclaim New Silk Road Visions

Japan and the USA, two major actors which initially refused to endorse Beijing's BRI vision or send leadership-level delegations to the 2017 and 2019 BRFs, had themselves some years earlier presented distinct proposals for reviving the ancient Silk Road networks. Their blueprints were unconnected and very much more modest than Xi Jinping's later, more extensive, proposals. Nonetheless, Japanese and US leaders did re-imagine the ancient Silk Road for their own national purposes, and they initially pursued these concepts with some, albeit short-lived, vigour. The contradiction between Japanese and US interest in establishing their own respective vision of a 'New Silk Road' and their well-expressed outrage at and counteraction vis-à-vis China's BRI blueprint was a notable feature of the history and evolution of the 'New Silk Road' discourse.

4.6.1 Tokyo Thinks up the New Silk Road

Japan's case, i.e. its engagement with the 'New Silk Road' construct, was instructive. After the Soviet collapse, Tokyo established diplomatic relations with the five Central Asian *Stan*s in 1992 and five years later, formally launched its 'Silk Road Diplomacy' targeting these states. Japan first focused on Kyrgyzstan, but the Kyrgyz government did not respond to Tokyo's offers of major projects, and soon, Japanese interest veered towards Kazakhstan. Uzbekistan, considered strategically significant, also attracted Japanese attention. Diplomatic forays began in 1992 when Foreign Minister Michio Watanabe visited Kazakhstan. Two years later, Kazakh leader, Nursultan Nazarbayev, went to Tokyo. However, oil-rich Kazakhstan evinced little enthusiasm for Japanese credit-financing and Uzbekistan, much more receptive, drew closer. Uzbek President Islam Karimov visited Japan in 1994, and by 1997, Tokyo had extended $500 m in grants and loans to Uzbekistan.[99]

---

[97] Ruta (2018, pp. 1–2).
[98] Ruta and Boffa (2018).
[99] Rakhimov (2014).

4.6 Japan and the US Proclaim New Silk Road Visions

Japanese aid to the *Stans* was also informed by an urge to match growing Chinese economic-commercial diplomacy. Interventions by eminent Japanese analyst Mifune Emi underscored these drivers. In two well-regarded volumes, *Bei-Chu-Ro Power Shift to Nihon* (2017) and *Chugoku Gaikou Senryaku* (2016),[100] Mifune sketched out Tokyo's perceptual landscape vis-à-vis China once Beijing's blueprint began taking shape across Eurasia. He described himself as 'neither a "panda hugger" (pro-China) nor a "dragon-slayer" (anti-China)'. That self-acclaimed detached objectivity placed Mifune at the centre of mainstream Japanese view of China in the context of Japan's locus in the world order being shaped by evolving great power dynamics. Describing Beijing's BRI design as a covert quest for global domination, Mifune stated, 'China is restructuring the international order by utilising OBOR, aiming to achieve Pax Sinica'. Rejecting Chinese insistence that BRI was fundamentally an economic integrative framework, he insisted 'the economic aspect' was 'just a part' of BRI.

Mifune quoted Xi Jinping as having stated that BRI went from 'the country-to-country community of common destiny to the regional community of common destiny' and even further to 'the community of common destiny for mankind', thereby 'uniting the development of China and the countries on the route'. Mifune thus saw BRI as 'an initiative to form a China-led international order, in other words, "the community of common destiny for mankind", therefore it pursues the world order of Pax Sinica'.[101] Mifune posited that a syncretic design, BRI synthesised up pre-existing Eurasian economic and connectivity projects, e.g. Kazakhstan's 'Bright Path', Russia's Eurasian Economic Union, ASEAN's Connectivity Master Plan, Turkey's 'Middle Corridor', Mongolia's 'Road to Development', the UK's 'Northern Powerhouse', the European Union's 'Juncker Plan', China's '16 + 1' collaborative grouping with Central and Eastern Europe and Poland's 'Amber Road' programmes.[102]

Mifune criticised 'China's new colonialism in Indian Ocean', specifically the construction and control of Sri Lanka's Hambantota Port. He and other Japanese academics described the MSR as an instrument of China, 'a land power, trying to rise as a sea power'. BRI's integrated information network, especially its proposed 'Space Information Corridor', would secure for Beijing 'cyber power and space power'. BRI would also further China's 'Made in China 2025' blueprint for joining 'the world's manufacturing powerhouse' by 2025, raising Chinese industry 'to the mid-level among the world's manufacturing powers' by 2035 and becoming the 'top of the manufacturing powerhouse' by 2045.

Finally, with BRI-partnerships, China was 'making its policy to be heard in the international community, or using it as a means to deter the policy or behaviour of the international community that China does not favour'.[103] BRI, in this view, was thus a sophisticated mix of geopolitical and geoeconomic drives aiming peacefully to secure global eminence. This perspective was widely shared among Japan's national security

---

[100] Mifune (2016, 2017).
[101] Mifune (2018).
[102] Mifune (2018).
[103] Mifune (2018).

community.[104] The divergence between this viewpoint and Tokyo's perceived early post-Cold War Eurasian interests could not be starker. Japan, after all, had been a major terminus of the ancient Silk Road with goods and travellers crossing the Sea of Japan from China and travelling back across Eurasia. Buddhism was arguably ancient Japan's most influential Silk Road import. The Imperial Shosoin Treasure Repository in historic Nara city, a centre of both Shinto and Buddhist faiths, houses valuable relics acquired over centuries from ancient Silk Road commerce.[105]

Japan's role in the accretion of the 'New Silk Road' construct, like China's, was a post-Soviet initiative. Prime Minister Ryutaro Hashimoto was the first major non-Chinese leader to echo Premier Li Peng's earlier proposal for establishing a 'New Silk Road' to revive trade, investment, travel and people-to-people contacts with renascent Central Asian polities. Generally negligent of former Soviet republics, Tokyo paid little attention to the region until 1997. Interest in stabilising a Eurasia convulsed by post-Soviet turbulence guided Hashimoto's geoeconomic approach. In an address to Japan's corporate leaders in July, Hashimoto outlined his vision shaping Japan's diplomacy vis-à-vis Russia, China 'and the "Silk Road" region'.

Recalling the post-Cold War context, Hashimoto described the period as 'the age of the so-called borderless society', which presented opportunities while posing challenges: 'by enhancing interdependence between countries and regions', economic globalisation rendered nation-state-centric approaches to problem-solving more complex. However, 'by expanding the potential loss resulting from a cut-off of exchanges with partners, such globalisation has the potential to contribute to maintaining peace'.[106] Having carefully observed the outcomes flowing from the European Union's Maastricht summit and the Euro-Atlantic G7 summit meeting in Denver which was expanded into the G-8 with Russia's inclusion, and noted the interest those governments evinced in Tokyo's post-Cold War priorities and preferences, Hashimoto asserted, Japan 'must introduce a new dynamism into our national foreign policy by forging a perspective of a Eurasian diplomacy viewed from the Pacific'.

Looking westward across the Eurasian continent, 'the perspective which now emanates from within us spans the Russian Federation, China and the Silk Road region, encompassing the Central Asian Republics of the former Soviet Union and the nations of the Caucasus region'. Appreciative of the growth China had recently achieved and the challenges it faced, Hashimoto believed that 'participation by China in international frameworks and strengthening of its status as a constructive partner in the international community will sharply advance openness and reform in China'. This was 'indispensable for the stability and prosperity of the Asian region'. Although Japan had some difficulties in its relations with both, 'the developments in these two great powers, Russia and China, now hold the key to the formation of an international order'.

---

[104] Report (2018d), Editorial (2017) and Kenichi (2012).

[105] Imperial Household Agency (Undated).

[106] Hashimoto (1997).

4.6 Japan and the US Proclaim New Silk Road Visions

Hashimoto pointed out that during the Cold War, Japan engaged with China in the economic-commercial realm and 'long since to stress to the US the legitimacy of our policy of active engagement in China'. It was 'time to strive even harder to build even more constructive relations with Russia and with China'. Linking and linked to Russia and China, stretched resource-rich Central Asian republics and the Caucasus, 'which we may call the Silk Road region', and which had 'great potential to serve as bridges, offering distribution routes within the Eurasian region'. Tokyo looked out to this landmass because 'Japan has deep-rooted nostalgia for this region stemming from the glory of the days of the Silk Road'.[107] This was why Hashimoto offered help with state-building to the renascent Central Asian polities.

Hashimoto directed Japanese diplomacy vis-à-vis the 'Silk Road region' towards political dialogue 'aiming to enhance trust and mutual understanding', economic collaboration, 'cooperation for natural resource development aiming to foster prosperity' and 'cooperation to build peace through nuclear non-proliferation, democratisation and the fostering of stability' across this vast region experiencing transformative transition. Hashimoto stressed the import of building relations with Russia, China and the 'Silk Road region' by avoiding 'situations in which a clash of national interests on a particular issue constrains wide-ranging exchanges'. A belief that 'conflict of interest and confrontation lie in our minds' led him to conclude that creative thinking would enable 'those responsible for managing confrontation to examine wider possibilities'. This visionary outline framed Japan's early 'Silk Road' initiatives.

The role of China, especially the quality of Japan–China relations, would define Tokyo's ability to realise this vision. Hashimoto's 1997 visit to China, after a stressful period, restored normalcy to bilateral ties. His successor, Keizo Obuchi, had to focus on the fallout from the 'Asian Economic Crisis' afflicting Japan's neighbourhood and collaborating with ASEAN, China, South Korea and multilateral financial institutions, especially the ADB, on arresting recessionary trends, stabilising volatile exchange rates, initiating growth-focused policies catalysing consumption and building up reserves to manage any future crises.[108] This priority informed the Jiang-Obuchi Joint Declaration issued at the end of Jiang's November 1998 visit to Japan. The two leaders also expanded economic cooperation, boosting Japanese investment in and transfer of technology and skills to Chinese manufacturing, specifying 33 items.

These included the construction of the Beijing-Shanghai High-Speed Railway, the first stage in the subsequently expanded network covering much of China, and the 'preservation of Silk Road cultural heritage sites'.[109] Even before Hashimoto's 1997 speech, Tokyo took a 'developmental' rather than 'geopolitical' approach to diplomacy vis-à-vis the 'Silk Road region'.[110] It began, as noted, by offering economic assistance to Kyrgyzstan, one of the smallest post-Soviet successor polities. However, Kyrgyzstan was unprepared to receive such aid, and Japan switched its attention

---

[107] Hashimoto (1997).
[108] Obuchi (1998), Okada (1998) and MoFA (1998).
[109] Obuchi and Jiang (1998) and MoFA (1999).
[110] Len et al. (2008, p. 11).

to Kazakhstan and Uzbekistan, offering loans in 1995. As Kazakhstan declined to borrow Japanese funds, Uzbekistan became Tokyo's Central Asian focus. When Hashimoto spoke of the 'Silk Road region', Japanese diplomats had already recognised the region's import in Tokyo's diplomacy vis-à-vis Moscow, Beijing and the Middle East, although nobody spelt out the linkages in 'concrete' terms.[111]

Just preceding Hashimoto's 'Silk Road' address, his future successor Obuchi led a 60-strong delegation of legislators, officials, businessmen and academics to Kazakhstan, Kyrgyzstan, Turkmenistan and Uzbekistan as part of Japan's 'Eurasian diplomacy'. On Hashimoto's instructions to draft his *Keizai Doyukai* address, MoFA summarised its views as 'Silkroad Diplomacy'. Hashimoto's resignation on domestic political disputes and Obuchi's succession slowed Tokyo's Central Asian initiatives. Still, in 1999–2000, Japanese embassies in Uzbekistan and Kyrgyzstan were enlarged by inducting JICA (Japan International Cooperation Agency) representatives. By then, Japan had become the largest source of ODA to Kazakhstan, Kyrgyzstan and Uzbekistan. In 2001, Tokyo hosted a donors' conference for Tajikistan, inviting President Emomali Rakhmanov to attend. In 2000–2008, Tokyo's yen-loans to Central Asia reached $2 bn and grant aid, $600 m, of which $260 m provided capacity-building technical assistance.[112]

After the USA launched military operations in Afghanistan in October 2001 following terrorist attacks on New York and Washington in September, UN-sponsored international aid to the region grew fast. Japan financed much of it. As several *Stans* deepened relations with Washington, opening military facilities to US forces for Afghan operations, Tokyo, too, gained. By this time, JICA had provided infrastructure-finance to Uzbekistan, home to the US air-base at Khanabad, laying fibre-optic cables, modernising airports, opening vocational training centres, building a railway-repair hub and installing a large power plant. In 2003, Tokyo proposed a 'Central Asia + Japan' forum to boost regional and multilateral policy-coordination. Foreign Minister Yoriko Kawaguchi visited the region to launch the Forum in August 2004. By then, Japan's ODA to the *Stans* reached 260bn yen ($2 bn+). Kawaguchi stressed the principles guiding Tokyo's 'Central Asia + Japan' forum in an address delivered in Tashkent[113]:

- Respect for diversity
- Cooperative competition
- Openness to third-party participation.

She and her regional counterparts focused on Central Asian peace and prosperity as major contributors to stability and progress in Eurasia and globally, reaching consensus on cooperative development of energy and transportation networks and water resources, and joint counterterrorism and anti-narcotic action.[114] Tokyo's aid in refurbishing regional railways culminated in Kawaguchi signing an agreement on

---

[111] Kawato (2008) in Len et al. (2008, p. 17).

[112] Kawato (2008) in Len et al. (2008, p. 19).

[113] Kawaguchi (2004) and Len (2005).

[114] MoFA (2005).

a $140 m project to build a 220 km railway line that would knit Southern Kazakhstan with other parts of that large territory.

Following the Andijan Incident in May 2005, during which Uzbek security personnel killed many unarmed protesters demonstrating after a violent prison-raid by armed militants, Tashkent responded to Western critique by asking Washington to end its military presence. Uzbekistan then signed a mutual security treaty with Russia instead. Deepening polarisation contributed to the *Stans* moving closer to Russia and China at the cost of American and Japanese influence. At the 2005 SCO summit, Hu Jintao offered soft-loans worth $900 m to SCO-partners. In an effort to rebuild trust, Tokyo hosted the 2nd Central Asia + Japan Foreign Ministerial meeting in June 2006. In August, Prime Minister Junichiro Koizumi visited Astana, Tashkent and Samarkand, reiterating Japan's interest in energy resources, but leaving Japanese corporations to pursue commercial operations without any direct official intervention.

In late-2006, after Shinzo Abe succeeded Junichiro Koizumi, Foreign Minister Taro Aso launched 'the Arc of Freedom and Prosperity' initiative covering Central Asia, the Caucasus, Balkans, Baltic States and the Middle East. He emphasised Japan's interest in promoting 'democracy, freedom, human rights, the rule of law, and the market economy', across the troubled arc[115] However, the initiative proved short-lived. After Aso was 'reshuffled' in August 2007, the 'Arc of Freedom and Prosperity' lost steam. Aso was only able to reinvigorate Japan's Silk Road initiative after he assumed national leadership in September 2008. His immediate focus as prime minister was, however, on addressing the impact of the *Great Recession* economic crisis, as Tokyo contributed substantially to both regional and multilateral efforts at building bulwarks against the recession further deepening. Looking ahead, in April 2009, Aso proclaimed an 'Asian Growth Initiative' to double Asia's total output by 2020.

Aso pledged to achieve this by 'encouraging region-wide development and expanded consumption' via 'the development of sub-regional infrastructure and industry as well as improvements in customs and other processes in a coherent manner, thereby enabling a smoother flow of people, goods and capital'.[116] Exemplifying the Ho Chi Minh City, Vietnam—Chennai, India shipping lane, an unwitting template for China's future MSR, Aso suggested building land–sea routes with smooth regulatory processes that could slash travel-time from 14 to 8 days. To this end, he proclaimed a 'Comprehensive Asian Development Plan' that would also counteract the downturn's regional impact. To fund his vision, he offered to prime local demand with finance: $20 bn in ODA, $20 bn for insuring infrastructure development, $5 bn for 'Leading Investment to Future Environment' and $22 bn for trade financing.[117]

Aso also proposed helping ASEAN to balance growth between the highly developed and wealthy Singapore, and ASEAN's poorer member-states, inviting leaders of Cambodia, Laos, Myanmar, Thailand and Vietnam to the first Mekong–Japan summit on Mekong Basin regional development. He next proposed building on the

---

[115] Aso (2006).

[116] Aso (2009a).

[117] Aso (2009a).

first Japan–China–RoK summit meeting held in December 2008, advancing trilateral cooperation via annual summits. To further stabilise post-Soviet Eurasia, Aso also boosted cooperation with Russia. Extending these initiatives, in mid-2009, he proclaimed 'The Initiative of a Eurasian Crossroads and the Concept of a Modern-Day Version of the Silk Road', pledging cooperation to 'bring the Eurasian continent together both north to south and east and west via this region'.

Aso's proposed 'North–South Logistics and Distribution Route', comprising roads and railways, would run from Central Asia through Afghanistan to the Arabian Sea. This would link to a horizontal 'East–West Corridor' running from Central Asia through the Caucasus to Europe, with new ports developed on the Caspian's shores. New infrastructure would 'unite resource-rich Central Asia and the Caucasus in one whole region' including Afghanistan and Pakistan. His blueprint, resonating with several connectivity plans being pursued by various UN organs for over a decade, preceded Beijing's OBOR proclamations by several years.

This was Japan's unwitting conceptual contribution to Beijing's subsequent BRI vision. Aso also returned to his idea of building land–sea connectivity using Japanese technology providing one-stop border-procedures to slash shipping times from Ho Chi Minh City to Chennai. In sum, Aso envisioned a series of routes along which 'people, goods and capital flow freely, traversing the entirety of the Eurasian continent beginning at the Pacific Ocean and ending in Europe. This could also be called a modern-day version of the Silk Road'.[118] Clearly, Aso pursued a vision that would knit Eurasia along a series of networks that would have resembled what the early stages of Xi Jinping's BRI framework looked like. But in September 2009, Japan's voters booted the LDP out, electing Yukio Hatoyama's Democratic Party of Japan (DPJ) to power. That dramatic result effected an atypically radical shift in broader Japanese policy.[119]

In Central Asia, though, Tokyo's activism underscored continuity. Its 'Silk Road Diplomacy' comprised political dialogue, economic cooperation, joint efforts at strengthening nuclear non-proliferation, and enhancing democratisation and stability. In 1997–2010, Tokyo delivered $600 m in grant aid to Central Asian states, with about $260 m going into capacity building. Over the same period, Japan gave the *Stans* ODA loans totalling $2 bn, most of which was spent on infrastructure, especially refurbishing and extending Soviet-era railways, modernising roads, bridges and airports, laying optical-fibre cables, repairing power plants, building vocational schools and water-facilities.[120] By providing financial and technical support to rebuilding, modernising and expanding Central Asia's inter-connectivity infrastructure, Japan had, by 2010, contributed significantly towards enabling the *Stans* to enhance their post-Soviet sub-regional economic-commercial integration while also boosting Central Asia's potential as a hub for future trans-Eurasian transport and communications networks.

---

[118] Aso (2009b).
[119] Editorial (2009).
[120] Rakhimov (2014).

4.6 Japan and the US Proclaim New Silk Road Visions

Table 4.2 Japan's major port-infrastructure investments in the IOR

Country	Project	Cost—US$
Bangladesh	Matarbari deep-sea port and power-station	3.7 bn
India	Mumbai trans-harbour transport link	2.2 bn
Myanmar	Dawei port and special economic zone	800 m
Myanmar	Yangon port container terminal	200 m
Madagascar	Toamasina port	400 m
Mozambique	Nacala port	320 m
Kenya	Mombasa port and related infrastructure	300 m

The DPJ hiatus unexpectedly saw Sino-Japanese tensions over the East China Sea (ECS) maritime territorial dispute spike. When the LDP was re-elected, Tokyo's historical China-rooted insecurity and competitive impulses, resonating with the US' 'Asia–Pacific Pivot/Rebalance', returned to the fore. As Beijing spread its post-financial crisis global investment portfolio, offering aid in infrastructure-construction, especially ports and power-stations, and laid the foundations of what later emerged as the SREB/MSR, Tokyo's more subtle and understated response mirrored its neighbour's initiatives. With quiet encouragement from Japan's 'Quad' partners, Tokyo secured the acceptance of its offers of assistance towards the building of substantial projects in several Asian and African states described as either established BRI-partners, or non-members keen to garner BRI's benefits.

Even before Japan launched its 'Free and Open Indo-Pacific Strategy' in April 2017 and with India, initiated the 'Asia–Africa Growth Corridor' in May, Japanese corporations, financed by officially sponsored organs, were already constructing major facilities along the Indian Ocean rim. The most substantial among these projects, as listed in Table 4.2, highlighted the often unnoticed and little-discussed scale of Japan's maritime ambitions and the spread of its Afro-Asian involvement across the IOR: [121]

In 2018, while launching Japan's official development assistance (ODA) White Paper, Shinzo Abe and his cabinet outlined Japan's ODA plans for the next few years. Tokyo's 'three pillars' reinforced the view that Japan would deploy economic resources to building 'quality infrastructure' across and linking the 'Indo-Pacific' and the AAGC spatial constructs while advancing an ideational mirror-image of and functional counterpoint to Beijing's BRI blueprint. Japan's enunciated principles—promotion of 'a rules-based order', enhancement of connectivity via 'high-quality' infrastructure and ensuring 'maritime law enforcement'—challenged China's connectivity infrastructure investment policy by suggesting BRI lacked those attributes and implicitly questioned the merit of BRI's MSR component specifically.

---

[121] Brewster (2018).

Although some observers saw Tokyo's policy framework convergent with BRI and China–Japan relations moving along an upward trajectory, others discerned strongly competitive tendencies.[122] Ambiguity aside, Tokyo having initiated a New Silk Road endeavour shortly after the Cold War ended, China's BRI/MSR undertaking energised Japan to mount a significant counter-initiative which promised, with BRI's own projects running in parallel, to rapidly develop the wider region's infrastructure stock, connectivity and pace of economic activity. Japan joined hands with India, the USA and Australia—its Quad partners—to pursue connectivity infrastructure investments which promised transparency, 'high quality' and an absence of 'debt traps'. This competitive response to BRI/MSR simultaneously widened and narrowed the perceptual and policy gulfs separating Japan and China.

With the two neighbours signing 50-plus joint-infrastructure agreements worth over $18 bn, including for 3rd-party projects, during Abe's October 2018 visit to Beijing,[123] this led to the China Development Bank (CDB), a key financer of major BRI projects, and the Japan Bank for International Cooperation (JBIC), Tokyo's principal ODA-funding agency, signing an MoU on joint-infrastructure development in 'third countries', without naming BRI. JBIC Governor Tadashi Maeda admitted to being 'surprised' by CDB counterparts 'unreservedly signing on to all' of Tokyo's principles—'transparency, inclusiveness, project viability, debt sustainability and rule of law'.[124] Sino-Japanese cooperation could transform BRI's future, but their bilateral grand-strategic tensions clouded the prospects. Just over a year after that fraternal exchange, Tokyo's defence white paper identified Beijing as the source of Japan's gravest insecurities,[125] illuminating the uncertainties flowing from *systemic transitional fluidity*.

4.6.2 Washington's New Silk Road Proposal

When the DPJ's unexpected electoral victory turned Taro Aso's 'Silk Road' regional connectivity plans into archival curiosities, Tokyo's American allies took up the case. Washington's accretive proposition began as a modest, hydrocarbon-fuelled programme which, however, originated elsewhere. Central Asian oil-and-gas reserves had been a key Soviet asset before 1992, when the *Stans* assumed ownership within their newly restored borders. Before then, as Mikhail Gorbachev steered the Soviet Union away from Cold War confrontations with the West, political fluidity reshaped the Soviet periphery. The older proposal to build a natural-gas pipeline linking Central Asian energy reserves to energy-starved South Asia, especially India, now began looking increasingly realistic.

---

[122] Zhao (2018b), Xinhua (2018), Fischetti and Roth (2018), Kohara (2018) and Furuoka (2018).
[123] Ren (2019).
[124] King (2019).
[125] Kono and Iwaya (2019).

4.6 Japan and the US Proclaim New Silk Road Visions

It acquired salience since 1989, shortly after Soviet forces withdrew from Afghanistan, enabling all parties to imagine a more benign future for Central Asia–South Asia relations. That was when the Iran–Pakistan–India (IPI) pipeline project was mooted. This 3000 km pipeline, costing $7.5 bn to build, would carry Iranian gas from its South Pars field across Pakistan's Balochistan and Sindh Provinces to India.[126] A spur-line would off-take some of the gas for the Pakistani industrial-commercial hub in Karachi. After years of talks which stalled owing to disputes over delivery-price, transit fees and confidentially articulated political-strategic differences, the IPI became mired in minutiae. But it energised those who viewed any expansion of Iran's influence as a threat. The result was a counter-proposal to build a pipeline linking Turkmenistan's large gas reserves to the Indian market via Afghanistan and Pakistan.

This 1800 km Turkmenistan–Afghanistan–Pakistan–India (TAPI) pipeline would cost $7.6 bn to build and when completed, would move a maximum of 33bcm of Turkmen natural gas annually for 30 years.[127] Discussions of US interest vis-à-vis post-Soviet Central Asian states grew to a crescendo in the late-1990s when public discussions precipitated legislative action. Congress amended the Foreign Assistance Act of 1961 to focus US aid on 'economic and political independence of the countries of the South Caucasus and Central Asia', while also enacting the 'Silk Road Strategy Act of 1999'. The latter Act, effective vis-à-vis Armenia, Azerbaijan, Georgia, Kazakhstan, Kyrgyzstan, Tajikistan, Turkmenistan and Uzbekistan, was explained on the strength of the US' strategic and geoeconomic imperatives[128]:

- The development of strong political, economic and security ties among countries of the South Caucasus and Central Asia and the West will foster stability in this region, which is vulnerable to political and economic pressures from the south, north and east.
- Building open market economies and open democratic systems across South Caucasus and Central Asia will provide incentives for international private investment, increased trade and other forms of commercial interactions with the rest of the world.
- Many South Caucasus states have secular Muslim governments seeking closer alliance with the USA and that have diplomatic and commercial relations with Israel.
- South Caucasus and Central Asia could produce oil and gas in sufficient quantities to reduce US dependence on energy from the volatile Persian Gulf region.
- US foreign policy and international assistance should be targeted to support the economic and political independence, democracy building, free market policies, human rights and regional economic integration across South Caucasus and Central Asia.

---

[126]Rousseau (2011).

[127]Kawawaki et al. (2012).

[128]US Senate (1999, Sec. 2).

The Silk Road Strategy Act, providing the legal framework for and the juridical basis of subsequent diplomacy and aid, stipulated US policy vis-à-vis countries of the South Caucasus and Central Asia be on the merit of US normative, grand-strategic and economic objectives[129]:

- promote and strengthen independence, sovereignty, democratic government and respect for human rights
- promote tolerance, pluralism and understanding and counter racism and anti-Semitism
- assist actively in the resolution of regional conflicts and to facilitate the removal of impediments to cross-border commerce
- promote friendly relations and economic cooperation
- help promote market-oriented principles and practices
- assist in the development of the infrastructure necessary for communications, transportation, education, health and energy and trade on an East–West axis in order to build strong international relations and commerce between those countries and the stable, democratic and market-oriented countries of the Euro-Atlantic Community and
- support US business interests and investments in the region.

Washington was engaged in bilateral diplomacy with these countries since the mid-1990s, developing policy-prescriptions relevant to each country as seen from the US perspective. Initial efforts generated bilateral investment treaties (BIT) crystallising the parameters, purposes and procedures for advancing and securing US investment in the region's struggling economies and establishing the frameworks for resolving any future disputes between US investors and recipient-state governments and local partners[130]:

- Armenia: BIT entered into force on 29 March 1996.
- Azerbaijan: BIT entered into force on 2 August 2001 and soon encountered several disputes between US investors and the Azeri Government.
- Georgia: BIT entered into force on 17 August 1997.
- Kazakhstan: BIT entered into force on 12 Jan 1994. Over the Treaty's first nine years, four US firms engaged in dispute with Kazakh authorities. One firm filed an appeal with a US court against Astana's refusal to license the export of Kazakh uranium, but its case was dismissed; a second firm disputed Astana's failure to repay debts owed to it but then negotiated a resolution before arbitration procedures were completed; a third firm obtained favourable arbitration outcome on Astana's expropriation of its oil-field concession before negotiating a resolution; a fourth firm claimed that Astana expropriated its real-estate development assets without paying appropriate compensation and moved the dispute to international arbitration.

---

[129] US Senate (1999, Sec. 3).

[130] Bureau of European and Eurasian Affairs (2003).

- Kyrgyzstan: BIT entered into force on 12 January 1994. A US telecommunications firm challenged the Kyrgyz government's alleged breach of terms on which the former had established a joint venture and moved to a hearing by OPIC.
- Tajikistan: Negotiations on a BIT were suspended in April 1993.
- Turkmenistan: Negotiations on a BIT were suspended in March 1998.
- Uzbekistan: The two governments signed a BIT on 16 December 1994. After the US government undertook not to enforce the Treaty until Tashkent completed economic reforms to ensure Uzbek policies did not violate the Treaty's terms, the US Senate advised the Executive Branch on its conditional consent on ratification. US–Uzbek relations significantly improved following the accession of President Shavkat Mirziyoyev to office, as highlighted during his May 2018 meeting with Donald Trump at the White House, but the BIT remained suspended.[131]

Parallel to Washington's economic and strategic trajectories converging in Central Asia, US multinationals and regional actors achieved a measure of consonance. On 15 March 1995, promoted by Argentina's energy-firm Bridas, Pakistan—whose economic prospects were hobbled by energy shortages, and Turkmenistan, Central Asia's poorest state despite its gas-wealth, signed a MoU to build the TAPI pipeline. This led to US giant Unocal's joint venture with Saudi oil-firm Delta to propose an alternative alignment. Unocal and Delta signed an agreement with President Saparmurat Niyazov on 21 October 1995. Unocal, leading an international alliance of energy firms, and the Turkmen government, incorporated the CentGas Consortium in Ashgabad on 27 October 1997 with the objective of building the TAPI pipeline.

The 1600 km pipeline would start at Turkmenistan's Dauletabad gas field, cross into North-Western Afghanistan, run parallel to the Herat–Nimruz–Kandahar Highway, cross into Pakistan's Balochistan Province, pass Quetta, Dera Ghazi Khan and Multan, finally reaching the Indian border at Fazilka. Costing $10 bn to build, it would carry 33bcm of natural gas annually.[132] Since the pipeline would have to cross Afghan territory, CentGas needed Taliban-ruled Afghanistan's consent.[133] Shortly, after CentGas was incorporated, the US Ambassador to Pakistan, Robert Oakley, underscoring strong US support for the TAPI project, moved from diplomatic service to the pipeline-builder's management. In January 1998, Kabul's Taliban Government selected CentGas over Bridas for the construction of the TAPI pipeline across Afghan territory and signed an agreement authorising CentGas to proceed. In June 1998, Russia's Gazprom sold its 10% stake in the undertaking, making the project largely US-led.

After *al-Qaeda* bombed the US embassies in Nairobi and Dar-es-Salam on 7 August 1998, and Washington accused Osama bin Laden of masterminding the attacks, Taliban leader Mullah Omar proclaimed support for bin Laden. Pipeline negotiations ended and in December 1998, Unocal withdrew from the consortium.[134] The project suffered the loss of US engagement for a dozen years when US and

---

[131] White House (2018) and UNCTAD, BITs—Uzbekistan (Undated).
[132] Report (2015b, 2016c).
[133] Report (1998).
[134] Croissant and Aras (1999).

Western attention veered towards more acute security concerns. Thereafter, regional interest became TAPI's prime mover.

In May 2002, leaders of Afghanistan, Pakistan and Turkmenistan met in Islamabad and agreed on jointly advancing the project. Oil and gas ministers from the three countries comprised of a steering committee to guide and monitor progress. At its first formal session in Ashgabat in July 2002, this supervisory committee requested the ADB to finance TAPI-related studies. In December 2002, the ADB extended the first tranche of such assistance. After the partner-governments asked the ADB to assume TAPI's secretariat functions in 2003, ADB launched its first technical assessment, examining TAPI's planned route, preliminary design, cost estimates and environmental impact analyses. ADB specialists conducted market studies ascertaining gas demand in Pakistan and Northern India, a risk analysis, and a risk mitigation study covering likely downstream difficulties for consumers in case of supply disruptions.

The ADB recommended construction of underground storage facilities in Pakistan to counteract unforeseen consequences of supply disruption, an inter-party agreement on host-government responsibilities, gas transportation and gas-sale-and-purchase, soliciting bids from private corporations interested in building the pipeline, and urged the parties to secure appropriate legal advice for formalising these agreements.[135] The US war in Afghanistan, however, prevented realisation of the TAPI blueprint. In December 2010, the Presidents of Turkmenistan, Afghanistan and Pakistan and India's Energy Minister met in Ashgabat to negotiate a framework for reviving the project.

They authorised the construction of the 1700 km, $7.6 bn pipeline, while the energy ministers signed a more detailed framework document. Despite leadership-level enthusiasm, the pipeline's route across Afghanistan's violence-prone Helmand and Kandahar Provinces and Pakistan's Balochistan Province raised questions which, certainly for the moment, remained unanswered. ADB President Haruhiko Kuroda called it a 'historic project' but admitted it would not be an 'easy' project to implement.[136] He proved right. The challenges confronting TAPI and the opportunity costs flowing from failed implementation led Afghanistan and Pakistan to explore alternatives with Tajikistan and the Kyrgyz Republic, both with surplus energy resources. This parallel endeavour followed the TAPI template.

In May 2006, ministers from the four governments met in Islamabad and agreed to explore electricity trading prospects. At a second meeting in October in Dushanbe, they resolved to fashion a Central Asia-South Asia Regional Electricity Market (CASAREM), by building the transmission and trading infrastructure capable of transmitting 1000 MWe to Pakistan and 300 MWe to Afghanistan. The project was named CASA-1000. In August 2008, they established an intergovernmental council to drive project implementation. As with TAPI, the ADB funded preliminary CASA-1000 feasibility studies. In 2011, the four partners asked the IFC to engage with project designing and implementation. In 2012, they signed bilateral agreements with the IFC enabling the latter to lead the selection of developers and operators

---

[135] Kawasaki (2008).

[136] AFP (2010).

for the project.[137] In 2018–2019, bids were solicited for the construction of key elements of the project.[138] Thus what began as a US initiative to link Central Asian and South Asian economies using natural gas as the vector, turned into a WBG-managed regional energy-market project.

In 2011, Washington formally launched its New Silk Road Initiative with the aim to reintegrate Afghanistan regionally by reviving traditional trade routes and rebuilding connective infrastructure destroyed in decades of warfare. Then-Secretary of State Hillary Clinton described the initiative as 'the economic side' of the US Afghan strategy, as, 'we all recognise that Afghanistan's political future is linked to its economic future—and in fact to the future of the entire region'. Her Deputy Secretary, William Burns, reinforced the emphasis on economic linkages securing the bases for future regional stability and prosperity.[139] The USA sought to link Afghanistan to Central Asia, Pakistan, India, 'and beyond', focusing on four 'key areas'[140]:

- Regional energy markets: Central Asia, 'a repository vast energy resources', could help South Asia's 1.6bn-plus population sustain growth. Directing some of the former's resources to the latter via Afghanistan 'would be a win–win'. The US supported the CASA-1000 power grid project with $15 m after the World Bank offered $526 m to the project in March 2014. Washington spent over $1.7 bn in building transmission lines, hydropower plants and associated features in Afghanistan since 2010 and also added 1000 MWe to Pakistan's power grid, helping 16 m people.
- Trade and transport: Improved trade and transit across Central- and South Asia required 'reliable roads, railways, bridges and border crossing facilities', as well as harmonised customs regulations, multilateral trade institutions and collaborative efforts against barriers. The USA built/rehabilitated over 3000 km of Afghan roadways, enabled Kazakh and Afghan accession to the WTO and supported formalisation of the 2010 Afghanistan–Pakistan Transit-Trade Agreement and the Kyrgyz–Tajik–Afghan Cross-Border Transport Agreement.
- Customs and border operations: Washington encouraged increased speed and efficiency of border crossings, and enhancing border security and frontier-regional governance to prevent weapons-, drugs-, other contrabands and human-trafficking. It noted that Central Asian intra-regional trade grew by 49% since 2009, average costs of border crossing across the region fell by 15% since 2011 and streamlined customs procedures at seven Afghan border-crossing points had reduced release time from eight days in 2009 to three-and-a-half hours in 2013, generating annual savings of $38 m.
- Businesses and people-to-people: The USA considered regional economic connectivity to signify more than 'infrastructure, border crossings and the movement of goods and services'. US encouragement of sharing ideas and expanding markets included opportunities for the youth, women and minorities, and boosting stability

---

[137] Khosla (2014).
[138] CASA-1000 project website https://www.casa-1000.org/. Accessed 8 Oct 2019.
[139] Clinton (2011), Gulamhusein (2011) and Burns (2014).
[140] DoS (2017).

and prosperity. Washington funded post-graduate scholarships for Afghan students across Central Asia, sponsored the Central Asia–Afghanistan Women's Economic Symposium, and South Asia Women's Entrepreneurship Symposium supporting thousands of businesswomen from the two regions, and financed regional trade conferences in Central- and South Asian cities generating trade deals worth over $15 m.

So, for several years before Xi Jinping proclaimed Beijing's BRI vision and launched China's 'trillion-dollar project of the century', the USA was engaged in fashioning its own New Silk Road programme, albeit one narrowly focused on a modest sub-regional scale. The US's grand-strategic objectives across Central Asia, indeed the Eurasian 'world island', were key drivers motivating the exercise. A perceived need to counteract the steady accretion of a collaborative and integrationist palimpsest called the SCO provided some of the impetus, while a comparable drive to draw the military campaign in Afghanistan to a close added urgency.

US Central Command (CENTCOM), responsible for managing and supervising the Afghan operations, under General David Petraeus, initiated some of the early blueprints for building a peaceful Afghanistan at the heart of a prosperous Central Asian-South Asian regional construct, as US analysts recorded.[141] Petraeus's successor, General James Mattis, focusing on immediate combat-related exigencies, shifted supervision of the New Silk Road Initiative to the DoS. Clinton and her team drove the initiative with diplomatic panache, but energy and urgency were dissipated.[142] As Washington's focused priorities shifted, the initiative became marginal.

4.6.3 The US' Lower Mekong Initiative

Washington also worked on another sub-regional collaborative enterprise before China's LMC mechanism took flight. In July 2009, Clinton met her counterparts from Cambodia, Laos, Thailand and Vietnam on the sidelines of a regional gathering in Phuket, Thailand. Establishing the Lower Mekong Initiative (LMI), the ministers agreed to deepen cooperation in managing the environment and boosting health, education and infrastructure development. Myanmar joined the group three years later. Although the original framework was expansive, US focus and resource allocations underscored three more specific goals[143]:

- Water management: Washington pledged to support lower riparian states in developing environmental protection and sustainable projects to 'help address future challenges' flowing from climate change. This included the development of a

---

[141] Cooley (2012), Kuchins (2013) and McBride (2015).
[142] Rosenberger (2017) and Standish (2014).
[143] DoS (2012).

predictive modelling tool called 'Forecast Mekong' to aid Lower Mekong countries in better appreciating climate change-rooted challenges and preparing adaptive policy-responses. The US Mississippi River Commission signed a 'sister-river' agreement with the Mekong River Commission; the partnership would help improve management of Lower Mekong basin trans-border water issues. Washington also aided projects designed to develop practices and regimes of sustainably exploiting water and forest resources.
- Education: The US Fulbright Programme and other educational projects funded more than 500 annual exchange programmes involving students and scholars from the region. The US government and private charities helped expand educational facilities and enrolment among school-age children in the region and develop rural broadband network coverage. US funding helped many local leaders to improve their English language skills enabling effective engagement with external interlocutors. Washington expanded its International Visitors' Programme covering the region to bring over education, health and environment-related practitioners and officials, helping them acquire best-practice knowledge from US bodies, and develop inter-bureaucratic ties.
- Health systems: US health-care assistance covered a catchment area directly affecting 2 m people and 'contributed to the 50% reduction in HIV/AIDS infection rate' in Cambodia. Washington boosted regional programmes to contain and counter pandemic outbreaks of influenza and supported Lower Mekong partners in tracking, diagnosing and treating multi-drug resistant forms of malaria and tuberculosis. In June 2010, the US and Vietnamese governments jointly sponsored the first US-Lower Mekong Health Conference on 'Transnational Cooperation to Respond to Infectious Disease Threats'. The event brought together regional and international health professionals to exchange information and coordinate policy-responses to regional health issues and establish a precedent and a template for the future.

Three years later, marking the 40th anniversary of US diplomacy 'as a friend of the ASEAN community', Washington released an update on US-ASEAN cooperation; US-LMI outcomes received particular attention. The US' LMI blueprint emphasised regional socio-economic and cultural integration, but the Fact Sheet underscored strategic and geopolitical intent. Washington pledged 'to commit substantial resources to LMI over the next three years through the Asia-Pacific Security Engagement Initiative (APSEI)'. The recently proclaimed APSEI was an 'integrated framework' engaging LMI states on 'current pressing bilateral and transnational issues and positions the US and its partners to sustain regional stability and support an inclusive regional economy in our shared future'.[144] Signalling 'the lasting nature of US engagement', Washington proclaimed 'LMI 2020', a blueprint incorporating LMI into APSEI, and pledging $50 m over three years to 'support a substantial expansion' of LMI programmes aimed at 'strengthening regional capacity to address transnational issues'. China was not named, but its ghost loomed Banquo-like over the proceedings.

---

[144]Office of the Spokesperson (2012).

Washington elaborated on its assistance, packaged via the USAID, PACOM, the latter's Asia-Pacific Centre for Security Studies, Harvard Kennedy School, and bodies from Japan, the RoK, Australia and New Zealand, towards improving the LMI-members' capacity in governance, education, specialised English language training, health, pandemic-monitoring, prevention and preparedness, quality pharmaceuticals production and supply, the environment, gender equality and women's empowerment. In short, the US' LMI vision was a comprehensive template for the region's socio-economic-cultural development. The instrumental deployment of PACOM's resources and support from the US key regional allies underscored the enterprise's fundamentally geopolitical character. Clinton's personal leadership in deepening the LMI endeavour gave ASEAN's newer and poorer members 'options to balance fast-expanding Chinese assistance',[145] which may have driven the initiative.

Clinton's successor, John Kerry, stressed the importance of protecting the Mekong Delta's residents from the ravages of encroaching climate change, especially rising sea levels pumping saline water into the Delta's rice paddies, Vietnam's bread basket.[146] The Trump Administration discarded much of its Obama-era inheritance, marginalising the LMI within the overall recrafting of its Asia policy. Nonetheless, building partly on the LMI legacy, Trump translated some of his campaign rhetoric into a broad campaign to constrain China's geoeconomic endeavours while setting out his 'vision for a free and open Indo-Pacific'. Less than a year into his presidency, addressing APEC leaders and business executives in Vietnam after a visit to Beijing, Trump slammed Chinese commercial practices in remarks presaging the imminent trade war:

'Countries were embraced by the WTO, even if they did not abide by its stated principles. Simply put, we have not been treated fairly by the WTO … The US promoted private enterprise, innovation and industry. Other countries used government-run industrial planning and state-owned enterprises'. Trump contrasted US practice to what he considered unacceptable behaviour: 'We adhered to WTO principles on protecting intellectual property and ensuring fair and equal market access. They engaged in product dumping, subsidised goods, currency manipulation and predatory industrial policies. They ignored the rules to gain advantage over those who followed the rules, causing enormous distortions in commerce and threatening the foundations of international trade itself'. Trump's 'Indo-Pacific Dream' demanded that 'all play by the rules, which they do not right now'.[147]

Trump warned, 'Such practices, along with our collective failure to respond to them, hurt many people in our country and also other countries … We can no longer tolerate these chronic trade abuses, and we will not tolerate them'. He stressed, 'We will no longer tolerate the audacious theft of intellectual property. We will confront the destructive practices of forcing businesses to surrender their technology to the state and forcing them into joint ventures in exchange for market access. We will address the massive subsidising of industries through colossal SOEs that put

---

[145] Bower (2012).

[146] Kerry (2014).

[147] Trump (2017).

private competitors out of business—happening all the time ... Those days are over'. Trump's new construct: 'we must uphold principles that have benefitted all of us, like respect for the rule of law, individual rights and freedom of navigation and overflight, including open shipping lanes ... So let us work together for a peaceful, prosperous and free Indo-Pacific'.[148]

Secretary of State Mike Pompeo fleshed out Trump's 'Indo-Pacific Dream'. 'Free Indo-Pacific' meant 'all nations, every nation, to be able to protect their sovereignty from coercion by other countries (and domestically,) good governance and the assurance that citizens can enjoy their fundamental rights and liberties'. As for an 'Open Indo-Pacific', the USA wanted 'all nations to enjoy open access to seas and airways. We want the peaceful resolution of territorial and maritime disputes'. Economically, 'open' meant 'fair and reciprocal trade, open investment environments, transparent agreements between nations and improved connectivity to drive regional ties'. Pompeo announced a $113 m initiative to advance energy ($50 m), the digital economy ($25 m), and infrastructure-building ($38 m) across the Indo-Pacific, with this 'strategic investment in the most competitive part of the world' aiming 'to catalyse' follow-up action by the US' corporate giants.[149]

Pompeo acknowledged that this amount was small compared to the region's need for new infrastructure and that only private capital could meet that demand. He assured his audiences that with US firms, 'what you see is what you get: honest contracts, honest terms and no need for off-the-books mischief'. Pompeo insisted, 'integrity in business practices is an essential pillar of our Indo-Pacific economic vision, and it is what each country in the region needs'.[150] Although Pompeo did not name China, his assertion of 'the US commitment to a free and open Indo-Pacific' covering the continuum from domestic dispensation to interstate transactions challenged much that China sought via Xi's 'Chinese dream' generally, and Beijing's BRI vision specifically.

The openly adversarial rhetoric defined the intensely competitive milieu. The USA thus led in drawing up cooperative developmental plans, including for certain infrastructure projects, using the 'New Silk Road' rubric and the Lower Mekong Initiative, before Beijing formalised its BRI blueprint. However, after Beijing launched BRI, the USA used these frameworks to build collective resistance and counteract China's expanding footprint.

Widely divergent scales, resourcing, political leadership and persistence distinguished US and Chinese efforts. While Beijing emphasised its economic, indeed geoeconomic focus, Washington's much more modest initiatives candidly advertised their geopolitical drivers. Irrespective of the transience of US efforts and the uncertain prospects of Chinese ones, however, successive US Administrations, and many US visionaries, scholars, technological innovators and corporations, along with counterparts from Europe, Japan, Russia, Australia and New Zealand, and various UN organs and Western-led multilateral institutions, could justifiably claim much credit for contributing to imagining what eventually emerged as China's globe-girdling BRI/OBOR vision.

---

[148] Trump (2017).

[149] Pompeo (2018).

[150] Pompeo (2018).

References

AFP (2010) TAPI gas pipeline: presidential push for lingering project. Express Tribune, 12 Dec 2010

AP (2019) Perennial presidential candidate Lyndon LaRouche dead at 96. WP, 13 Feb 2019

ASEAN (2002) Feasibility study for the missing links and spur links of the SKRL. MoFA/IAI, Vientiane, 28 June 2002, p 3

Askary H, Ross J (2017) Extending the New Silk Road to West Asia and Africa: a vision of an economic renaissance. Schiller, Washington

Aso T (2006) Arc of freedom and prosperity: Japan's expanding diplomatic horizons. MoFA/Japan Institute of International Affairs, Tokyo, 30 Nov 2006

Aso T (2009a) Overcoming the economic crisis to rekindle a rapidly developing Asia: remarks at the 15th international conference on the future of Asia. PMO, Tokyo, 21 May 2009

Aso T (2009b) Japan's diplomacy: ensuring security and prosperity—speech at the Japan Institute of International Affairs. PMO, Tokyo, 30 June 2009

Azizov U (2017) Regional integration in Central Asia: from knowing-that to knowing-how. J Eurasian Stud 8:128

Bastos P (2018) Exposure of Belt and Road economies to China trade shocks. WBG, Washington, June 2018

Beets M, Merry-Baker M, Billington M, Kokinda R, Small D, Ross J (eds) (2018) The New Silk Road becomes the world land-bridge: a shared future for humanity, vol II. Schiller, Washington

Belt and Road Portal (2019) The BRI: progress, contributions and prospects. Leading Group for Promoting the BRI, Beijing, 22 Apr 2019

Boffa M (2018) Trade linkages between the Belt and Road economies. WBG, Washington, May 2018

Bower E (2012) US moves to strengthen ASEAN by boosting the LMI. CSIS, Washington, 24 July 2012

Bradsher K (2013) Hauling new treasures along the Silk Road. NYT, 20 July 2013

Brewster D (2018) Japan's plans to build a 'free and open' Indian Ocean. Interpreter, 29 May 2018

Bureau of European and Eurasian Affairs (2003) Assessments required by the Silk Road Strategy Act of 1999: report to congress. DoS, Washington, Jan 2003

Burns W (2014) Remarks on economic connectivity in Central Asia. DoS/Asia Society, New York, 23 Sept 2014

Chatzky A (2019) China's Belt and Road gets a win in Italy. CFR, New York, 27 Mar 2019

Chen J (2016) Becoming an international stakeholder. CD, 15 Mar 2016

Chen W (2017) A positive view of China as responsible world stakeholder. CD, 30 Oct 2017

Chen D, Feng S (2016) The Russia-India-China trio in the changing international system. China Q Int Strateg Stud Shanghai Inst Int Stud 2(4):431–447

Clinton H (2011) Remarks at the New Silk Road ministerial meeting. DoS, New York, 22 Sept 2011

Committee on Transport (1998) Emerging issues and developments at the regional level: transport, communications, tourism and infrastructure development. ESCAP, Bangkok, 23 Jan 1998

Cooley A (2012) Great games, local rules: the new great power contest in Central Asia. OUP, New York, pp 3–15, 30–50

Cornell S, Engvall J (2017) The EU and Central Asia: expanding economic cooperation, trade, and investment. Central Asia-Caucasus Analyst, 6 Sept 2017

Croissant M, Aras B (1999) Oil and geopolitics in the Caspian Sea region. Santa Barbara, Greenwood, p 87

Cui T (2019) Why the US shouldn't sit out the BRI. Fortune, 23 Apr 2019

Ding G (ed) (2009) Hillary Clinton visits China: US 'smart power'. PD, 20 Feb 2009

Directorate General for International Cooperation and Development (2017) International cooperation and building partnerships for change in developing countries: Central Asia—transport. European Commission, Brussels, p 1

References

DoS (2012) Lower Mekong Initiative. Washington. https://www.state.gov/p/eap/mekong/. Accessed 24 May 2018

DoS (2017) US support for the New Silk Road. https://2009-2017.state.gov/p/sca/ci/af/newsilkroad/index.htm. Accessed 24 May 2018

Editorial (2009) The vote that changed Japan. Economist, 3 Sept 2009

Editorial (2011) China committed to peace and growth. PD, 7 Sept 2011

Editorial (2017) Japan and 'one belt, one road'. Japan Times, 24 June 2017

ESCAP (2015) Intergovernmental agreement on the Trans-Asian Railway network. UN, Bangkok, 24 Nov 2015

ESCAP (Undated-a) About ESCAP. http://www.unescap.org/about. Accessed 1 Mar 2018

ESCAP (Undated-b) Asian Highway. http://www.unescap.org/our-work/transport/asian-highway/about. Accessed 2 Mar 2018

ESCAP (Undated-c) History. http://www.unescap.org/about/history. Accessed 1 Mar 2018

Fischetti A, Roth A (2018) Xi and Abe guarantors of Chinese-Japanese entente. EAF, 7 Sept 2018

Freund C, Ruta M (2018) BRI: overview. WBG, Washington, 29 Mar 2018

Funk K, Cho H, Dominguez R, Barbour P (2012) Country partnership strategy for the PRC. WBG, Beijing, 11 Oct 2012

Furuoka F (2018) Japan's new aid strategy builds bridges with the BRI. EAF, 16 Mar 2018

Garlick J (2016) China's role in nuclear safety shows it is a responsible player. CD, 31 Mar 2016

General Secretariat (2017) Council conclusions on the EU strategy for Central Asia. Council of the EU, Brussels, 19 June 2017

Gulamhusein H (2011) US secretary of state Hillary Clinton discusses New Silk Road initiative. Aga Khan Development Network, Dushanbe, 22 Oct 2011

Hashimoto R (1997) Address to the Japan Association of Corporate Executives. PMO, Tokyo, 24 July 1997

High Representative for Foreign Affairs and Security Policy (2018) Connecting Europe and Asia: building blocks for an EU strategy. European Commission, Brussels, 19 Sept 2018

High Representative for Foreign Affairs and Security Policy (2019) EU-China: a strategic outlook. European Commission, Brussels, 12 Mar 2019, p 1

Hillman J (2018) The rise of China-Europe railways. CSIS, Washington, 6 Mar 2018

Hodgkinson P (1999) Development of the Trans-Asian Railway: TAR in the southern corridor of Asia-Europe routes. ESCAP, New York

Horvath B (2016) Identifying development dividends along the BRI: complementarities and synergies between the BRI and the SDGs. UNDP, New York, pp 5–11

Imperial Household Agency (Undated) The Shosoin repository. https://www.kunaicho.go.jp/e-about/shisetsu/shosoin01.html. Accessed 8 Oct 2019

Information Office (1996) Symposium on economic development along new Euro-Asia continental bridge opens in Beijing. UN, Beijing/New York, 9 May 1996

Information Office (2011) China's peaceful development. State Council, Beijing, 6 Sept 2011, ch 3

Information Office (2016) Wuhan Asia-Europe logistics. Government of Hubei, Wuhan, 22 Apr 2016

Iseri E (2009) The US grand strategy and the Eurasian heartland in the 21st century. Geopolitics 14(1):26–46. Feb 2009

Jeremy S (2017) Unit-45 holds the key to container trade. Supply Chain. https://www.supplychaindigital.com/brochure/unit-45. Accessed 20 May 2018

Juncker J (2016) Commission delegated regulation (EU) 2017/849. European Commission, Brussels, 7 Dec 2016

Kaeser J (2018) Keynote speech on Belt and Road International Summit by Siemens President and CEO. Siemens, Beijing, 6 June 2018

Karimov I (1994) Remarks at meeting with PRC premier Li Peng in Li P (1994) China's Basic Policy Towards Central Asia. Beijing Review, Vol.37, No.18, May 2-8, 1994, pp.18-19

Kawaguchi Y (2004) Adding a new dimension—Central Asia plus Japan: policy speech at the University of World Economy and Diplomacy. MoFA, Tashkent/Tokyo, 26 Aug 2004

Kawasaki F (2008) TAPI natural gas pipeline project (phase II): technical assistance completion report. ADB, Manila, 16 Dec 2008, pp 1–3

Kawawaki F, Chansavat B, Chenoweth J (2012) TAPI natural gas pipeline project, phase 3. ADB, Manila, p 1

Kenichi I (2012) On the way Japan's Foreign and Security Policy should be: testimony to budget committee of the diet house of representatives. JFIR, Tokyo, 2 Mar 2012

Kerry J (2014) Remarks at the friends of the Lower Mekong ministerial meeting. DoS, Naypyitaw, 9 Aug 2014

Khosla S (2014) Project information document: CASA-1000/PIDA2581. WBG, Washington, 27 Mar 2014

Kim J (2017a) The BRI: building bonds across Asia, Europe and beyond. WBG, Washington, 13 Oct 2017

Kim J (2017b) The BRI: building bonds across Asia, Europe and beyond. WBG, Washington, 12 Oct 2017

King A (2019) No appetite for a 'new Cold War' in Asia. EAF, 9 June 2019

Kohara M (2018) Japan-China relationship is on the mend. JT, 16 May 2018

Kono T, Iwaya T (2019) Defense of Japan 2019. MoD, Tokyo, Sept 2019, pp i–ii

Kuchins A (2013) Great games, local rules: book review roundtable. Asia Policy. No. 16, NBR, Washington/Seattle, July 2013, pp 175–178

LaRouche L (1990) The productive triangle: locomotive for the world economy. Schiller Institute, West Berlin (in German)

LaRouche H (1997) Eurasian land-bridge: a new era for mankind. Exec Intell Rev 24(19):21–25. 2 May 1997

LaRouche H (2017) The New Silk Road is changing the world: US must join in 2018. Schiller, Washington, 28 Dec 2017

LaRouche H (2018) The New Silk Road is shaping strategic affairs. Schiller, Washington, 5 Apr 2018

LaRouche Political Action Committee (2018) The US joins the New Silk Road: a Hamiltonian vision for an economic renaissance

Leesburg (Undated) https://larouchepac.com/20151229/us-joins-new-silk-road, pp 1–5. Accessed 19 May 2018

Len C (2005) Japan's Central Asian diplomacy: motivations, implications and prospects for the region. China Eurasia Forum Q 142. Nov 2005

Len C, Uyama T, Hirose T (eds) (2008) Japan's Silk Road diplomacy: paving the road ahead. Nitze School of Advanced International Studies, Washington

Li P (1994) China's basic policy towards Central Asia. Beijing Rev 37(18):18–19. 2–8 May 1994

Li H (2009) Genuinely ready to see China's rise? PD, 17 Nov 2009

Li Y (ed) (2017) The review of developments in transport in Asia and the Pacific. ESCAP, Bangkok, p 32

Liu S, Rohland K, Nehru V (eds) (2013) China 2030: building a modern, harmonious, and creative society. WBG/NDRC, Washington/Beijing

Macguire E, Anderson B (2013) 'Silk Road' railways link Europe and Asia. CNN, 27 June 2013

Mackinder H (1904) The geographical pivot of history. Geogr J 23:422

Maitra R (2013) The alliance of India-Russia-China. Paper presented at conference on the 2nd American Revolution: developing the Pacific and ending the grip of empire. Schiller, Los Angeles, 2 Nov 2013

Malik K (ed) (2008) UNDP at work in China: annual report. UNDP, Beijing, June 2008, p 24

McBride J (2015) The new geopolitics of China, India, and Pakistan: building the New Silk Road. CFR, New York, 22 May 2015

MEA (2019) 16th meeting of the foreign ministers of Russia, India and China in Wuzhen, 27 Feb 2019. GoI, New Delhi, 22 Feb 2019

Media Centre (2016) Geodis completes first rail transport for Kaporal between China and France. Geodis, Paris/London, 5 Sept 2016

References

Media Office (2017) OBOR: starting shot for intermodal train from China to Europe. Hupac, Korla/Chiasso, 26 May 2017

Meng J (2010) High-tech companies go west. CD, 6 Nov 2010

Mifune E (2016) Chugoku Gaikou Senryaku (Chinese diplomatic strategy). Kodansha Sensho Metier, Tokyo

Mifune E (2017) Bei-Chu-Ro power shift to Nihon (US, China, Russia's power shift and Japan). Keishu Shobo, Tokyo

Mifune E (2018) Participation in the dialogue with the world: Eurasia 2025. Jpn Forum Int Relat (JFIR) E-letter 11(3), Issue 67, 20 June 2018

MoFA (1998) Sixth ASEAN + 3 summit. Hanoi, 15–17 Dec 1998

MoFA (1999) Diplomatic bluebook 1999. Tokyo, ch 1, p 3

MoFA (2005) Joint statement: 'Central Asia + Japan' dialogue/foreign ministers' meeting—relations between Japan and Central Asia as they enter a new era. Astana/Tokyo, 28 Aug 2005

Mohieldin M (2017) BRI: a global effort for local impact. In: Conference on connecting cities for inclusive and sustainable development, UNIDO, Vienna, 26 Sept 2017

NDRC (2007) National office for promoting the rising of central part of China was established. State Council, Beijing, 10 Apr 2007

Network Information Security Group (2019) EU coordinated risk assessment of the cybersecurity of 5G networks. European Commission, Brussels, 9 Oct 2019, pp 21–22

Obuchi K (1998) Four initiatives for Japan-ASEAN cooperation toward the 21st century. MoFA, Tokyo, 14 Dec 1998

Obuchi K, Jiang Z (1998) Japan-China Joint Declaration on building a partnership of friendship and cooperation for peace and development. MoFA, Tokyo, 26 Nov 1998

Office of the Spokesperson (2012) Fact sheet on LMI. DoS, Washington/Jakarta, 13 July 2012

Okada M (1998) Press conference: visit of Prime Minister Keizo Obuchi to Hanoi for the ASEAN summit meeting. MoFA, Hanoi, 15 Dec 1998

Pan M (2017) High-speed rail project finally gets on track. CD, 22 Dec 2017

Peyrouse S (2009) Business and trade relationships between the EU and Central Asia. EU-Central Asia Monitoring, June 2009, pp 5–7

Pijl K (ed) (2015) Handbook of the international political economy of production. Edward Elgar, Cheltenham, ch 5, pp 76–97

Pomfret J (2010) World's workshop heads to inland China. Reuters, Zhengzhou, 25 Aug 2010

Pompeo M (2018) America's Indo-Pacific economic vision: remarks at Indo-Pacific business forum. DoS, Washington, 30 July 2018

Rabitz C (2009) Friedrich Schiller's works have withstood the trials of revolution. DW, Bonn, 10 Nov 2009

Rakhimov M (2010) Internal and external dynamics of regional cooperation in Central Asia. J Eurasian Stud 1(2):99–101. July 2010

Rakhimov M (2014) Central Asia and Japan: bilateral and multilateral relations. J Eurasian Stud 5:78

Ren X (2019) Keeping China-Japan relations afloat. EAF, 24 Jan 2019

Report (1994) Chinese leader wants Silk Road renewed. UPI, Moscow, 19 Apr 1994

Report (1998) Country-report: Turkmenistan. EIU, London, p 24

Report (2010) Rising labor costs trigger industrial relocation. CD, 6 July 2010

Report (2015a) China economic update, June 2015. WBG, Beijing, 2 July 2015, p 1

Report (2015b) Turkmenistan starts work on gas link to Afghanistan, Pakistan, India. Reuters, Mary, 13 Dec 2015

Report (2016a) Silk Road to expand reach via Chengdu-Europe express rail. CD, 17 Nov 2016

Report (2016b) Khorgos Gateway goes live with Navis N4 terminal operating system. Navis, Oakland, 8 Nov 2016

Report (2016c) TAPI gas pipeline: infographic. ADB, Manila, 8 Apr 2016

Report (2017a) Schiller Institute plans to bring America into the 'New Silk Road' paradigm this spring. Schiller, Washington, 20 Mar 2017

Report (2017b) Kazakhstan's Khorgos dry port aiming to handle 0.5 m TEU per year by 2020. Centre for Transport Strategies, Kiev, 21 July 2017

Report (2018a) The world is turning towards the New Silk Road. Schiller, Washington, 22 Aug 2018

Report (2018b) New China-Europe freight train route launched. Xinhua, Urumqi, 20 Jan 2018

Report (2018c) China's pursuit of Europe trade links powers freight rail surge. Bloomberg, 7 Mar 2018

Report (2018d) Japan takes the lead in countering China's BRI. SCMP, 10 Feb 2018

Report (2019a) Xi, Merkel, Macron and Juncker meet in Paris. DW, 26 Mar 2019

Report (2019b) China-Europe freight train cargo value surges by 106 percent. Xinhua, Beijing, 22 Apr 2019

Roberts D, Meyer H, Tschampa D (2012) The Silk railroad of China-Europe trade. Bloomberg, 21 Dec 2012

Rosario D (2019) EU foreign investment screening regulation enters into force. European Commission, Brussels, 10 Apr 2019

Rosenberger L (2017) The rise and fall of America's New Silk Road strategy. Economic Monitor, 12 May 2017

Rousseau R (2011) Pipeline politics in Central Asia. Foreign Policy in Focus, 24 June 2011

Rudd K (2008) China a 'constructive partner' in trade talks. PD, 8 Aug 2008

Ruta M (2018) Three opportunities and three risks of the BRI. WBG, Washington, 4 May 2018

Ruta M, Boffa M (2018) Trade linkages among Belt and Road economies: three facts and one prediction. WBG, Washington, 31 May 2018

Saling S (2016) UNDP and China to cooperate on BRI. UNDP, New York, 19 Sept 2016

Sand V (2015) Press release: Kalmar delivers port equipment to Kazakhstan's premier in-land dry port. Cargotec, Paramount/Helsinki, 18 Aug 2015

Schiller Institute (Undated) From productive triangle to Eurasian land-bridge. https://www.schillerinstitute.org/russia/rus_eal_chronology.html#triangle. Accessed 23 Jan 2018

Severo R (2019) Lyndon LaRouche, cult figure who ran for president 8 times, dies at 96. NYT, 13 Feb 2019

Shepard W (2016a) How those China-Europe 'Silk Road' trains first began. Forbes, 29 June 2016

Shepard W (2016b) How a company called Unit 45 revolutionized China-Europe 'Silk Road' rail transport. Forbes, 8 Nov 2016

Shepard W (2017a) Belt and Road's new ports stimulate the development of innovative transport technologies and faster logistics. SCMP, 28 Mar 2017

Shepard W (2017b) These 8 companies are bringing the 'New Silk Road' to life. Forbes, 12 Mar 2017

Smith K (2017) China-Europe rail freight continues to soar. Int Railw J. 18 Apr 2017

Standish R (2014) The US' Silk Road to nowhere. FP, 29 Sept 2014

Starr S (ed) (2007) The New Silk Roads: transport and trade in Greater Central Asia. Johns Hopkins, Washington

Starr S, Cornell S (2015) The EU and Central Asia: developing transport and trade. Central Asia-Caucasus Analyst, 10 Dec 2015

Suzuki W (2019) China's Belt and Road hits a speed bump in Kazakhstan. Asian Review, 24 Apr 2019

Tedjarati S (2017) What is the BRI? Honeywell News, Morris Plains, 8 Sept 2017

Togt T, Montesano F, Kozak L (2015) From competition to compatibility: striking a Eurasian balance in EU-Russia relations. Clingendael, The Hague, Oct 2015, pp. 306

Trump D (2017) Remarks at APEC CEO summit. White House, Da Nang, 10 Nov 2017

Tsuruoka D (2017) US multinationals angle for inside track in B&R push. Asia Times, 1 July 2017

UNCTAD, BITs—Uzbekistan (Undated) http://investmentpolicyhub.unctad.org/IIA/CountryBits/226#iiaInnerMenu. Accessed 29 May 2018

References

Urban W (2018) The 'New Silk Road': opportunities for European companies? Sberbank Europe, Vienna, 6 Feb 2018

US Senate (1999) Silk Road Strategy Act of 1999. In: 106th congress, Washington, 10 Mar 1999

Wang Z (2016) Honeywell, big supplier to BRI. CD, 1 Nov 2016

White House (2018) The US and Uzbekistan: launching a new era of strategic partnership. Washington, 17 May 2018

Xinhua (2013) Chinese experts see US-DPRK antagonism as root cause of nuke test. PD, 17 Feb 2013

Xinhua (2018) China, Japan should consolidate momentum of improvement in ties: premier Li. PD, 13 Sept 2018

Xu S (1997) The new Asia-Europe land bridge: current situation and future prospects. Jpn Railw Transp Rev 30–33. Dec 1997

Zhao H (2018a) China will be a responsible stakeholder in Arctic affairs. CD, 22 Feb 2018

Zhao H (2018b) Japan and China crossing tracks in Southeast Asia. EAF, 15 Sept 2018

Zhou Y (2009) Pfizer looks inland for less costly talent pool. CD, 30 Nov 2009

Zhou W (2013) Rail route to Europe improves freight transport. CD, 13 Sept 2013

Zoellick R (2005) Whither China: from membership to responsibility? DoS/National Committee on US-China Relations, New York, 21 Sept 2005

Zoellick R (2007) China 'can help counter' slowdown. PD, 18 Dec 2007

Chapter 5
Case Study 1: The China–Pakistan Economic Corridor

5.1 China-Pakistan Economic Corridor, a Belt and Road 'Flagship'

In May 2013, during a visit to Pakistan, Chinese Premier Li Keqiang proposed the establishment of a 1864-mile/3000 km-long China–Pakistan Economic Corridor (CPEC), linking Kashgar in Xinjiang to Karachi and Gwadar ports in Southern Pakistan via a network of roads, railway, pipelines and digital connectivity, enabling enhanced exchanges of goods, capital, technology, information and people between the two economies and societies, generating much mutual benefit, and promising to dramatically accelerate Pakistan's progress.[1] Two months later, during Pakistan's Prime Minister Nawaz Sharif's return visit to China, the two governments signed a MoU, formally launching CPEC. In May 2015, when President Xi Jinping visited Pakistan, members of his delegation and their hosts signed agreements on Chinese contractors and suppliers building 51 infrastructure—and related projects across Pakistan. These would be largely financed with Chinese loans and investments worth $46bn. Most of Beijing's funding of Balochistan Province's Gwadar Port was converted into grants.[2]

Lauding the substantial expansion and deepening of the well-established Sino-Pakistani collaborative relations which CPEC represented, Xi Jinping noted: 'We should use China-Pakistan Economic Corridor (CPEC) to drive our practical cooperation with focus on Gwadar Port, energy, infrastructure development and industrial cooperation so that the fruits of its development will reach both all the people in Pakistan and the people of other countries in our region'.[3] Xi's emphases indicated focused attention on energy security. Then-Prime Minister Nawaz Sharif, a driving force behind the CPEC construct, resonated with Xi's perspective on both the bilateral and regional implications of the enterprise: 'Pakistan and China are iron

---

[1] Qamar (2017).
[2] Jorgic (2017a), Gul (2017) and Report (2015).
[3] Xi (2017).

brothers, and friendship with China is the cornerstone of Pakistan's foreign policy. CPEC is a game changer which will transform the lives of the billions of people of the region'.[4]

Pakistan's political establishment was polarised on many points, including CPEC, although not on Pakistan-China relations per se. Nawaz Sharif's principal domestic political opponent, leader of the Pakistan Tehreek-e-Insaf (PTI) party, Imran Khan, was initially critical of CPEC's key aspects, and vocally condemned these on the campaign trail. However, after he won the 2018 general elections which led him to succeed Sharif as prime minister, Khan acknowledged, 'China gives us a huge opportunity through CPEC, to use it and drive investment into Pakistan. We want to learn from China how they brought 700 million people out of poverty'.[5]

The roads, railways, digital connectivity links, power plants and industrial units included in the CPEC projects portfolio would be built commercially by Chinese firms, to be ready within 15 years. They would create around 700,000 jobs and could add 2.5% to Pakistan's annual GDP growth rate.[6] The publication of CPEC's financial and scheduling details was the highlight of Xi's visit, which coincided with confirmation that China would sell eight new submarines to the Pakistani Navy, with half to be assembled at the Karachi Shipyard and Engineering Works in Pakistan, using Chinese design and technology, nearly doubling Islamabad's subsurface assets, underscoring the import of security linkages in Sino-Pakistani ties.[7]

For China, CPEC would connect it to markets in Asia, Africa, and Europe, reduce shipping distances and time and, by bypassing the Malacca Strait 'choke-point', enhance its economic, specifically energy, security.[8] As CPEC's scale and speed of implementation gained momentum, with its funding figures from Chinese sources rising several times to reach $62bn in 2017, the potential of it nearing $100bn by completion in 2030 beckoned.[9] In 2019, credit rating agency Moody's analysts anticipated that CPEC implementation could contribute between 9% and 10% to Pakistan's GDP in FY2018–2019; their Deloitte counterparts estimated CPEC would contribute around 2.5% to Pakistan's GDP growth rate.[10] Optimism was tempered by Islamabad's acknowledgement that in 2018–2019, Pakistan's GDP had only grown by 3.3%, the lowest rate of expansion in nine years.[11] CPEC's boost would take longer to take effect.

In June 2018, on the 5th anniversary of CPEC's launch, Beijing and Islamabad pointed to Pakistan's FY2017–18 GDP growth climbing to 5.8%, compared to the previous FY's 3.8%, as evidence of the CPEC vision's validity and value. The brightest symbol of Pakistan's fragile economy potentially being transformed to a state of

---

[4]Report (2017a).
[5]Khan (2018a).
[6]Bilal (2018) and Report (2016a, pp. 1–2).
[7]Gady (2016), Macfie et al. (2015) and Panda (2015).
[8]Tunningley (2017), Report (2016a, p. 2) and You (2007).
[9]Siddiqui (2017) and Research Snipers (2018).
[10]Malik (2019) and Deloitte (undated).
[11]Mu (2019).

steady growth came in the form of power supply. Until 2015, some areas suffered up to 20 h daily without electricity while nearly 70% of Pakistan experienced 12–14 h of power cuts. WBG analysts found the economic cost of Pakistan's power-sector distortions reached $17.69bn (6.53% of GDP) in FY2015, and consumer subsidies comprising fiscal costs were $2.15bn (0.80% of GDP). In FY2015, lack of reliable electricity-supply cost the economy $12.87bn (4.75% of GDP), and in 2017, almost 51m rural Pakistanis (26% of the population) had no access to electricity.[12] By early June 2018, however, cuts had been eliminated across nearly three-fourths of Pakistan and significantly reduced elsewhere.[13] By the end of 2017, around half of the 10,000 MWe added to Pakistan's national grid came from CPEC-listed projects.[14] Much of the credit rested with two new CPEC-listed coal-fired power plants, each producing 1320 MWe, at Sahiwal in Punjab, and Port Qasim in the commercial hub of Karachi, respectively.

The two plants were designed to generate 18bn KWh of electricity annually, addressing electricity needs of eight million families. A number of other conventional and renewable power-projects, being built to collectively eliminate power shortages and enable the country's industrial, agricultural, and transport-and-communications sectoral growth to accelerate, headed towards completion. In 2019–2020, CPEC implemented six projects which would add 6910 MWe to Pakistan's national grid.[15] Energy production and distribution lay at the core of the CPEC blueprint. The corridor implemented 15 'priority projects', 'actively promoted' four other projects and encouraged the preliminary examination of two 'potential energy projects'. This focus on eliminating Pakistan's energy deficit and meeting the country's growing electricity needs was highlighted in the CPEC's long-term plan (LTP) published after the 7th meeting of the China-Pakistan Joint Coordination Committee (JCC) in late 2017.

Resonating with the goals driving Pakistan's Vision-2025 development plan, the jointly developed LTP reviewed progress in implementing projects over CPEC's first five years and established a framework for advancing Pakistan's industrialisation, by 2030. Given CPEC's transformative promise for Pakistan, for China–Pakistan relations, and for the impact CPEC's realisation could have on the prospects for BRI's wider SREB and MSR components, CPEC's champions in Pakistan and China, and critics elsewhere, described CPEC as a BRI 'flagship'.

Senior Pakistani diplomat, Sartaj Aziz, was among many officials who underscored CPEC's leading role within Beijing's BRI firmament: 'By linking China with Arabian Sea and the Persian Gulf, CPEC will optimise trade potential and enhance energy security of China, Pakistan and our wider region'. This 'flagship project of China's BRI' would 'directly benefit three billion people inhabiting China,

---

[12] Fan (2019).
[13] Bhatti and Liu (2018).
[14] Iqbal (2017).
[15] EIU (2019).

South Asia, Central Asia and the Middle East'.[16] In Pakistani and Chinese narratives, CPEC's seven 'salient features', i.e. connectivity, energy, trade and industrial parks, agricultural development and (rural) poverty alleviation, tourism, 'cooperation in areas concerning people's livelihood and non-governmental exchanges' and financial cooperation, would positively transform Pakistan and its neighbourhood.

The minister heading the Pakistani team leading the Sino-Pakistani endeavour summarised the enterprise's objectives: 'CPEC will greatly speed up the industrialisation and urbanisation process in Pakistan and help it grow into a highly inclusive, globally competitive and prosperous country capable of providing high-quality life to its citizens'.[17] CPEC's LTP articulated both 'hardware' and 'software' approaches. The connectivity projects, with three components—a highway network,[18] a railway network,[19] and a fibre optic project,[20] would expand 'the flow of goods, information and people across regions', a flow defining the core of the CPEC vision. The organisers of China's Hainan-based annual Boao Forum for Asia, in their 2018 annual report, stated, 'CPEC, a flagship project under the BRI, has not only improved local infrastructure but also is extending toward Afghanistan, reducing poverty, the hotbed of terrorism, and bringing better prospects for local people's lives'.[21]

Laudatory reviews by Pakistani and Chinese voices were unsurprising. However, their apparent assumption of CPEC's self-evident and universally admirable features was immediately and strongly challenged by critics in India, a country bordering both China and Pakistan, and one whose relations with both have historically been adversarial and, occasionally, openly hostile. One Indian critic noted, 'The $50bn corridor that connects Kashgar in western China with Gwadar port in Pakistan, spanning 3000 km, is one of the pain points in ties between India and China'. India's envoy in Beijing, Gautam Bambawale, diplomatically insisted: 'One of the norms is that the project should not violate the sovereignty and territorial integrity of a country. Unfortunately, there is this thing called the CPEC which violates India's sovereignty and territory (sic) integrity'. His conclusion: 'Therefore, we oppose it'.[22] Ambassador Bambawale's superiors in Delhi were less diplomatic: 'The so-called 'China-Pakistan Economic Corridor' violates India's sovereignty and territorial integrity. No country can accept a project that ignores its core concerns on sovereignty and territorial integrity'.[23]

---

[16] Xinhua (2017).

[17] Iqbal (2017).

[18] For more information, you can find the 'CPEC Highway Network Map' here: http://www.cpec.gov.pk/map-single/1 (last accessed October 16th, 2019).

[19] For more information, you can find the 'CPEC Highway Network Map' here: http://www.cpec.gov.pk/map-single/1 (last accessed October 16th, 2019).

[20] For more information, you can find the 'CPEC Fibre Optic Project Map' here: http://www.cpec.gov.pk/map-single/3 (last accessed October 16th, 2019).

[21] Boao Forum for Asia (2018) and ZD (2018).

[22] Achom (2018).

[23] Official Spokesperson (2018).

According to one Indian analyst, 'Beijing's seeming unwillingness to substantially address Indian sensitivities', including CPEC's alignment crossing 'Pakistan-occupied Kashmir, has allowed Indian strategic concerns to crystallise into disaffection and suspicion toward the BRI'.[24] India's increasingly intimate strategic partner in their shared anxiety over China's expanding geopolitical footprint, the USA, too, reflected profound unease over Beijing's BRI framework, including its 'CPEC flagship'. Legislators shared grave concerns expressed by senior executive branch officials. Senator Charles Peters, a member of the US Senate Armed Services Committee (SASC), while questioning then-Secretary of Defence James Mattis on US policy vis-à-vis Washington's long-running military campaign in Afghanistan, posited, 'The One Belt One Road strategy seeks to secure China's control over both the continental and the maritime interest, in their eventual hope of dominating Eurasia and exploiting natural resources there, things that are certainly at odds with US policy'.[25]

Under CPEC, highway networks were either being built anew or refurbished, expanded and improved. The port of Gwadar in southern Balochistan, the terminus of CPEC's infrastructural spine, was being developed as a commercial, industrial and transport hub. The principal road-link between China's Xinjiang Autonomous Region in the north, and Pakistan's Punjab, Khyber Pakhtunkhwa, Sindh and Balochistan provinces to the south, crossed the Karakoram mountain range of the western Himalayas at the Khunjerab Pass. Thus, CPEC's blueprint essentially followed the decades-old Karakoram Highway (KKH) alignment that linked Pakistan's national highway network to its Chinese counterpart via National Highway (NH) 35, between Islamabad and Kashgar, almost 800 miles apart.

Construction of the KKH began in 1966, following the September 1965 India–Pakistan war over the disputed and divided region of Jammu & Kashmir. It was opened to commercial traffic in 1986. In addition to offering the trans-Himalayan neighbours a land-link, the KKH symbolised Beijing's strategic backstop in the defence of an insecure Pakistan vis-à-vis India. More directly impactful for Pakistani citizens, a new 392 km Multan-Sukkur Motorway, CPEC's longest road-project, would, for instance, allow traffic to flow between Punjab and Sindh at 120 km per hour, cutting travel time between the two manufacturing hubs from 10 h to four.[26] CPEC's highway network would be complemented with a railway counterpart which built on, modernised and expanded Pakistan's creaky rail-network. These refurbished-cum-newly built road-and-rail networks would virtually transform Pakistan's terrestrial transport facilities. And this was just the beginning.

CPEC's digitalisation project included a cross-border optical fibre cable linking Pakistan and China, enabling the promotion of 'technologies of the fourth industrial revolution'. Pakistan's industrialisation would benefit from new industrial parks and 'special economic zones' (SEZs) built 'all over the country'. Agriculture, hitherto the backbone of Pakistan's economy, would receive new technologies, e.g. biological breeding and drip irrigation, to boost productivity and efficiency, with the focus on

---

[24] Jacob (2017).
[25] Peters (2017).
[26] Bhatti and Liu (2018).

small farmers. In the 'software' realm, Islamabad and Beijing enhanced inter-central bank monetary cooperation, adopting bilateral currency swaps for financing CPEC projects, rather than relying on 'any third-party currency', i.e. the US dollar.

Regulatory resonance was designed to remove bottlenecks, expedite the transfer of goods, capital, technology and information, and hasten project implementation. Inter-state coordination would be supplemented with cultural cross-fertilisation of ideas via exchanges of scholars, students and tourists, whose flow would comprise 'an integral component of the corridor'. Planned 'coastal tourism' would complement these efforts. Coastal leisure-resorts were built or planned along three routes: Karachi-Bandar-Keti, Ormara-Sonmiani, and Jhala Jhao-Gwadar-Jiwani, to boost Balochistan's economic prospects, improve the experience of Pakistani visitors and foreign tourists, and widen Sino-Pakistani people-to-people exchanges.

Pakistan's Ministry of Planning, Development and Reforms (MPDR), the NDRC's CPEC partner, claimed that when completed in 2030, CPEC's energy-related projects would double Pakistan's power-generating capacity. This would enable the growth of small-and-medium enterprises, and major production units, brightening prospects for Pakistan's economic development. CPEC's conjunction of enhanced grid-delivery and terrestrial transport-and-communications expansion improved jobs and income generation. By mid-2018, Chinese firms implementing CPEC projects had created 'over 100,000 jobs' for Pakistanis.[27]

CPEC marked a transition in China-Pakistan relations. Pakistani analysts described it as a framework that took Sino-Pakistani 'strategic and economic partnership to a new height'. CPEC promised to stabilise Pakistan's highly challenged economy. For China, CPEC's Xinjiang-Gwadar alignment would 'serve as an alternative to the Malacca straits' for 'time- and cost-effective' transportation of energy, enabling China to enhance its 'energy security by reducing reliance on the Malacca route'. This route, after all, was 'already a possible flashpoint of blockade by the US or Indian navy'. They noted that much potential benefits notwithstanding, CPEC 'increased Indian mania' because it could 'counter hegemonic designs of India in the Indian Ocean, Arabian Sea and the Persian Gulf'.[28]

Since the consolidation of Sino-Pakistani ties following the 1962 Sino-Indian border war, linkages coloured interactions in strategic and geopolitical hues. China's assistance to Pakistan, instanced by large-volume transfers of military *materiel*, construction of the KKH, physically forging a Chinese stake in Pakistan's territorial integrity, and Beijing's clandestine aid to Pakistan's nuclear weapons programme, counteracted Pakistani perceptions of insecurity vis-à-vis India and reinforced Beijing's regional strategic leverage. China also invested in Pakistan's energy and mining sectors. CPEC, reflecting a resonance between the partners' interests, mirrored a culmination of trajectories defining bilateral ties cemented over decades.

Islamabad was keen to promote growth, private-sector investment, employment-generation, export-trade, transfer of technology and skills, and secure benefits from the relocation of Chinese manufacturing. Beijing, in turn, was anxious to relocate

---

[27] Editorial (2018a), Bhatti and Liu (2018) and Chen et al. (2018).
[28] Butt and Butt (2015).

5.1 China-Pakistan Economic Corridor, a Belt and Road 'Flagship'

surplus manufacturing capacity by expanding overseas production bases, establishing and securing new markets, and forging energy cooperation with partners with a view to ensuring energy security.[29] Sino-Pakistani economic cooperation, however, long preceded CPEC. The partners signed their free trade agreement (FTA) in 2006, with the FTA taking effect in 2007. Nonetheless, some Chinese observers discerned risks to timely and full implementation of many CPEC projects. They feared a lack of focus and complementarity could hobble the enterprise. Although bilateral trade volumes grew, they paled beside China's commerce with other regional partners, including those with which Beijing's relations were fraught, e.g. Delhi, Hanoi and Manila.

CPEC dramatically increased Chinese FDI flows into Pakistan's power generation, construction, transport and communications, and financial sectors, but little of it went into light manufacturing or small-and-medium enterprises (SME), arguably the core of Pakistan's economic dynamism. Official data nevertheless indicated that of the $207m in FDI Pakistan received in November 2017, for instance, $206m was Chinese.[30] Total FDI flows into Pakistan rose from $3.45bn in 2017 to $3.79bn in 2018[31]; much of this came from China. As Pakistan faced a severe resource crunch while negotiating concessionary help from the IMF and Saudi Arabia in 2018–2019, Beijing promised to deposit $2.5bn with Pakistan's central bank, raising to $4.5bn in commercial loans China advanced to Pakistan in 2018–2019. Beijing also gave $1bn in grants for helping Pakistan's education, health, vocational training, drinking water and poverty alleviation projects over 2019–2021.[32] By the autumn of 2019, 11 CPEC projects had been completed and another 11 were being built; of the target of adding 11,110 MWe to the national grid, 10,000 MWe had 'been achieved'.[33]

Beijing's 'all-weather friendship' notwithstanding, Chinese economists cautioned Islamabad against taking CPEC's largesse for granted, urging Pakistani officials to 'convince (Chinese) private investors' of 'investment opportunities in Pakistan', and recommending the establishment of 'an FDI Advisory Board that shall promote' a 'new image of the country'. One urged Islamabad to address 'security issues and risks, hard infrastructure challenges, especially SEZ-specific constraints like energy, roads to SEZs etc.,' suggesting Pakistan eliminate 'soft infrastructure challenges' like 'corruption, rule of the law, coordination among institutions, inadequate capacity and cultural biases'. Although 'long-term trust at the government level' was 'solid', the two private sectors needed 'more mutual dialogue and exchanges', allowing 'the peoples' to 'get to understand each other'.[34] CPEC's completion with Chinese funding looked less than watertight if Islamabad avoided significant reforms first.

---

[29] Jia (2018).
[30] Sajjad (2017).
[31] Report (2019b).
[32] Gul (2019).
[33] Yao and Shah (2019).
[34] Jia (2018).

5.2 China–Pakistan Economic Corridor Deepens Threat Perceptions

Most external critics aimed their guns at CPEC's northern and southern termini, i.e. Beijing-built infrastructure traversing Gilgit-Baltistan in Northern Pakistan directly south of the Sino-Pakistani border, and the newly built deep-water port at Gwadar in coastal Balochistan on the Arabian Sea. Overseas anxiety seemed to intensify as the original Chinese financed CPEC budget of $46bn first rose to $55bn and then, $62bn.[35] Indian and US critique of CPEC traversing disputed territory, i.e. Gilgit-Baltistan—under Pakistani administration since 1947 but also claimed by India, has been noted. Gwadar, however, elicited wider concerns.

Notwithstanding uncertainty over the prospects of the Chinese-funded and -built Gwadar Port, Beijing built a school and infirmaries and granted around $500m to build an airport ($230m), potable water facilities ($130m), a hospital (100m), and a technical and vocational training college ($10m) there.[36] By mid-2018, China Overseas Ports Holding Company (COPHC), operating the port on a Pakistani licence, reported the presence of variously specialising 20 corporations in the Gwadar Free Zone, with aggregate investments exceeding $460m.[37] Gwadar's unusually high level of Chinese grant-aid was suspected to reflect Beijing's purported geopolitical interests in dominating the Persian Gulf-Indian Ocean SLoC, deepening Indo-US anxiety vis-à-vis China and its semi-alliance with Pakistan.[38]

5.2.1 India's Critique of the Corridor

Indian criticism of CPEC, including in statements by Prime Minister Narendra Modi and senior officials, focused on the status of Gilgit-Baltistan, a former principality that came under Pakistani control at the time of the 1947 Partition on the initiative of a British-Indian military officer. Like all disputed territories, Gilgit-Baltistan elicited divergent Indian and Pakistani narratives, each supporting the validity of one's own claims while rebutting the other's.[39]

Indian practitioners noted that of the 222,236 km^2 expanse of the original Kashmir state which the British nominally 'sold' to Jammu's *Maharaja* Gulab Singh in March 1846 after defeating Kashmir's Sikh rulers, Delhi administered 101,437 km^2. The remainder was under Pakistani and Chinese control. Pakistan-administered 'Azad (free) Kashmir' comprised 13,297 km^2; Gilgit-Baltistan, in Indian view 'a part of the

---

[35] Siddiqui (2017).
[36] Jorgic (2017b).
[37] Bhatti and Liu (2018).
[38] Vickery (2018), Sen (2018), PTI (2017a), Conrad (2017), Jorgic (2017b) and Fazil (2016).
[39] Asif and Yang (2019), Singh (2013), Lambah (2016, pp. 227–237), Singh (2016), Ahmad (2016), Siddiqa (2017, pp. 108–123) and Akhund (2018).

Indian State of Jammu and Kashmir', had an area of 72,496 km^2; 42,685 km^2, including 5180 km^2 'illegally ceded to China by Pakistan in 1963', was under Beijing's control.[40] Islamabad's decision to maintain Gilgit-Baltistan as an autonomous region outside Azad Kashmir 'proved' Gilgit-Baltistan's status was 'tragic, ambiguous and undefined'.[41]

In 1881, following the post-Mutiny transfer of the East India Company's South Asian holdings to the crown, colonial authorities established an Agency in Gilgit, separate from Kashmir. After constitutional changes in 1935 granted Indians partial self-governance, the British took the Gilgit Agency on a 60-year lease to guard against long-feared Russian/Soviet threats. On 30 July 1947, the departing British transferred Gilgit-Baltistan to Kashmir's Maharaja Hari Singh, who appointed Brigadier Ghansara Singh its governor. After a Pakistan-endorsed march into Indian-occupied Kashmir by Pashtun tribal *lascar* militias in defence of a Muslim-Kashmiri uprising, Hari Singh acceded the Muslim-majority Kashmir to India. Responding to local sentiments, Major William Brown of the Gilgit Scouts, the senior British official present in the region, declared for Pakistan, detained Brigadier Singh and secured Gilgit-Baltistan.

In early November, while war raged across western Kashmir, Major Brown raised the Pakistani standard at Gilgit-Baltistan's administrative headquarters in Skardu. A fortnight later, the Pakistan-appointed Political Agent, Sardar Mohammad Alam, took charge. Sir George Cunningham, Governor of the North-West Frontier Province, later renamed Khyber Pakhtunkhwa, instructed Brown to maintain order across Gilgit-Baltistan. In 1948, King George VI awarded Major Brown the 'Most Exalted Order of the British Empire' for his action in 1948, legitimising the area's inclusion in Pakistan. This partly irregular action by an Indian Army officer, seconded to the regional paramilitary militia, taken against the backdrop of post-Partition turbulence roiling principalities ruled by minority-community princes, established Gilgit-Baltistan's unique status as Pakistan's 'Northern Areas'. In August 2009, the Gilgit-Baltistan Empowerment and Self-Governance Order renamed the region 'Gilgit-Baltistan'.[42]

There was no challenge among the region's populace to Gilgit-Baltistan's incorporation into Pakistan, but the latter, anxious to maintain its claim to Jammu & Kashmir in its entirety, could not logically integrate either Azad Kashmir or Gilgit-Baltistan into the Pakistani federation without accepting Indian offers of converting the 'Line of Control' drawn up along successive post-war ceasefire lines, and thereby giving up its broader claims. Gilgit-Baltistan was, therefore, administered as a special territory directly by Pakistan's federal government, rather than as part of either Azad Kashmir or Khyber Pakhtunkhwa. Pakistan repeatedly amended Gilgit-Baltistan's

---

[40] Lambah (2016, p. 227) and Report (2016b).
[41] Lambah (2016, p. 227).
[42] Report (2016b); Singh (2016) and Lambah (2016, p. 227).

legal, constitutional and administrative arrangements but failed to resolve contradictions inherent in balancing between sustaining contested claims on the one hand and ensuring stability and control on the other.[43]

The launch of CPEC, Indian protests against Chinese presence within 'disputed territory', growing local demands for greater rights and responsibilities, and Chinese anxiety to secure this key way-station for CPEC's transport-and-digital corridor, generated fluidity in the Gilgit-Baltistan discourse but offered no easy escape from the contradictions. Still, a review of recent records of Indian responses to Sino-Pakistani trade and infrastructure collaboration across Gilgit-Baltistan raised questions. This went back to the early 1960s aftermath of India's border war with China, which inflicted profound material and psychological trauma on the Indian elite. Delhi's national security establishment and the penumbra of *cognoscenti* opinion thereafter identified China as the source of the most acute threats to India's strategic interests.[44]

Indian analysts and friendly external observers posited, Sino-Pakistani collaboration, originating in the mid-1950s, was rooted in a 'purely strategic' calculation by Beijing and Islamabad to jointly thwart Indian well-being. A senior Indian practitioner insisted, 'China's agenda in Pakistan, whether vis-à-vis India or Afghanistan, has been purely geopolitical'. In his view, 'China's policy has been driven by power political criteria, and economics has counted for very little'.[45] In the sensitive nuclear-arms field, where dialectics drove the dynamic, Delhi saw Sino-Pakistani cooperation in an Indo-centric perspective: China's 'destructive engagement with Pakistan' and 'continued technology and weapon transfers to Pakistan are intended to provoke conflict between India and Pakistan'. China 'seeks to increase the capabilities of Pakistan in the hope that Pakistan might deliver a disturbing threat to India'.[46] That Pakistan might contemplate exercising agency in its own national security calculations appeared to escape this analyst almost entirely.

Another Indian commentator asserted, 'The irrefutable reality is that India's military adversaries, China and Pakistan are nuclear weapons states conjugally tied in an unholy strategic nexus pointedly targeting India ever since 1962'. A 'complicating strategic reality' was that the USA was 'traditionally inclined to tilt towards Pakistan and China whenever the chips are down against India'.[47] Kashmir-focused insecurity, fundamental to mutually inconsistent Indian and Pakistani national mythologies, betrayed incendiary escalatory potential between the nuclear-armed rivals in early 2019 when first India, and then Pakistan, conducted substantial aerial assaults on or across swathes of Kashmir in 'adverse possession'. Indian claims of destroying 'massive' militant bases in Pakistan were rebutted, and Pakistani ordnance caused little damage in India, but both claimed successes in aerial combat, denying losses. Pakistan captured an Indian pilot who was, in a 'peace gesture', repatriated to a hero's

---

[43] Siddiqa (2017, pp. 112–114).

[44] Ali (2017c), Sharma (2013), Ganguly and Pardesi (2009), Ali (1999), Vajpayee (1998) and Hansen (1967).

[45] Menon (2016a) and Small (2015).

[46] Mishra (2001).

[47] Kapila (2013).

welcome; Indian air-defence may have shot down their own helicopter in error.[48] Regional restraint eventually calmed passions, but India's constitutional revision transforming the disputed state of Jammu & Kashmir into two separate 'Union Territories' directly under Delhi's administration, and violent suppression of Kashmiri protests, betrayed persistent volatility.[49]

That is where China got involved. Delhi and its foreign friends posited that the core of India's insecurity resided in perceived Chinese threats to the latter's strategic autonomy; Beijing's 'all-weather friendship' with Pakistan, manifest in its aid to Islamabad's military strength and, now, in building CPEC, must be confronted. That imperative shaped Delhi's policy responses, including in its nuclear arsenal-building, and external balancing endeavours.[50] This became clear in October 2019 when, just before leaving for an 'informal summit' with Prime Minister Modi in India, Xi Jinping received Imran Khan and, on Kashmir, assured him, 'China supports Pakistan to safeguard its own legitimate rights and hopes that the relevant parties can solve their disputes through peaceful dialogue'.[51] Delhi retorted, 'Jammu & Kashmir is an integral part of India. China is well aware of our position. It is not for other countries to comment on the internal affairs of India'.[52] India viewed CPEC's construction in that regional context.

5.2.2 Western Anxiety

Indian views resonated with broad strands of Western thinking. Influential bodies of opinion identified China as the source of long-term, strategic, challenges to Western values and global pre-eminence and urged partners to mount concerted counteraction designed to thwart Chinese 'covert designs' and thereby sustain a modified version of the status quo.[53] Minority views only marginally leavened this competitive, even adversarial, dynamic.

A few Western commentators believed China did not seek to fundamentally revise the US-led order but only refine it to better accommodate its growing interests. Others saw dangers of domestic turbulence flowing from the discordance between China's increasingly open economic growth model and its closed political dispensation.[54] However, their voices, while adding flavour and variety to the 'China threat' discourse, remained marginal to the mainstream of 'first world' perceptions of China. This imbalance coloured the cultural palimpsest on which much China policy was

---

[48] Pastricha (2019) and Safi and Zahra-Malik (2019).
[49] Ratcliffe (2019), Tharoor (2019) and PTI (2019a).
[50] Kapur (2018), Bennett (2017), Report (2017b), Sheth (2015) and Report (2012).
[51] Report (2019a).
[52] Kumar and PTI (2019).
[53] Amedeo (2018), DeAeth (2018), Brown et al. (2018), Editorial (2017a), Orchard (2017), Fallon (2017) and Blackwill and Tellis (2015).
[54] Rudd (2018), Goldstein (2018), Sloan (2018) and Lampton (2018).

grounded. That was the prism through which China–Pakistan collaboration, and particularly CPEC, was perceived by its critics, especially those in the USA.

RAND analysts, while acknowledging CPEC's benefits to both partners, if and when the 'nearly 60 road, rail, port, power generation, communications and industrial zone projects' were completed over CPEC's 15-year build-up period, saw many risks. Most lay in 'local politics across weak frontier economies plagued by poor transportation and communications, conflicting trade and technical standards, corruption and terrorism'. Security challenges threatened implementation. By late 2017, notwithstanding efforts by a newly raised Army Special Security Division (SSD) comprising nine composite infantry battalions and thousands of Civil Armed Forces soldiers organised in six wings, Baloch separatists had killed more than 50 workers and several Chinese personnel who were building CPEC projects in Balochistan.[55]

A former US intelligence practitioner with regional experience, too, highlighted challenges to CPEC's prospects flowing from 'Pakistan's questionable ability to repay loans', and 'security threats posed by separatist insurgency, terrorism, and organised crime'.[56] In this view, CPEC's 'future prospects are uncertain'. Additionally, 'In the grand scheme of things, CPEC is but a modest part of the larger BRI'. Although CPEC's scale and scope were 'immense', they nonetheless 'pale in comparison to what Beijing hopes to do' with BRI. The latter 'boasts the potential' to involve 64 partner-countries home to nearly 4.5bn people and generating 'around 40% of the global economy'. Despite its relatively minor status within BRI, CPEC's 'overall significance … cannot be denied'. CPEC involved 'the world's likely next superpower pouring billions of dollars in infrastructure investments into a volatile, nuclear-armed country'; its outcome would be 'an early bellwether for how successful China can be in its ambitious pursuit of what may be the largest and most expensive trade link the world has ever seen'.[57] This was the geoeconomic lens through which US analysts examined CPEC.

Western and Western-aligned efforts to apply pressure on China and Pakistan explained Sino-Pakistani anxiety. Whereas China had been subjected to such difficulties since mid-1989, Pakistan's longstanding concerns over India and Afghanistan may have deepened its insecurity. In August 2017, while proclaiming his Afghanistan strategy, Donald Trump strongly criticised Islamabad: 'Pakistan often gives safe haven to agents of chaos, violence and terror. The threat is worse because Pakistan and India are two nuclear-armed states, whose tense relations threaten to spiral into conflict'. Trump's equating US goals in Afghanistan and Pakistan, focusing on nuclear security, deepened concern: 'America's interests are clear. We must stop the resurgence of safe havens that enable terrorists to threaten America. And we must prevent nuclear weapons and materials coming into the hands of terrorists'.[58]

Trump proclaimed a new US approach to Pakistan: 'We can no longer be silent about Pakistan's safe havens for terrorist organizations, the Taliban, and other groups

---

[55] Gishkori (2016), Dossani and Erich (2017), Rafiq (2017), Crabtree (2018) and Report (2018a).
[56] Kugelman (2017, pp. 15–17).
[57] Kugelman (2017, pp. 25–27).
[58] Trump (2017).

that pose a threat to the region and beyond'. He acknowledged Pakistani suffering and counterterrorism efforts and cooperation, 'but Pakistan has also sheltered the same organizations that try every single day to kill our people'. His outrage was apparent: 'We have been paying Pakistan billions and billions of dollars, at the same time they are housing the very terrorists that we are fighting'. Trump warned Islamabad, 'But that will have to change. And that will change immediately'.[59] Washington suspended reimbursing several hundred million dollars accrued by Pakistani 'alliance support operations', and other defence assistance.

To drive the point home, Trump added, 'Another critical part of the South Asia strategy for America is to further develop its strategic partnership with India'. Given Pakistan's congenital duel-unto-the-death over national legitimacy with, and history of insecure belligerence vis-à-vis India, Trump's description of Pakistan's nemesis as 'the world's largest democracy and a key security and economic partner' confirmed the Pakistani elites' worst suspicions. Trump's vow regarding India that 'We are committed to pursuing our shared objectives for peace and security in South Asia and the broader Indo-Pacific region',[60] delivered a pointed threat to Pakistan. Islamabad's residual faith in Washington's friendship likely evaporated with this speech and subsequent US action. Pakistan's critics welcomed Trump's remarks.[61]

The US commander of NATO and allied forces in Afghanistan equated 'the behaviour of Pakistan, Russia, and Iran' in aiding Taliban and other Afghan militants challenging Kabul's authority and battling US-led forces.[62] This perspective ignored Pakistan's losses in blood and treasure—nearly 80,000 civilian and military dead, and over $120bn-sustained in 2001–2018 in counterterrorism toll.[63] Some Western observers questioned the merit of Trump's Pakistan policy in advancing the USA's geopolitical interests in and around Pakistan. A US academic, considered a hardliner on Pakistan, warned, the Trump Administration's 'growing pressure on Pakistan … will have the effect of forcing Pakistan into an ever-tighter embrace of China'.[64] In contrast, Beijing stressed the 'great sacrifices' and 'important contributions' Islamabad had made in the fight against terrorism. The difference could not be clearer.

A Chinese spokeswoman told reporters, 'We believe that the international community should fully recognize Pakistan's anti-terrorism efforts'. She added, 'We are happy to see Pakistan and the United States carry out anti-terror cooperation on the basis of mutual respect, and work together for security and stability in the region and world'. As for the USA, 'We hope the relevant US policies can help promote the security, stability and development of Afghanistan and the region'.[65] Islamabad insisted it had, via hard military campaigns, eliminated terrorist safe havens and

---

[59] Trump (2017).

[60] Trump (2017).

[61] Thakur (2018), Roggio (2018) and Mangaldas (2018).

[62] Miller (2018).

[63] Nawaz (2018).

[64] Malik and Jennings (2017).

[65] Report (2017c).

could not 'be a US ally and be put "on trial" at the same time'. It wanted Washington to view USA–Pakistan relations through a bilateral prism, not one tinted with US goals in Afghanistan or India.[66] Sino-Pakistani labours built CPEC on that anxious landscape.

5.3 China–Pakistan Economic Corridor's 'Long-Term Plan'

In November 2017, with a high-powered Chinese delegation visiting Islamabad, the two partners proclaimed CPEC's Long-Term Plan (LTP).[67] The LTP was driven by China's goal to 'advance the western development strategy, promote economic and social development in Western China, accelerate the B&R construction, give play to China's advantages in capital, technology, production capacity and engineering operation, and promote the formation of a new open economic system'. Pakistan, for its part, sought to 'fully harness' its 'demographic and natural endowment' by enhancing 'its industrial capacity through creation of new industrial clusters, while balancing the regional socioeconomic development, enhancing people's well-being, and promoting domestic peace and stability'.[68]

The LTP defined CPEC as 'a growth axis and a development belt' incorporating a 'comprehensive transportation corridor and (Sino-Pakistani) industrial cooperation' as 'the main axis', and 'concrete economic and trade cooperation, and people-to-people exchange and cultural communications as the engine'. CPEC's physical blueprint comprised large infrastructure-building, industrial development and 'livelihood improvement' projects aimed at 'socio-economic development, prosperity and security' along CPEC's alignment across China's Xinjiang Uygur Autonomous Region and Pakistan's entirety. The Corridor would link key nodal points in the cities of Kashgar, Atushi, Tumshuq, Shule, Shufu, Akto and Tashkurgan Tajik in China with Gilgit, Peshawar, Dera Ismail Khan, Islamabad, Lahore, Multan, Quetta, Sukkur, Hyderabad, Karachi and Gwadar in Pakistan.[69]

The LTP described CPEC's 'core' and 'radiation' zones as the 'one belt, three axes and several passages'. The 'one belt', representing CPEC's core zone, stretched from Kashgar, Tumshuq, Atushi and Akto county in Xinjiang, to Pakistan's Gilgit-Baltistan, Azad Kashmir, federal capital Islamabad, and parts of Khyber Pakhtunkhwa, Punjab, Sindh and Balochistan provinces. The 'three axes', i.e. horizontal alignments, connected six nodes: Lahore-Peshawar, Sukkur-Quetta, and Karachi-Gwadar. 'Passages' linked nodal points Islamabad, Karachi and Gwadar via trunk highways and railways. Vertically, CPEC combined five functional zones[70]:

---

[66] Afzal (2017) and Nawaz (2018).
[67] Wang and Iqbal (2017).
[68] CPEC Coordinator (2017, p. 9).
[69] CPEC Coordinator (2017, p. 4).
[70] MPDR (2017a).

5.3 China–Pakistan Economic Corridor's 'Long-Term Plan'

- Xinjiang foreign economic zone
- Northern border trade logistics and business corridor and ecological reserve
- Eastern and Central Plain economic zone
- Western logistics corridor business zone, and
- Southern coastal logistics business zone.

CPEC converged China's regional development plans with Pakistan's medium-term Vision 2025 programme.[71] It also pursued the formation of a 'new international logistics network and industrial layout' around major transport-infrastructure, 'elevate the status' of South and Central Asian countries within the global economy, and advance regional economic integration through growing trade, international economic and technological cooperation, and personnel exchange.[72] The LTP sequenced CPEC's implementation in three phases which radiated its growth and developmental effect from the 'core' to the Central and South Asian regions[73]:

- Phase-1 (2015–2020) would concentrate on 'basically addressing' major bottlenecks hindering Pakistan's socio-economic development, following which 'CPEC shall start to boost the economic growth along it for both countries'.
- Phase-2 (2020–2025) would strive to ensure that 'CPEC building' was 'basically done', the industrial system 'approximately complete', major economic functions activated 'in a holistic way', people's livelihood 'significantly improved', regional development 'more balanced', and all the goals of 'Vision 2025 achieved' by the end of this phase.
- Phase-3 (2025–2030) would seek to 'entirely accomplish' CPEC building, establish the 'endogenous mechanism for sustainable economic growth', effect CPEC's role in 'stimulating economic growth in Central Asia and South Asia into holistic play', and grow South Asia into 'an international economic zone with global influence'.

The LTP identified seven 'Key cooperation areas' that CPEC building would focus on[74]:

- Connectivity
- Energy-related fields
- Trade and industrial parks
- Agricultural development and poverty alleviation
- Tourism
- People's livelihood and non-governmental exchanges, and
- Financial cooperation.

---

[71] MPDR (2014a).
[72] MPDR (2017b, p. 9).
[73] MPDR (2017b, p. 10).
[74] MPDR (2017b, pp. 14–22).

5.3.1 Connectivity

Considered one of the highest priorities within the CPEC blueprint, connectivity included transportation hardware that could drive socio-economic development across the CPEC catchment area, widening and deepening the capacity for movement of people and goods between China and Pakistan on the one hand, and throughout Pakistan, on the other. It also included the construction of information network infrastructure, boosting the exchange of ideas and information between China and Pakistan, and also across Pakistan. CPEC launched many individual and inter-connected projects within these two categories.

Integrated transport systems: CPEC's new and renovated transport system would include major highways and railways, as well as metropolitan rapid transit systems in key nodes. The most substantial projects.

- Construction and development of Kashgar–Islamabad, Peshawar–Islamabad–Karachi, Sukkur–Gwadar, and Dera Ismail Khan–Quetta–Sorab–Gwadar road infrastructure
- Modernisation and capacity expansion of Pakistan's railway trunk lines, especially ML-1 linking Peshawar–Lahore–Karachi, the track along whose entire length would be doubled, and building new lines and related supportive infrastructure
- Implementing the Gwadar City Master Plan, and construction and development of Gwadar city and port, with an integrated transport system; improving port infrastructure, construction of the East Bay Expressway and Gwadar International Airport, and enhancing the Gwadar Free Zone's competitiveness
- Strengthening China-Pakistan cooperation in technical training for implementing CPEC and developing sustainable transportation.

Information network infrastructure: China and Pakistan would boost information connectivity, jointly construct and operate communication, broadcast and TV networks, and synchronise construction of information, road and railway infrastructure.

- Constructing optical fibre cable connectivity linking China and Pakistan, and building 'the backbone optical fibre networks' in Pakistan
- Upgrading Pakistan's networking facilities, including its national data centre and the country's second submarine cable landing station
- Expediting Pakistan's adoption of the Chinese-designed Digital Terrestrial Multimedia Broadcasting (DTMB) standard
- Promoting Pakistan's ICT-enabled development, e.g. e-government, e-commerce, border electronic monitoring, and 'safe city' construction
- Enhancing Pakistan's information industry by building IT industrial parks and IT industry clusters
- Increasing personnel exchange programmes boosting the Pakistani workforce's technical skills, establishing technical training centres across Pakistan, and expanding Pakistan's ICT human resource base.

5.3.2 Energy

CPEC would establish power stations and distribution grids across Pakistan, using oil, gas and coal, building major thermal-power, hydropower, coal gasification, and renewable power generation projects and strengthen Pakistan's power transmission capacity:

- Enabling China and Pakistan to boost and diversify hydrocarbon supplies, with consideration of building refineries and storage depots along CPEC's alignment
- Optimising sourcing and use of coal, enhancing research on and use of Pakistan's own coal reserves, developing gasification technologies and augmenting coal mining
- Promoting river planning, preparing projects and building hydropower stations to improve renewable power content within Pakistan's energy mix
- Developing Pakistan's wind and solar energy generation capacity and integrating such power with its national energy portfolio
- Strengthening construction and expansion of Pakistan's high-voltage power grids and transmission and distribution networks
- Developing and expanding Pakistan's industrial capabilities for manufacturing energy-related equipment, including for renewable energy generation and distribution.

5.3.3 Trade and Industry

CPEC would secure Chinese cooperation in advancing Pakistan's trade and manufacturing capabilities, expand Sino-Pakistani exchanges and enhance bilateral trade liberalisation:

- Expanding Pakistan's textiles and garments industries, quality-control, value addition, and portfolio of high value-added products, and using export-processing to boost their contribution to Pakistan's economy
- Establishing SEZs in all provinces and regions, and linking these to China's Kashgar Economic and Technological Development Zone and Caohu Industrial Park
- Advancing Pakistan's industrial capability from assembling imported components to the production of these locally, and encouraging Chinese firms to invest in improving Pakistan's manufacturing energy-efficiency
- Expanding Pakistan's industrial base to include chemicals, pharmaceuticals, agrochemicals, engineering goods, iron and steel, light fabrication, home appliances and construction materials, to meet domestic and export demands
- Enhancing the exploration, extraction and utilisation of Pakistan's mineral resources, and establishing mineral-processing zones and industries

- Jointly developing industrial parks to deepen bilateral economic and trade cooperation while ensuring industrial concentration
- Optimising logistics, business-to-business cooperation, investment opportunities and trade structures, focusing on ports and transport-logistics, notably by developing the Gwadar free zone and linking Pakistan's free zones to Chinese equivalents
- Launching standardised data exchange pilot projects to support customs supervision areas linking special zones and industrial parks located along CPEC's alignment.

5.3.4 Agricultural Development and Poverty Alleviation

CPEC would enable the partners to cooperate on agro-personnel training, technical exchanges, biological breeding, agro-production, processing, storage and transportation, infrastructure construction, disease prevention and control, water resource utilisation, land development, ICT-enabled agriculture, and marketing farm-products to build an agro-industry:

- Upgrading agricultural infrastructure, building water-saving modern agricultural zones, strengthening drip irrigation technology, and developing low-to-medium yield land to boost productivity
- Expanding collaboration in crop farming, livestock breeding, forestry and foodstuff cultivation, and aquatic and fisheries production
- Improving handling, storage, shipping and marketing of agro-products, and expanding Pakistan's forestry, horticulture, fisheries and livestock sectors
- Enhancing Pakistan's water resource management, developing pastoral areas and deserts and, to these ends, introducing remote-sensing technology
- Boosting Pakistani production and utilisation of pesticides, fertilisers, machinery and support services, including agricultural education and research.

5.3.5 Tourism

CPEC would develop trans-border tourism facilities and attractions along and across Sino-Pakistan borders, and along CPEC's transport routes, especially in Southern Pakistan:

- Examining prospects for developing tourism, and building resorts along China-Pakistan borders, and along CPEC's highway and railway routes across Pakistan
- Improving cross-border tourism services in public information, transportation, security and protection towards building a developed trans-frontier tourism industry

5.3 China–Pakistan Economic Corridor's 'Long-Term Plan'

- Developing a coastal leisure-and-vacation route in Southern Pakistan, linking urban centres in Keti Bandar, Karachi, Sonmiani, Ormara, Jhal Jhao, Gwadar and Jiwani
- Reviewing the potential of fashioning a '2 + 1+5' tourism spatial structure[75] in Southern Pakistan, with the two centres at Karachi and Gwadar Ports, an axis along the future coastal tourism belt linking the two ports, and five tourist zones built around Jiwani and Gwadar, Jhal Jhao, Ormara, Sonmiani and Keti Bandar.

5.3.6 People's Livelihood and Non-governmental Exchanges

CPEC would strengthen collaboration and communications among Pakistani and Chinese NGOs, deepen people-to-people friendship, improve people's livelihood prospects and enhance service provision in cities lying astride CPEC's alignment:

- Utilising global and Chinese urbanisation concepts in developing nodal-municipalities, especially in public transport, water supply and drainage systems; utilising China's strength in technology, equipment and finance to pilot livelihood projects
- Expanding training programmes for Chinese and Pakistani central, regional, local and party officials in CPEC areas; selecting meritorious Pakistani students for further education and for effecting cultural exchanges at universities in Xinjiang
- Strengthening cooperative socio-economic development, improving Pakistani vocational training, enhancing design and R&D activities at Pakistan's educational institutions, deepening research collaboration among Pakistani and Chinese institutes, especially in technology-transfers and technical training of Pakistani nationals at Xinjiang's Academy of Central Asia Regional Economic Cooperation
- Expanding medical services and upgrading existing medical facilities in CPEC regions
- Exchanging cooperative experiences of public social welfare for the Gwadar region and enhancing public support for CPEC
- Cooperatively planning and implementing Pakistan's water resource development, utilisation, conservation and protection projects, and managing flood and drought prevention and disaster relief.

5.3.7 Financial Cooperation

CPEC would aid both China and Pakistan in boosting their 'financial reform and opening up' efforts, establishing 'multi-level cooperation mechanisms' and policy coordination, innovating financial products and services, and controlling financial risks:

---

[75]MPDR (2017b, p. 19).

- The partners would deepen cooperation between their central banks and regulatory agencies, coordinate financial regulations, implement and expand bilateral currency-swap agreements, promote trade and investment settlements using their two currencies to reduce third-party currency demands, and promote orderly but free flow of capital using the Cross-Border Inter-Bank Payment System
- China supported Pakistan's cooperation with AIIB and the partners would open their markets to each other's financial institutions, encouraging these to finance CPEC projects, including with externally sourced loans; they would fashion a cross-border credit system and promote export credit, project financing, syndicated loans, trade-finance, lease-finance, investment banking, e-banking, cross-border RMB business, financial markets and asset management
- China and Pakistan would open and develop securities markets, jointly provide multi-currency direct financing of Pakistan's central and provincial government projects, boost cooperation among stock exchanges, and support their enterprises and institutions in financing CPEC projects
- Pakistan would draw on the experience of the Shanghai and other Chinese Pilot Free-Trade Zones in building the Gwadar Port Free Zone and explore offshore RMB financial business; the partners would deepen financial cooperation between their free-trade zones and explore the formation of 'a RMB backflow mechanism'.[76]

5.4 The Corridor's Investment and Financing Mechanisms

The LTP set out investment and financing mechanisms and supporting measures for funding CPEC projects from initiation through completion. These were based on China giving 'full play to its advantages in investment and financing' on the principles of joint investment, joint construction and shared benefits. The parties would 'create necessary commercial conditions' for establishing projects 'according to the market-oriented principle', fashioning a 'reasonable cost and revenue sharing mechanism', thus securing 'reasonable commercial returns' from the projects 'for all shareholders'.[77] CPEC's funding-investment plans outlined several sources:

- Government funds: States bore 'primary responsibility' for funding 'public welfare projects'. China offered grant, interest-free loans, concessional loans and preferential buyer's credit; Pakistan's central and provincial governments assumed 'some investment and financing responsibilities' with budgeted funds and bond-capital. To reduce costs and protect investors' rights, both parties extended credit support[78]
- Direct investment: Both governments encouraged Chinese SOEs, private corporations and private funds of other entities to invest in CPEC projects while also

---

[76] MPDR (2017b, pp. 20–21).
[77] MPDR (2017b, p. 24).
[78] MPDR (2017b, p. 24).

5.4 The Corridor's Investment and Financing Mechanisms 195

urging Pakistan's private capital to invest in such projects, and in private financial institutions investing in Pakistani infrastructure projects
- Indirect financing: The partners would deepen 'strategic cooperation' among their policy banks, development finance institutions and commercial banks, resolve CPEC project-funding issues, and support the Silk Road Fund, and China-Eurasia Economic Cooperation Fund in financing CPEC projects
- International financial institutional loans: China and Pakistan invited the WBG, the ADB, the AIIB and other multilateral and international financial institutions to extend long-term concessional loans for financing CPEC projects
- Other innovative investment and financing methods: Pakistan's federal and provincial governments, financial institutions and businesses would be encouraged to secure RMB financing in China including Hong Kong, and other offshore RMB centres; the partners would encourage Chinese and Pakistani market players to obtain funds from international markets and from within Pakistan, and invest in CPEC projects.

5.5 The Corridor's Projects in Pakistan

The LTP white paper issued by the Pakistani government set out details of these and related mechanisms with which China and Pakistan would secure, invest and protect financing for CPEC projects from diverse sources. By mid-2018, those responsible for implementing the CPEC blueprints claimed significant successes. However, the list of projects implemented variously under the CPEC rubric was long.

5.5.1 Energy Projects

The LTP listed a total of 19 individual projects, with details of location, output, and costs, along with two potential projects, whose prospects appeared unclear. Projects were classified as 'priority projects', 'actively promoted projects' and 'potential energy projects'. Priority projects are listed in Table 5.1:
The LTP's four 'Actively-promoted projects' were

- Kohala, Azad Kashmir—1100 MWe hydropower station—completion in 2025
- Rahim Yar Khan, Punjab—1320 MWe imported coal-fired power plant
- Cacho, Sindh—50 MWe wind farm
- Jhimpir, Thatta, Sindh—50 MWe Western Energy wind farm.

Table 5.1 CPEC LTP priority projects

Project location	Project summary	Ready status
Port Qasim, Karachi	2 × 660 MWe coal-fired power plants	25 Apr 2018
Naran, Khyber Pakhtunkhwa (KPK)	Suki Kinari hydropower station	Dec 2022
Sahiwal, Punjab	2 × 660 MWe coal-fired power stations	28 Oct 2017
Bahwalpur, Punjab	1000 MWe solar park	30% op in Aug 2016
Karot, Punjab	720 MWe hydropower station	Dec 2021
Matiari, Sindh-to-Lahore, Punjab	660 kV HVDC transmission line	Dec 2019
Thar Block I, Sindh	6.8 MTPA coal mine; 2 × 660 MWe plants	Dec 2019
Thar Block II, Sindh	4 × 330 MWe lignite-fired power plants	10 July 2019
Thar Block II, Sindh	3.8 MTPA surface coal mine	Dec 2018
Thar Block VI, Sindh	1320 MWe coal-fired power station	Dec 2023
Gharo, Thatta, Sindh	Hydro China 50 MWe wind farm	5 Apr 2017
Jhimpir, Thatta, Sindh	100 MWe UEP wind farm	16 June 2017
Jhimpir, Thatta, Sindh	50 MWe Sachal wind farm	11 Apr 2017
Jhimpir, Thatta, Sindh	2 × 50 MWe Three Gorges wind farms	9 July 2018
Gwadar, Balochistan	300 MWe coal-fired power plant	Dec 2022
Hub, Balochistan	1320 MWe coal-fired power plant	17 Aug 2019

The LTP also listed two 'Potential energy projects' to be considered later on:

- Phandar, Gilgit-Baltistan—80 MWe hydropower station
- Karakoram International University, Gilgit-Baltistan—100 MWe hydropower station.

5.5.2 Major Infrastructure Projects

The LTP set out plans to build/rebuild/modernise five multi-lane highways and three railway projects across Pakistan. In late 2019, these were at various stages of progress.

Major roads and highways included

- Karakoram Highway (KKH) Phase II Thakot-Havelian section—120 km—completion Mar 2020
- N-35 KKH Thakot–Rajkot remainder—136 km

- Peshawar–Karachi motorway Multan–Sukkur section—392 km—completion Aug 2019
- N-30 Khuzdar-Basima road—110 km
- N-50 D.I.Khan/Yarik-Zhob upgradation Phase I—210 km.

Major railway projects included

- Expansion/reconstruction of ML-1: Peshawar-Rawalpindi-Gujranwala-Lahore-Sahiwal-Khanewal-Bahawalpur-Rahim Yar Khan-Rohri-Nawabshah-Hyderabad-Karachi—1872 km—completion 2022
- Havelian dry port—450 m TEU annual capacity
- Pakistan Railway—capacity building, expansion and staff training.

5.5.3 Mass-Transit Rail Projects

CPEC included substantial urban mass-transit light railway projects for the four provincial capital cities:

- Karachi Circular Railway—completion scheduled for Oct 2020
- Lahore Orange Line—completion July 2019
- Greater Peshawar Region Mass Transit
- Quetta Mass Transit.

5.5.4 Special Economic Zones

China and Pakistan planned to establish nine Special Economic Zones (SEZ) focusing on specific products across Pakistan for accelerating industrialisation and boosting trade:

- Moqpasand SEZ, Gilgit-Baltistan (marble/granite, iron ore/steel, foodstuffs, leather)
- Mirpur SEZ, Azad Kashmir (mixed industrial and manufacturing units)
- Allama Iqbal Industrial City, Faisalabad (textiles, steel, pharmaceuticals, chemicals, engineering, food processing, plastics, agro-implements)
- ICT Model Industrial Zone, Islamabad (steel, foodstuffs, pharmaceuticals, chemicals, printing and packaging, light engineering)
- Rashakai Economic Zone, Nowshera, KPK (foodstuffs, fruits, textiles)
- Mohmand Marble City, KPK (marble, stoneware)
- Bostan Industrial Zone, Balochistan (chromite, ceramics, cooking oil, foodstuffs/fruit processing, agro-machinery, pharmaceuticals, electrical appliances, cold-storage)

- China SEZ, Dhabeji, Sindh
- Pakistan Steel Mills Industrial Park, Port Qasim, Karachi (chemicals, pharmaceuticals, steel, automotive parts, printing and packaging, garments).

5.5.5 Social Sector Development Projects

CPEC included four social developmental programmes

- People-to-people exchanges via annual cross-border cultural events
- Inter-sectoral transfer of knowledge, especially on urban/rural development
- Establishment of the Pakistan Academy of Social Sciences with links to CASS
- Education sector knowledge transfer through consortium of Business Schools.

5.5.6 Other Major Projects

Many of the CPEC projects were included in Pakistan's own Public Sector Development Programmes (PSDP) for FY 2017–2018. Additionally, CPEC included three 'other projects':

- Cross-border optical fibre cable (820 km linking Khunjerab Pass to Islamabad via Gilgit-Baltistan, KPK, Punjab; provides 3G/4G services in Northern Pakistan, enhanced ICT security with an alternative fibre-route)—activated in July 2018; commercially open in February 2019[79]
- Digital Terrestrial Multimedia Broadcast pilot project—completed in May 2018
- Pakistan's Meteorological Department's early warning system parallel to WBG project.

5.6 The Corridor's Early Harvest Projects

Shortly after CPEC's launch, China invested $14bn in 30 early harvest projects. By the autumn of 2016, project implementation had made some progress. The Chinese embassy in Islamabad reported that 16 projects were already under construction and were steadily moving forward.[80] Later updates similarly provided information on progress made in project implementation.[81]

---

[79] Nasim (2019).
[80] Zhao (2018).
[81] Press Release (2018).

5.6 The Corridor's Early Harvest Projects

5.6.1 Khyber Pakhtunkhwa Province

- Joint feasibility study for upgrading ML1
- Upgradation of ML1
- Establishing Havelian Dry Port
- KKH II—Havelian-Thakot
- KKH III—Rajkot-Thakot
- N-50 Highway—D.I.Khan-Quetta
- Suki Kinari Hydropower Project
- Optical fibre cable—Khunjerab-Rawalpindi; opened in July 2018.

5.6.2 Punjab Province

- Optical fibre cable—Rawalpindi-Khunjerab; opened in July 2018
- Haier-Ruba Economic Zone II
- Lahore-Karachi Motorway (Multan-Sukkur section).
- Joint feasibility study for upgrading ML1
- Upgradation of ML1
- Sahiwal coal-fired power plant
- Rahimyar Khan coal-fired power plant
- Karot hydropower plant
- Lahore Metro Train Orange Line
- Lahore-Matiari power transmission line
- Faisalabad-Matiari power transmission line
- Quaid-e-Azam solar park, Bahwalpur.

5.6.3 Sindh Province

- Matiari-Lahore power transmission line
- Matiari-Faisalabad power transmission line
- Port Qasim Engro power plant
- Thar Block II power plant and surface coal mine
- Dawood wind farm
- Jhimpir wind farm
- Sachal wind farm
- China-Sunec wind farm
- Upgradation of ML1.

5.6.4 Balochistan Province

- N-30 Khuzdar-Basima Highway
- N-50 D.I.Khan-Quetta Highway
- Hubco coal-fired power plant
- Gwadar-Nawabshah LNG terminal and pipeline
- Gwadar Eastbay Expressway—completion in Oct 2020
- Gwadar Eastbay Expressway II
- Gwadar power plant
- Gwadar New International Airport
- Gwadar Smart Port City master plan—completed in Aug 2019
- Gwadar Port terminal upgrade, including breakwater and dredging
- Gwadar City wastewater treatment plan
- Gwadar primary school
- Gwadar Hospital upgrade
- Gwadar Technical and Vocational College
- Gwadar Free Zone freshwater supply.

5.6.5 Additional 'New Projects'

After announcing province-wise distribution of CPEC projects, and discussions with Chinese counterparts in late-2016, the CPEC Secretariat, located within the Government of Pakistan's MPDR, published a list of 'new' projects, indicating an expansion of CPEC's project portfolio:

- Gilgit-Shandor-Chitral-Chakdara link road—Gilgit-Baltistan/KPK
- Mirpur-Muzaffarabad-Mansehra link road—Azad Kashmir/Punjab
- Chiniot iron ore mining, processing and steel mills complex—Punjab
- Keti Bunder Sea Port development—Sindh
- Naukandi-Mashkhel-Panjgur Road linking with M8/N85 highways—Balochistan
- Quetta-Pat Feeder Canal water supply scheme—Balochistan.

5.7 The China-Pakistan Economic Corridor's Terminal Challenges

CPEC's 'flagship' status notwithstanding, implementation faced early hiccups. Shortly after the LTP was issued, Beijing suspended funding for three road projects

5.7 The China-Pakistan Economic Corridor's Terminal Challenges

until mutually agreed 'new guidelines' took effect. These were the KKH III—Rajkot-Thakot section, the Khuzdar-Basima Highway, and the Dera Ismail Khan-Zhob section of the N50 DI Khan-Quetta Highway.[82] Funding suspension indicated differences over details, which delayed progress. A fortnight earlier, Pakistan had withdrawn a $14bn dam project from CPEC. A senior official told Pakistani legislators, 'Chinese conditions for financing the Diamer-Bhasha Dam were not doable and against our interests'. He informed Parliament that Chinese terms included operating and maintaining the dam on completion and taking ownership of it. This being 'unacceptable', Pakistan decided to fund the dam with its own resources.[83]

Given Islamabad's resource constraints, this proved challenging, but Indian observers saw the move variously as a 'diversionary tactic', a 'crisis situation', and a cause for 'despair'.[84] When Imran Khan's Pakistan Tehreek-e-Insaf (PTI) party won the July 2018 general elections defeating the two traditional major parties, some conjectured that he might, in the interest of greater transparency, slow CPEC-related investment down. In the event, Khan assured everyone: 'the Corridor will receive wide support from all sectors of Pakistani society … the Corridor has a positive effect on the development of Pakistan. I believe that the Corridor construction in the medium and long term will be firmly guaranteed and will continue to be implemented'.[85]

Policy differences, funding challenges and political turbulence cast a shadow on CPEC, but the impact of regional and *systemic* anomie flowing from CPEC's northern and southern termini appeared much more severe. CPEC's extremities, in Gilgit-Baltistan and Balochistan's Gwadar Port, respectively, raised grave external concerns. Repeated Chinese and Pakistani assurances that CPEC was an economic endeavour to improve the socio-economic lot of Pakistan's impoverished populace, and that it had no geopolitical content, did not allay Indian and Western anxiety. The evolution of CPEC's termini, especially the decades of unopposed operation of the China–Pakistan Karakoram Highway (KKH), the basis of CPEC's northern alignment, and the 1950s-era US engagement with the conceptual origins and development of Gwadar's deep-sea port, suggested recent histrionics were partly overblown, even inconsistent.

5.7.1 The Corridor's Karakoram Highway-Rooted Beginnings

The 800-mile-long KKH, 'the world's highest transnational roadway' linking Kashgar in Xinjiang with Islamabad via Havelian in Northern Pakistan, was 'completed' in 1978–1979. It had taken two decades of difficult engineering work across the

---

[82]Raza (2017).
[83]Rana (2017).
[84]Chaudhury (2018) and Singh (2018).
[85]CPEC Secretariat (2018) and Report (2018b).

Karakoram mountain range by thousands of Pakistani and Chinese engineers, technical personnel and roadbuilders.[86] With rockfalls, mudslides and avalanches caused by both tectonic plate friction and years of blasting of the mountains often shutting down higher stretches, and seasonal floods threatening its lower reaches, work never really ended, but the KKH was opened to commercial traffic in 1982. Increasing trade between proto-allies China and Pakistan, especially since the October-November 1962 China–India border war, in which India was humiliated[87] notwithstanding US military and intelligence support, was the public explanation of this costly—in both lives and treasure—civil-engineering enterprise.

Running southward from Kashgar, the KKH snaked its way across western Xinjiang, linking valleys separating peaks of the Sarykol range which formed a junction between the Pamir and Kunlun mountain ranges. At the 4700 m-high Khunjerab Pass slicing the Karakoram range, the roadway crossed into Gilgit-Baltistan, winding across troughs before reaching the upper Indus Valley just east of Gilgit. The KKH then followed the Indus, past the Nanga Parbat massif beyond which the Indus breaks out of its headwaters. It then swung to a different alignment, running southward towards Havelian and another 52 miles on to Islamabad. Many critics noted the strategic implications of a modern highway linking the two near-allies against the backdrop of deepening tension between India on the one hand, and China and Pakistan on the other.[88] And yet, in mid-1978, approaching the KKH's official opening, Delhi made 'only a formal protest, just for the record'.[89] That *pro forma* protest reflected policy for decades.

The KKH originated before the Sino-Indian War. In 1958, teams from the Pakistan Army Corps of Engineers, later forming Pakistan's Frontier Works Organisation (FWO), began building an all-weather road between Gilgit and Swat. In 1966, following the 1965 India-Pakistan war over Kashmir, Pakistan and China decided to establish a road link for speedy delivery of Chinese *materiel* to Pakistani forces, while demonstrating Beijing's commitment to Islamabad. The Gilgit-Swat road was the KKH's nucleus. Initially, FWO teams built a 400 km shingle road from Thakot to Khunjerab Pass, with temporary bridges spanning myriad streams, ravines and gorges. The section was completed in February 1971. After Pakistan's war that year with India over Bangladesh, resulting in the loss of Pakistan's eastern wing, the partners replaced all temporary bridges with permanent ones, extending the KKH's Pakistani section from Thakot to Havelian, and linked it to the Havelian-Islamabad motorway.

Construction of the mostly dual-lane KKH entailed 21 million m^3 of earthwork using 80,000 tons of cement, and rock blasting using 8000-plus tons of explosives. Pakistani and Chinese engineers deployed an average of 15,000 men, 1200 vehicles

---

[86] Jacobs (2013), Azam (2012), Singh (1981, p. 18) and Topping (1979).
[87] TNN (2012), Report (2014), Abdi (2017) and Jain (2018).
[88] Topping (1979), Singh (1981, pp. 18–26) and Sering (2012).
[89] Ram (1978).

and 1000 pieces of varied plant to the project over its duration. Officially, 404 individuals lost their lives and a similar number suffered serious injury.[90] Other sources cited 'official figures' claiming more than 800 Pakistanis and 82 Chinese were killed in blasts, rockfalls, avalanches and falls into gorges.[91] On 18 June 1978, President Zia-ul Haq and Vice Premier Geng Biao formally inaugurated the motorway. Initially, only Pakistan's national security establishment drew the KKH's benefits. Civilian traffic began flowing in 1982; foreigners gained access in 1986.

Terrestrial commerce and tourism along the KKH grew gradually. The motorway's alignment placed Northern Pakistan at the heart of energy-rich but landlocked Central Asian Republics' quest for access to the Arabian Sea. This led to the May 2004 trade agreement among China, Pakistan, Kazakhstan and Kyrgyzstan on exchanges of commodities, energy-resources and other products, using the Pakistani dry port at Sost on the KKH. But the narrow, often unstable, stretches of 'the highest international highway' restricted the size and volume of goods traffic. Anticipating future growth, China and Pakistan began widening and improving the mountainous motorway. To ensure the KKH could handle round-the-year flows of trucks carrying 40-foot containers, they invested much time, effort and resources.

The scale of the challenge became apparent when, in January 2010, a massive landslide in Hunza buried 20 people, blocked the Hunza River, flooded villages, orchards and pastures, displaced thousands, submerged a 19 km section of the KKH and created the Attabad Lake. Until an alternative could be devised, trucks carrying cargo and containers had to be ferried across the newly formed lake, drastically slowing movement. In July, Beijing and Islamabad signed an agreement on widening and upgrading the KKH with the aim of devising an alignment bypassing the Attabad Lake and increasing KKH's haulage capacity threefold.

The project doubled the number of road-links at Khunjerab Pass from four to eight. A spillway installed on the lake in 2012 drained some of the overflow, revealing submerged villages. Engineers from the FWO's Chinese partner, China Road and Bridge Corporation (CRBC), an infrastructure construction SOE, elected to drill a tunnel into the mountains to fashion a new path away from the lake. For this, they had to haul in heavy equipment from Xinjiang and build some plants locally. In addition to boring a tunnel, and widening the existing lanes, they had to strengthen the load-bearing capacity of over a hundred bridges. CRBC engineers led the physical building project, but employed several thousand Pakistani personnel in various capacities while training others to become technical staff and skilled workers.[92] The $400m work was well-advanced and maturing before CPEC's formal launch.[93]

Long after the KKH reopened to commercial traffic, nature often played havoc, with heavy rains triggering landslides that shut down movement for days.[94] The challenges of sustaining this key artery at the northern bounds of the CPEC network

---

[90] FWO Pakistan https://www.fwo.com.pk/45-completed-projects/343-kkh. Accessed 7 July 2018.
[91] Kazim (2012).
[92] Huaxia (2015).
[93] Khan (2014) and Kazim (2012).
[94] APP (2016).

will clearly need constant care. Pakistani views of the KKH specifically and Sino-Pakistani relations generally, were overwhelmingly positive. Conversely, Indian commentary on the theme was invariably negative. However, Indian critique of the KKH, and the strength of Sino-Pakistani bonds it represented, did not feature prominently in the core of the Indian national security discourse.

This changed after Xi Jinping proclaimed his BRI vision and, subsequently, Li and Xi unveiled the CPEC initiative with their Pakistani counterparts. It was then that Indian analysts focused on what in their view were the highly malign geopolitical and strategic object, design and implications of the KKH, and the CPEC blueprint being fashioned around it.[95] Occasionally, elite outrage boiled over into calls for action against Sino-Pakistani 'aggression'.[96] Officially, Delhi made more nuanced statements, but only after CPEC's alignment was formally published. This suggested it was CPEC's transformative scale, not what the KKH could deliver to Pakistan, seen in the context of Indian anxiety vis-à-vis China, which made a difference sufficient for Delhi to register its suddenly stern objections. In this, India found comforting convergence in the views of its Trump Administration allies.

Pakistan's steady integration into China-Central Asia trade and infrastructure connectivity networks in the early 2000s generated dynamics reflecting regional developmental aspirations. These resonated with Beijing's Eurasian drive.[97] The evolution of this accretive palimpsest of interests, and an expanding range of choices, preceded the proclamation of Beijing's BRI blueprint. That convergence laid the foundation on which China and Pakistan found it logical to initiate the CPEC endeavour. Early in Donald Trump's presidency, reflecting a darkening China perspective clouding Washington, his then-Secretary of State, Rex Tillerson, and former Secretary of Defence, James Mattis, formally slammed the Chinese-Pakistani near-alliance generally, and CPEC in particular. They also inducted India into the discourse, characterising Delhi as a contrasting, and positive, force for good.

Tillerson complained that China and Pakistan threatened the global order which had secured peace and prosperity, while India was rising in a responsible fashion within the order. Mattis noted that CPEC's alignment, crossing disputed territory, raised questions about the intent and legitimacy of the enterprise. While Indian commentators found much to commend in Washington's endorsement of Delhi's critique of BRI/CPEC, China and Pakistan protested vigorously.[98] Simultaneously reflecting and reinforcing ambient dynamics of strategic fluidity at both *systemic* and *sub*-systemic levels, CPEC, especially its termini, thus emerged as a lightning rod for tensions dividing the US-Indian and Sino-Pakistani dyads in the context of US-Chinese competition. Geopolitics and geoeconomics were thus intensely enmeshed.

---

[95] Dutt and Ninan (2015), Gupta (2015) and Joshi (2015).
[96] Krishnamurthy (2018).
[97] Azam (2012) and Hays (2009).
[98] Report (2017d), PTI (2017b), Report (2017c, 2017e), Babar (2017) and Menon (2016b).

5.7.2 Gwadar's Evolutionary Emergence

Controversy was not, however, restricted to CPEC's northern terminus; its southern end-point, the Port of Gwadar on Balochistan's Arabian Sea coast, stirred deeper passions. As with the KKH, Gwadar's critics apparently represented national security interests of the same states which were allegedly threatened by CPEC's construction. The first phase of China-aided port-construction in Gwadar, completed in 2008, included a multi-purpose terminal. Under CPEC, a new 4.2 km jetty and a 1.5 km breakwater to protect it, were to be built. Officials listed 12 separate projects at Gwadar directly linked to the CPEC blueprint. The first 10, costing a total of $796m, would be financed almost entirely with Chinese grants. Funding needs of the remaining schemes were under discussion at CPEC Secretariat in 2018–2019, when Pakistani and Chinese official statements indicated their scale and priority. The mix of ambition, imagination, and China's grant-financing, precipitated much external angst.

The evolution of the Gwadar deep-sea port project betrayed political-economic irony. Shortly after Pakistan joined the USA's Middle-Eastern alliance system designed to defend Gulf oil fields from anticipated Soviet threats, in 1954, Washington assigned the US Geological Survey (USGS) to conduct a detailed survey of Pakistan's Balochistan coast. The USGS's lead-surveyor, Worth Condrick, recommended that Pakistan build a deep-sea port around the fishing village of Gwadar, 481 km west of Karachi, 72 km east of the Pakistan-Iran border, and 320 km from Oman's Cape al-Hadd.[99] Geography, marine-geology and Pakistan's economic needs, as well as the USA's alliance-defined exigencies and perspective, informed Condrick's recommendations, which were promptly endorsed by Washington.

At the time, Gwadar and surrounding areas belonged to Oman. Omani accounts say that in 1792, in the context of close political-economic ties between Oman and Balochistan, the Khan of Kalat, ruler of Balochistan, Mir Naseer Khan, gifted Gwadar to Oman's Sayyid Sultan bin Ahmed, for reasons not made clear.[100] Makrani Baloch soldiers helped consolidate and expand the Omani principality over the next 150 years. The recruitment of Baloch men into the Omani army, specifically the Sultan's bodyguards, proved particularly sensitive to the newly created Pakistan, notably because of Makran's proximity to Pakistan's capital, Karachi. Protests by Prime Minister Liaquat Ali Khan and his successors led to Britain suggesting that Pakistan pay £4m in compensation to Oman for Gwadar's return. An agreement was reached in 1957 and Karachi paid £3m for Gwadar's reversion to Pakistan's Balochistan in September 1958.[101]

Pakistan's first military administration under Ayub Khan pursued both economic and military development. World Bank-aided efforts led in 1964 to Karachi using the USGS report in identifying Gwadar as a future deep-sea port. However, it was only in 1988 that Pakistan, then absorbed in a US–Saudi-led covert campaign against Soviet

---

[99] Niazi (2017) and Anwar (2014).
[100] Ministry of Heritage and Culture (2010) and Kanwal (2018, p. 2).
[101] Anwar (2011, p. 98).

forces in Afghanistan, began work on building the Gwadar Port. The fishing village was converted into a small harbour in 1988-1992 with local investment worth Rupees 1623m. However, ships traversing the Persian Gulf-Arabian Sea-Indian Ocean route took no notice. In 1992, the Karachi Port Trust initiated a $200m deep-sea port project in Gwadar with 60% local currency investment; the remainder would be raised abroad. The project was to be completed by 1994. British consultants and their local partners conducted a feasibility study in 1993; in 1993–1994, Pakistan's 8th FYP included Gwadar Port's development as an 'essential element'.[102]

To attract foreign investment, Islamabad floated a build-operate-transfer (BOT) package in November 1994, but it elicited little interest. In December 1995, Pakistan approved plans to build the deep-water port with its own resources, but financial constraints precluded action. In 1997, Prime Minister Nawaz Sharif ordered a fresh review of Pakistan's maritime trade and infrastructure concerns. The review highlighted the need, among other things, to swiftly develop the Gwadar deep-sea port. However, after Pakistan followed India in conducting nuclear tests in May 1998, US-led economic sanctions prevented implementation. In May 1999, a year after the nuclear tests, China offered financial and technical assistance to the deep-sea port project but, after Pakistan's covert occupation of the Kargil Mountains in disputed Kashmir, and the Indian–Pakistani warfare it triggered, the offer apparently withered.

While visiting Beijing in May 2001, President Pervez Musharraf secured Premier Zhu Rongji's pledge to revive the offer. Chinese engineers surveyed Gwadar for several months, submitting a 'master plan' in October. The ground-breaking ceremony was scheduled for December 2001, but US combat missions, including aerial operations against Taliban-controlled Afghanistan mounted from Pakistani airbases in Balochistan, precluded progress. In March 2002, a Pakistani delegation visiting Beijing negotiated a framework agreement for the construction of a deep-sea port in Gwadar. On 22 March, Musharraf and Vice Premier Wu Bangguo presided over Gwadar's ground-breaking ceremony. Musharraf expressed hopes that Gwadar would transform Pakistan's economy and brighten the region's future.[103]

Pakistan's defence establishment recognised that the USA and 'key regional players' e.g. India and Iran, could view 'Chinese involvement in the Gwadar port project with suspicion'. Besides, 'any potential Chinese long-term military presence in the region' would pose 'complications for Pakistan'.[104] Still, the country's economic and strategic imperatives overrode these concerns. Phase-1 of the Chinese-funded and -built project included the completion of three multipurpose berths, while Phase-2 envisaged the construction of nine more berths, an approach channel, storage terminals, warehouses, trans-shipment and industrial facilities.[105] As work started, officials began expressing hopes that Phase-1 would be completed early, initially by March 2005 and, later, by Christmas 2004.

---

[102] Rizvi (2008).
[103] Musharraf (2005).
[104] Hassan (2005).
[105] Anwar (2011, p. 100).

5.7 The China-Pakistan Economic Corridor's Terminal Challenges

While work progressed, however, project costs rose from $248 to $298m, and the completion date was moved to September 2005.[106] Gwadar's build-up led to the construction of the 700 km Karachi-Gwadar Makran Coastal Highway linking the two ports via the Ormara and Pasni harbours, thereby reducing travel time from 48 h to seven. Construction of two roads, one linking Gwadar to Quetta and Chaman on the Iranian border, and the second to Khuzdar in eastern Balochistan, and preliminary work for connecting Gwadar to Pakistan's rail-network at a cost of $1.25bn, too, were begun. Several factors, including a bomb-blast killing three Chinese technicians in 2005, delayed Phase-1's completion. In late 2006, the project was sufficiently advanced that the Gwadar Port Authority (GPA) invited bids for managing the newly built port and ancillary facilities, and leading Phase-2 implementation.

In early 2007, Pakistan gave Singapore's PSA a $750m, 40-year, concession as Gwadar's management-partner. PSA would invest $5bn–$8bn in developing the port over the life of the contract while Gwadar generated estimated revenues of $17bn–$31bn.[107] The contract proved short-lived. The concession stipulated that GPA transfer a tract of land on which PSA would build some agreed facilities, but local critics, including Balochistan's Governor, petitioned Pakistan's Supreme Court to stop transferring control over Pakistani land to a foreign firm. In December 2010, the court issued a stay order preventing work. That, and the volatile security situation, kept PSA from meeting its investment pledges. As Pakistan's Supreme Court held up a verdict for over 18 months, GPA sought alternatives. Musharraf's successor, Asif Zardari, on one of his frequent visits to Beijing in mid-2012, discussed Gwadar's future with Chinese leaders. GPA then began negotiating the PSA concession's termination.[108]

In January 2013, Islamabad finalised the termination process and transfer of Gwadar's operational responsibilities to COPHC,[109] which began building out key elements of Gwadar's Phase-II development plan. Satisfied with its performance over four years, and in view of Gwadar's role within CPEC, in April 2017, Pakistan granted COPHC a 40-year lease covering both the port and the adjacent free-trade zone. Over the life of the lease, COPHC would retain 90% of the revenue from Gwadar's maritime operations, and 85% from the zone's earnings. Six months later, analysts noted that with CPEC implementation gathering pace, Gwadar's annual cargo-handling capacity would reach a million tons by the end of 2017. Medium-term plans would raise it to 13m tons, transforming Gwadar into 'South Asia's biggest shipping center'. CPEC's LTP envisaged that by 2030, its capacity would reach 400m tons.[110] The project's significance was seen as:

- Gwadar potentially offered the PLAN 'a convenient base of operations near the Strait of Hormuz'
- As CPEC's southern terminus, Gwadar was 'the project cargo port' for the vision

---

[106] Rizvi (2008).
[107] Khan (2012) and Aziz (2007).
[108] Khan (2012).
[109] Reuters (2013).
[110] Gul (2017) and Report (2017f).

- When planned pipelines were laid, Gwadar could help ship 1m bpd of oil overland from regional suppliers to Xinjiang, allowing 'China to bypass the Strait of Malacca'.[111]

5.7.3 Gwadar's Chinese-Funded Projects

In addition to the Chinese-built Phase-I, which PSA took over briefly, various Pakistani agencies, e.g. Karachi-based shipping and harbour authorities, the Pakistani Navy, and PSA itself, had constructed several installations in and around Gwadar by the time the port and ancillary facilities became the responsibility of COPHC and its subcontractors. Once Gwadar was designated the key terminus of the CPEC alignment, a large number of new projects, as well as old ones being adapted and enlarged for meeting their CPEC objectives, were listed as parts of Gwadar's developmental scheme. The consolidated project portfolio attracted substantial Chinese funding, most of it in the form of grants:

- New Gwadar International Airport ($230m): to be built 26 km east of Gwadar City; grant agreement signed in May 2017; preliminary work began in 2018; groundbreaking ceremony presided over by Prime Minister Imran Khan in March 2019
- East Bay Expressway ($140.60m): Six-lane expressway alongside a provisional 30-m wide railway corridor will connect Gwadar Port to N-20 Makran Coastal Highway through Gwadar Free-Trade Zone, bypassing Gwadar City, and avoiding urban traffic disruption; construction began in 2017—completion scheduled for 2020
- Potable water treatment, supply and distribution facilities ($130m): designed to meet expected water supply and sewage disposal/treatment demand in Gwadar City, Port and Free-Trade Zone to 2030–2050; includes water supply from Swad Dam and a new desalination plant being built on the coast[112]
- Breakwaters construction ($123m): COPHC engineers presented the design for the 1.5 km breakwater in 2017 to be built after Pakistani authorities modified/approved it
- Pak–China Friendship Hospital ($100m): Phase-2 included construction of a 50-bed state-of-the-art hospital and residential blocks on a 68-acre plot, with plans for another five 50-bed medical blocks to be built in Phase-3
- Free-Trade Zone development ($32m): Free-trade zone on 2280 acres, industrial zone on 3000 acres, and export-processing zone on 1000 acres; access roads, internal roads, utilities, power supply, security and ancillary services

---

[111] Report (2017f).
[112] Project Director (Water) (2017).

5.7 The China-Pakistan Economic Corridor's Terminal Challenges

- Dredging of berthing areas and channels ($27m): Additional terminals and berthing facilities extending to a maximum of 10 km; continual maintenance dredging provisions
- Technical and Vocational Institute ($10m): Enabling Gwadar's local population to enhance their technical and vocational skills, improve employment prospects and contribute to and benefit from Gwadar's development
- Gwadar Smart Port City master plan ($4m): Contract signed with China's Fourth Harbour Design Institute in May 2017; completed in August 2019
- Bao Steel Park, Petrochemicals, Stainless Steel and other industries
- Gwadar University, in collaboration with a Chinese university
- Livelihood—fisheries, boat-building and -maintenance services.

The Smart Port City Master Plan saga underscored practical challenges facing Gwadar's planned transformation. While policy-level Pakistani–Chinese cooperation was detailed and thorough, inter-bureaucratic coordination in finalising details, and collaboration among Chinese SOEs implementing the project and their Pakistani counterparts proved complex. First drafted in outline in August 2017, the Master Plan went through repeated reviews, refinements and revisions by the China Communications Construction Company (CCCC), its holding firm, the First Harbour Engineering Company (FHEC), and executives at the Gwadar Development Authority (GDA). Land procurement from private Pakistani owners further convoluted the process and delayed completion. However, CPEC managers insisted that the Master Plan's submission had been accomplished within its planned two-year schedule.[113]

Large flows of Chinese investment transforming a small fishing town on coastal Balochistan precipitated Indian and US suspicions that Gwadar could emerge as a PLAN base. Chinese activity in a barren locale, originally home to fewer than 100,000 people, that overlooked 'some of the world's busiest oil and gas shipping lanes', troubled those already unhappy with Beijing and Islamabad. Revised plans expected cargo-handling capacity, with five berths deepened to a depth of 20 m, to rise from 1.2m tonnes in 2018 to 13m tonnes by 2022. Beijing's plan for Gwadar airport included a 7000 m runway, but Pakistan asked for a 12,000 m runway capable of handling both airliners like the Airbus A380 Superjumbo, and large military aircraft.[114] Chinese finance lavished on Gwadar came atypically as grants; and the scale and speed of Gwadar's development, taken together with already deepening China-rooted anxiety, ruffled even more foreign feathers.

Chinese officials sought to allay Indian anxiety by noting Gwadar's commercial drivers—enabling Beijing to open up new cargo routes to the Middle East and Africa—behind the deep-sea port project, and offering to discuss Delhi's complaints over CPEC's alignment crossing disputed territory. However, India's response offered little hope that CPEC generally and its two Pakistani termini specifically would be seen in a neutral, unsentimental, light by Indian elites.[115] Gwadar's naval potential

---

[113] Khan (2019).
[114] Jorgic (2017a).
[115] Dasgupta (2016).

caused the deepest concerns. Quoting Chinese military analysts, e.g. Zhou Chenming, media reports noted the possibility of Gwadar eventually becoming a PLAN facility. Chinese and Pakistani officials either denied such plans or remained taciturn on the issue, but China-focused suspicions generated much critical commentary.[116]

The Chief of India's Naval Staff, Admiral Sunil Lanba, reflected these fears: 'In future, if PLA Navy ships operate from Gwadar, it will be a matter of concern; we will have to think of ways to mitigate the challenge'.[117] Lanba disclosed countervailing plans to reinforce Indian naval presence all along the IOR stretching from the Sunda and Lombok Straits in the east to Africa's east coast. Delhi supplemented this with its geoeconomic engagement with Iran, rebuilding and expanding the Chabahar Port on the Persian Gulf in Gwadar's vicinity, and committed $400m to constructing a railway line connecting Chabahar to Zahedan near the Iran-Afghan border. Given Indian suspicions of Sino-Pakistani collaboration and Pakistan's refusal to grant India transit access to Iran and Afghanistan, Delhi bypassed Pakistan via its Iran-Afghanistan linkage with Central Asia.[118]

Against that dialectic backdrop, Islamabad's grant of a 40-year lease to develop, maintain and operate Gwadar Port to COPHC in April 2017 appeared to confirm the fears of CPEC's critics. COPHC would also develop a 2281-acre free-trade zone near the port. The lease required COPHC to implement Gwadar's development plans and allowed it to earn 91% of the revenue from Gwadar's terminal and marine operations, and 85% of the revenue from the free-trade zone's operations.[119] These terms led Indian and US critics to infer that Gwadar's development was driven by China's interests, not Pakistan's.[120] The fact that apart from land and some labour, Pakistan obtained almost the entire financing and other resources needed to build Gwadar from China, and needed to repay its debts, did not carry weight in the critics' views. They seemed convinced that Islamabad lacked agency and was a mere tool used for advancing Beijing's geostrategic designs for securing regional domination.[121]

5.8 The USA's Corridor Critique

US-based analysts refined and advanced this argument. Notwithstanding Chinese and Pakistani insistence that given its extant, planned and potential transportation, communications and energy-linkages, Gwadar was 'a symbol of regional peace and prosperity', and that it had 'no strategic or political aims against a third country',[122] Western and Indian concerns coloured the discourse. Five years before CPEC was

---

[116] Habib (2018), Kanwal (2018), Rajagopalan (2018) and Gertz (2018).
[117] Chan (2018), Ahmed (2018) and Report (2017g).
[118] PTI (2016).
[119] PTI (2017c).
[120] Shakil (2017) and PTI (2018a).
[121] Rolland et al. (2019).
[122] Gul (2017).

5.8 The USA's Corridor Critique

launched, an Arlington-based think tank focused on maritime issues, the Centre for Naval Analysis (CNA), examined China-Pakistan relations and their implications for US interests[123]:

- 'Traditional geopolitical interests' underpinned Sino-Pakistani relations
- 'Similar geographical and historical concerns' vis-à-vis India made the two countries 'natural partners'
- China's 'growing economic equities with India' could subtly affect Sino-Pakistani ties
- Shared concerns over terrorism could also potentially affect China–Pakistan relations
- Pakistan's efforts to advance Chinese counterterrorism interests could generate domestic friction in Pakistan
- Economic relations, driven by China's need for energy and natural resources, and Pakistan's infrastructure needs, encompassed land and sea transport, nuclear and hydropower projects, and mining
- Help with Gwadar deep-sea port, the KKH and planned railway-lines promised to open up China's West to seaborne trade and accelerate its development
- China's growing investment in Pakistan increased its vulnerability to 'extremist threats'
- A 'strong military component' to Sino-Pakistani relations made the PLA 'a key player' in China's Pakistan-related decisions
- The USA and China shared 'important existential concerns' in Pakistan, but Beijing's 'predilection' for bilateralism limited US options; still, some coordination was possible
- Better understanding of Sino-Pakistani relations required 'further and broader study'.

Then-Secretary of Defence, James Mattis, widely considered an erudite and thoughtful former soldier, a philosopher-General, and one of the few 'adults' in the Trump Administration, illuminated the USA's grand-strategic anxiety and regional concerns vis-à-vis China's BRI/OBOR blueprint, and its consequences for US interests, and those of its allies: 'In a globalized world, there are many belts and many roads, and no one nation should put itself into a position of dictating "One Belt, One Road". That said, the OBOR also goes through disputed territory, and that in itself shows the vulnerability of trying to establish that sort of a dictate'.[124]

Other US analysts noted Pakistani appreciation of CPEC's economic, political and security challenges, including those facing Gwadar's planned development. After Islamabad established a 15,000-strong Special Security Division (SSD) in 2016 to guard CPEC projects and Chinese personnel, two US observers insisted, 'the primary obstacle to the CPEC's full implementation is security'; they believed the SSD would 'be stressed by the task at hand'. Like India, the USA, too, would 'have its own long-term concerns' as CPEC represented 'the leading edge of China's expanding

---

[123] Dumbaugh (2010).
[124] Mattis (2017).

access to, and likely influence within, Eurasia'. US efforts to compete with China for Pakistani favours, however, was 'best avoided, as it would be costly, unwinnable, and almost certainly counter-productive'. Instead, Washington should 'advance its own set of politically sustainable goals in Pakistan', focusing on 'terrorism, nuclear proliferation, and the war in Afghanistan'.[125]

Others underscored challenges confronting China and Pakistan in implementing CPEC, especially those rooted in Pakistan's domestic circumstances: 'local politics across weak frontier economies' afflicted with 'poor transportation and communications, conflicting trade and technical standards, corruption and terrorism'. Power shortages, insecurity and the medium- to long-term consequences of project finance could slow down implementation. Major regional and extra-regional players have been 'more competitive than cooperative', with the USA, Japan and India presenting BRI alternatives. Pakistan, a 'functioning democracy with a relatively educated bureaucracy and professional workforce', could face these 'daunting' challenges with some confidence. In a barb targeting Washington's own faltering 'New Silk Road' undertaking, they asserted, 'a lot can still go wrong' with CPEC and BRI, but CPEC projects 'already up and running send a clear message' to all Asia-Pacific actors: 'If you're not a serious player, you can't win'.[126]

Notaries of Indo-US insecurity claimed, 'In China's grand strategy, Gwadar is an important foothold that is part of its String of Pearls strategy for the Indo-Pacific'. Beijing's adherence to 'Mahan's theory of sea dominance' drove the PLAN's rapid expansion. Beijing 'clearly aims to dominate the Indo-Pacific'. Delhi responded by negotiating and signing agreements with the UAE, Oman, Djibouti, Seychelles and France for access and joint naval exercises counteracting PLAN's 'formidable challenge'.[127] But this would not be enough. The Quadrilateral Security Dialogue, or the 'Quad' binding the USA, India, Japan and Australia, 'should counter China's strategic outreach by networking with other like-minded countries' to fashion 'cooperative security frameworks to ensure a free, open, prosperous, and inclusive Indo-Pacific region', a prospect allegedly threatened by Sino-Pakistani collusion.[128]

This anxiety triggered fears that India and the USA had 'been distressingly ambivalent' vis-à-vis China's BRI. 'They have neither been willing to participate to a degree necessary to assure' that BRI adhered to the 'peaceful purposes the Chinese declare', nor were they willing and able to 'constitute their own maritime infrastructure initiative'.[129] This clearly competitive narrative revealed the geopolitical palimpsest on which efforts to implement the CPEC blueprint in particular, and the BRI vision more generally, proceeded. So, dialectics of mutually reinforcing angst-filled aspirations dividing CPEC's champions and detractors merited critical scrutiny.

---

[125] Markey and West (2016).
[126] Dossani and Erich (2017).
[127] Kanwal (2018, p. 4).
[128] Kanwal (2018, p. 1).
[129] Vickery (2018).

5.9 China-Pakistan Economic Corridor's Geopolitical Ecology

CPEC is being implemented against the backdrop of *systemic transitional fluidity*, flowing primarily from a widely perceived intensification of Sino-US strategic competition, itself resulting from status quo-oriented resistance to outcomes of relative power shifting from the USA to China and, to a lesser extent, other actors.[130] This dynamic, the backdrop against which Pakistan and China have been collaborating in pursuing both geoeconomic and strategic goals, evolved steadily since the unstated termination of the tacit Sino-US Cold War-era anti-Soviet alliance in 1989–91, when the *international security system* experienced rapid transformation.

CPEC will not only forge infrastructural, transportation, communication and energy links, and move resources across Sino-Pakistani borders, but will also colour perceptions of neighbours India, Iran and Afghanistan, and the *system-manager* (and China's *systemic*-level rival), the USA. Pakistan and China have complex relationships with these states. If all six actors felt perfectly secure and comfortable with each other, then peace and harmony would prevail across CPEC's physical alignment and in the world of conceptual constructs. If all six actors felt totally insecure vis-à-vis each other, conflict would rage constantly. Reality resides in the volatile middle of these extremes. Each action initiated by an actor affecting the perceived interests of others elicits a reaction which, in turn, drives a counteraction. Given the number of variables and the fluidity of interactions, the likely result is a very complex dialectic dynamic.

China and Pakistan forged a relatively stable tacit-alliance since the 1962 India–China war, but their relations with India and the USA experienced varied degrees of mutual insecurity which coloured their operational plans, force-posture, deployment patterns, elite-rhetoric and popular perceptions. Interests converged or diverged to varying degrees within *strategic* constructs,[131] several of which characterised the landscape on which China and Pakistan implemented CPEC:

Positive/benign dyads: China–Pakistan, India–Afghanistan, USA–India, USA–Afghanistan, China–Afghanistan, China–Iran
Negative/complex dyads: China–USA, Pakistan–India, China–India, Pakistan–Afghanistan, Pakistan–Iran, Iran–USA, Iran–Afghanistan,
Positive/benign triangles: USA–India–Afghanistan, China–Iran–Afghanistan
Negative/complex triangles: Pakistan–China–India; Pakistan–Iran–Afghanistan; Pakistan–India–Afghanistan; USA–China–Pakistan, USA–Pakistan–India.

Given largely resonant Pakistani and Chinese strategic perspectives and CPEC's symbolic import to both, it will likely be implemented to a broadly successful conclusion. Since the partners share an interest in timely and complete execution of agreed CPEC-related projects, an examination of the opportunities thrown up by and

---
[130] Haas (2018), Layne (2018), Steinbock (2017), Leonardo (2017), Moody (2016), Lai (2012, 2016) and Nye (2011).
[131] Ali (2017b).

challenges confronting CPEC can illuminate the processes, forces and tendencies shaping CPEC's alignment and its milieu.

5.9.1 The Corridor's Multi-layered Implications

Pakistan has been described as 'One of the most important countries in this initiative (BRI), perhaps the most important'.[132] Pakistani and Chinese commentary asserted, when fully implemented, CPEC promised to transform Pakistan's economy into a productive powerhouse that sits astride the BRI, and helps to boost China's economy and attain some of its national objectives. CPEC projects, if successfully completed, could dramatically alter Pakistan's intra-state and inter-state circumstances. Using CPEC as the case with which to study the opportunities, challenges and likely outcomes of China–Pakistan collaboration, and BRI generally, precipitates four levels of analyses:

1. State-level: how CPEC projects modify state-policy, elite perceptions and priorities in resource-allocation, leadership attention and effort, and intra-state power dynamics among key stake-holding factions within both Pakistan and China.
2. Bilateral level: how CPEC's implementation colours the Sino-Pakistani relationship, deepens inter-dependencies, and recasts mutual reliance in the pursuit of the two partners' respective geoeconomic and/or geopolitical objectives.
3. *Sub-systemic* level: How CPEC's implementation affects Pakistani and Chinese relations with India, Afghanistan and Iran; how the latter respond to deepened Pakistan–China bonds and the 'permanence' of the resulting physical infrastructure and policy overhang, and how China and Pakistan counter-respond in a cyclical, dialectic dynamic; and whether the resultant fluidity secures for two partners long term gains, or not.
4. *Systemic* level: With China increasingly seen as the USA's pre-eminent rival for regional (and potentially global) influence, and India a staunch member of the US-led anti-Chinese coalition built around the Quad, alongside South Korea, several ASEAN-members, and the USA's NATO allies Britain and France, how CPEC shapes the coalition's perceptions and policy-priorities vis-à-vis China and Pakistan, and if the latter can defend their own interests with diplomatic, economic and deterrent action.

External observers may be better-positioned to examine CPEC's third- and fourth-order implications, but only analysts with access to Sino-Pakistani policy documents and primary data-sets can knowledgeably address first- and second-order concerns. Their ability to make granular and nuanced appreciations of CPEC-related issues is unlikely to be matched by outside observers. To advance understanding of CPEC's

---

[132]Farr (2017).

geoeconomic and geopolitical impact on the region and beyond, meaningful interventions demand a synthesis of the four levels of analyses. Only such an examination could generate comprehensive inferences illuminating CPEC's consequences. Academic accuracy and policy-clarity demand such an endeavour.

5.9.2 Preliminary Questions

CPEC's champions in Pakistan and China see the scheme in an overwhelmingly positive light. If they did not, they would not have proposed it in the first place. However, logic aside, not everyone in the region and beyond shares their enthusiasm. Senior officials in Delhi, Kabul and Washington have openly expressed grave concern over and strong disapprobation of aspects of CPEC. They have articulated views directly challenging those of their Chinese and Pakistani counterparts. The contrast among these competitive narratives could not be starker.

Given the degree of divergence, until CPEC is fully functional as a leading BRI component, it would be presumptuous to claim meaningful understanding of the impact, implications and consequences of deepening Sino-Pakistani collaboration on the *sub-systemic* and *systemic* landscapes. Meaningful conclusions would demand an empirical study of CPEC's outcomes when it is fully operational. Answering these questions would be key to such analyses:[133]

5.9.3 First-Order Analytical Issues

- Which sectors of Pakistani and Chinese economies are receiving priority attention via CPEC project-planning and resource-allocation?
- How does that correspond to Pakistan's own development plans and priorities presented in Islamabad's Vision 2025 and the 11th Five-Year Plan blueprints?[134] Does either need to change to reconcile divergences or eliminate duplications?
- What is the balance of benefit on CPEC's completion among Pakistan's elite factions, and among Pakistan's provinces, as well as on elite-masses distributive dynamics?
- Will CPEC's implementation strengthen Pakistan's state-consolidation endeavours, boosting national unity and identity across demographic and ethnic divides, or can gulfs further widen, weakening the state? If the answer is the latter, then, as CPEC projects are built, mitigation needs to be planned and effected.
- CPEC being a medium- to long-term enterprise, how can its steady, accretive, implementation be protected from the vicissitudes of periodic party-political, democratic process-borne turbulence, an essential function of healthy politics?

---

[133] Ali (2017a).
[134] MPDR (2014b) and Planning Commission (2013).

- Given the internal-security situation across swathes of Khyber Pakhtunkhwa and Balochistan provinces, can Pakistan ensure security all along CPEC's length and breadth? Where are the main gaps, and how can these be cost-effectively filled?
- Considering CPEC's long-term, nation-wide and strategic implications, do Pakistani leaders perceive a need to build a knowledgeably supportive consensual constituency? What steps are being taken to that end? How effective are these proving to be? What more needs to be done, and how differently?

5.9.4 Second-Order Analytical Issues

- Out of total CPEC project-funding, what are the proportion and amount of credit? What are the terms and conditions? Do these help Pakistan's long-term economic interests? Which elite factions are making these calculations? And on what bases?
- Given Islamabad's fiscal straits, have the partners shown consideration to addressing Pakistan's fiscal-and-monetary imperatives once repayments start gathering pace?
- Is Beijing cognisant of Pakistan's absorptive capacities, institutional, structural and leadership-focused traditions, practices and aspirations, and are the two parties educating each other sympathetically? Does Sri Lanka's Hambantota Port experience offer a cautionary tale? Can the negativities of that experience be prevented here?
- China has shown concern over the presence, arming, training and operations of Uighur Muslim militants/separatist rebels from China's Xinjiang in Pakistan's former-FATA districts. After bloody counter-militant operations by Pakistani forces, these districts were integrated into Khyber Pakhtunkhwa. As CPEC links Xinjiang and Southern Pakistan via KPK, how will the partners resolve these concerns to mutual satisfaction?
- China is the more capable CPEC partner. Given the negative reactions in Pakistan's neighbourhood, how will Islamabad and Beijing ensure a long-term commitment to this tacit-alliance, dispel regional anxiety, and meet CPEC's inevitable challenges?
- Pakistan's relationship with China, deepened by CPEC, will likely be transformed. How appreciative is Beijing of Islamabad's need for help with managing the transition to deepened linkages, assuaging sovereignty concerns and securing qualitative shifts?
- Is there a shared framework of understanding that CPEC will integrate Pakistan into Beijing's BRI networks, and China's pathway to global eminence? Does Pakistan grasp its locus within CPEC's geoeconomic and geopolitical end-state?
- Pakistan is receiving substantial material and intangible assistance whereas China's gains are both economic and grand-strategic. Will CPEC subordinate Pakistan's agency as a Chinese protectorate? What are Pakistan's medium-term cost-benefit calculi?

5.9 China-Pakistan Economic Corridor's Geopolitical Ecology   217

- However, transformative CPEC proves to be, it will likely face myriad challenges. Do Pakistan and China have fall-back options? If they do, what are these?

5.9.5 Third-Order Analytical Issues

- Pakistan's integration into the BRI network, seen variously but critically by India, Iran and Afghanistan, is generating often negative elite-responses. Have Pakistan and China planned diplomatic and other measures to counteract such negativity?
- India has been notably truculent. India's adversarial history vis-a-vis Pakistan and its grand-strategic frictions with China[135] suggest CPEC/BRI will entrench this dynamic. What responses are the two partners deploying to alleviate India's anxiety?
- Pakistani leaders accused India and the USA of 'hatching conspiracies' against CPEC.[136] Pakistan's detention of two alleged Indian intelligence officers, one of them in Balochistan, his dramatic televised 'confession' of a subversive-paramilitary mission, and his controversial military trial both reflected and reinforced Pakistani insecurity vis-à-vis covert Indian operations in Balochistan.[137] This level of CPEC-rooted mutual insecurity imposes costs. How seriously are Pakistan and China taking this *sub-systemic* threat with *systemic*-level linkages?
- Given the foregoing, is the moribund SAARC now dead?[138] Can the SCO, which now includes both Pakistan and India, serve as a replacement, or will deepening China–Pakistan–Indian dynamics damage the SCO's collective prospects, too?[139]
- China enjoys better relations with Iran and Afghanistan than does Pakistan.[140] Can the latter seek Beijing's good offices to allay the former's concerns over Pakistan being strengthened with Chinese support?

5.9.6 Fourth-Order Analytical Issues

- As *systemic transitional fluidity* pits China and the USA in intensifying rivalry, and Beijing fashions geoeconomic counter-moves against Washington's geopolitical coalition-building, do the partners recognise CPEC's *systemic,* and *sub-systemic,* fallout?[141]

---

[135] Smith (2013).
[136] Report (2018c), Tanzeem (2018) and PTI (2017d, 2017e).
[137] PTI (2019b) and Report (2017h).
[138] APP (2018).
[139] PTI (2018b), Report (2018d) and Zeb (2018).
[140] Gul (2018), PTI (2018c) and Kemenade (2009).
[141] PTI (2017).

- With the *systemic core* and the *dominant system* polarised between US- and China-led camps, and Pakistan opting for integrative collaboration with Beijing, what are the likely pressures to afflict Pakistan as CPEC progresses towards completion?
- What measures are Pakistan's policy-elites considering and/or are adopting to counteract the negative outcomes flowing from these challenges?
- As CPEC transforms Pakistan as a political-economic actor, will it need to transform its policy frameworks, objectives and institutions to match parallel changes affecting the *system*?
- Is it necessary to mobilise and sustain national consensus on the cost-benefit choices being made, so that national support strengthens Pakistan's ability to cope with short- to medium-term costs in pursuit of longer-term benefits at all of the above levels?

The answers to these questions are not self-evident. However, as Pakistan and China proceed with implementing CPEC's LTP, these have acquired salience. How these are addressed could determine CPEC's long-term success. And that could foreshadow the prospects for Pakistan's future, BRI's outcomes, and China's emergent global role.

5.10 The Corridor's Insecurity Challenges

Across much of its temporal-spatial presence, CPEC faced myriad security threats. In August 2019, after Indian military firing across the intra-Kashmir ceasefire line killed three Pakistanis, officials evacuated around 50 Chinese staff working on a dam at the confluence of the Jhelum and Neelum rivers.[142] Militants, repeatedly assaulting Pakistani and Chinese personnel working on various projects, posed a graver threat. Diverse assailants driven by myriad motives attacked Chinese workers over many years. The impact of these strikes varied. In one attack, close to Gwadar in 2017, for instance, gunmen on motorbikes shot dead 10 road-builders, wounding two others.[143] In August 2018, a suicide-bomber from the separatist Baloch Liberation Army (BLA) blew up an explosive-laden vehicle next to a bus carrying Chinese staff to Balochistan's Saindak copper and gold mine, injuring five Chinese workers, while the bomber died.[144]

Saindak, Pakistan's biggest mine, was not, however, a part of CPEC. A more lethal attack occurred in Karachi in February 2018 when a Chinese employee of the COSCO Shipping Lines was shot dead in a 'targeted attack'.[145] In the most brazen assault, BLA gunmen struck the Chinese consulate in Karachi in November 2018.

---

[142] AFP (2019).
[143] Anadolu (2017).
[144] Yousufzai (2018).
[145] Reuters (2018).

5.10 The Corridor's Insecurity Challenges

Gunfire exchanges killed three militants, two policemen and a Pakistani father-and-son duo trying to pick up visas.[146] No Chinese were hurt. Even earlier, in 2014–2015, assailants killed 44 and wounded more than 100 workers—mostly Pakistanis, working on Chinese-funded projects.[147] Beijing has invested in many Pakistani projects outside CPEC, and Chinese staff engaged in several have faced similar threats. An engineer working at a hydropower plant near Islamabad went missing in December 2017, presumably abducted by unknown criminals.[148]

Dangers facing Chinese staff reached such levels that in October 2017, the Chinese embassy in Islamabad, reporting a threat against the ambassador himself, requested additional security.[149] Baloch separatists, angered by Beijing's military support for Islamabad, and Pakistan's robust counter-insurgency operations, were active across Balochistan, threatening CPEC. Myriad militant groups targeted Chinese staff working across Pakistan, implementing both CPEC-related and unrelated projects. In 2019, according to one estimate, the number of PRC citizens working in Pakistan reached 20,000, almost evenly divided between CPEC and non-CPEC projects. With the LTP approaching its second phase, numbers would rise significantly. In addition, Chinese nationals visiting Pakistan on short-term visas hit 70,000.[150] Their visibility and vulnerability would likely rise as CPEC implementation hit its stride in the 2020s.

Uighur, Uzbek and assorted devotees based in tribal districts along the Pakistan–Afghan border considered China an oppressive power occupying Xinjiang. Indigenous extremists, united under the 'Tehrik-e-Taliban Pakistan' (TTP) umbrella, saw Pakistan's secular establishment and its Chinese allies as a godless cohort. This perspective crystallised after a military assault on radical Islamists barricaded in Islamabad's Red Mosque in July 2007. Clerics and seminary-students, slamming foreigners and secular Pakistanis for their 'nudity', had demanded the imposition of 'shariah' law. When activists abducted Chinese beauticians accused of prostitution, Beijing urged action. After talks failed, Special Forces assaulted the complex, clearing it after 36 h of combat. During this period, armed tribesmen blockaded the KKH, protests rocked cities and militants mounted an abortive attack on the presidential aircraft. Six months later, 40 clerics commanding 40,000 fighters, formed the TTP in South Waziristan.[151]

CPEC inherited this legacy. Pakistani analysts examined developments from varying points of view. One noted, 'Whoever controls the Indian Ocean controls Asia', and 'extra-regional powers should stay out of what they would like to call the "India's Ocean"'. Such a mindset animating Pakistani elites was premised on the belief that 'India is indeed a challenge for both Pakistan and China'.[152] Pakistan's Navy

---

[146] Hadid and Sattar (2018).
[147] Orchard (2018).
[148] Dorsey (2018).
[149] Report (2017i, 2018e).
[150] Basit (2018).
[151] Hussain (2017), Walsh (2007) and Baker (2007).
[152] Tarar (2017).

made 'adequate arrangements', adopting 'a multi-pronged approach'. This included strengthening Gwadar Ports' physical security, conducting coastal exercises and security patrols, enhancing maritime domain awareness and engaging provincial and federal law-enforcement agencies, to ensure port security, vessel security and security of the sea-lanes. The Navy deployed a Pakistan Marines Force Protection Battalion. For affording CPEC activities and facilities maritime security, the Navy stood up Task Force-88. Islamabad raised electricity surcharges to provide for CPEC security.[153]

In June 2015, Pakistan's Ministry of Defence Production ordered four 600-ton and two 1500-ton patrol vessels from Chinese shipbuilders, with several to be built in Pakistan, specifically for this purpose.[154] Securing CPEC's maritime terminus was not enough, however. The protection of CPEC's terrestrial expanse, sites, assets and personnel, especially the maximum expected 30,000 Chinese workers, was a responsibility shared by the federal and provincial administrations. The SSD's 9000 military and 6000 paramilitary troops[155] alone shouldered CPEC-protection responsibilities in Balochistan's underdeveloped expanse, but in Pakistan's other provinces, regional administrations dedicated additional personnel to assisting SSD operations. Punjab, Pakistan's most populous and arguably politically the most significant province, established a 10,000-strong Special Protection Unit. Khyber Pakhtunkhwa committed 4200 policemen and Sindh, 2600 policemen, to help the SSD.[156] Still, as Chinese staff continued facing threats, the authorities were constantly reminded of the scale of the challenge. After a BLA raid on a luxury hotel in Gwadar took eight lives, Pakistan announced plans to raise a new Division-sized special formation for CPEC-security purposes.[157]

Pakistan's political vicissitudes added to insecurity. In mid-2018, dramatic electoral changes in the country's leadership brought to power a strong critic of Pakistan's dynastic political establishment, Imran Khan, whose campaign remarks questioned CPEC's sustainability. In his inaugural address, however, Khan stressed Pakistan's dire 'economic crisis': 'China gives us a huge opportunity through CPEC, to use it and drive investment into Pakistan. We want to learn from China how they brought 700 million people out of poverty'. Khan also sought ideas on governance: 'The other thing we can learn from China is the measures they have taken against corruption, how they have arrested more than 400 ministers there'.[158] Scepticism at home and abroad notwithstanding, Khan told journalists, 'In the future, the corridor will receive wide support from all sectors of Pakistani society. The construction of the

---

[153] Editorial (2017b), Tarar (2017) and Khan (2016).
[154] Tarar (2017).
[155] AFP (2017).
[156] Legarda and Nouwens (2018).
[157] PTI (2019c).
[158] Khan (2018b).

5.10 The Corridor's Insecurity Challenges

corridor continues to advance'.[159] Khan saw CPEC as a ray of hope in a dismal socio-economic landscape.[160]

In mid-July 2018, the usually critical Dawn, a newspaper, offered a positive assessment lauding the launch of the 'Pak-China Optical Fibre Cable' project after two years of work. The project comprised 820 km of underground cable running from Rawalpindi to Karimabad, and 172 km of aerial cable from there across landslide-prone areas to the Khunjerab Pass, where the cable linked to China's communications network. The project provided 26 high-capacity microwave links, connecting nine nodal-centres between Khunjerab and Rawalpindi, as backup. In addition to dramatically expanding telecoms facilities in Azad Kashmir and Gilgit-Baltistan, the project offered an alternative route to Pakistan's digital links to Europe, the Middle East and Africa, via China.[161] The link boosted Pakistan's ICT connectivity while reducing dependence on the Arabian Sea submarine cable, vulnerable to possible hostile interdiction.

CPEC completed its 'digital corridor' well within schedule. Islamabad would no longer have to rely on European countries, the USA or India for the exchange of 'sensitive information', a particular concern as the submarine cable digital links had been installed by a consortium including Indian parties considered 'a security risk'. Alternative routes for Internet traffic, cheaper tools for faster and more reliable e-governance, and advanced linkages with and within Pakistan's remote northern areas would advance Pakistan's developmental interests. CPEC would further help Pakistan connect to the world via new transport, communications, industrial processing, warehousing and port facilities being built in Gwadar. Dawn's editors hoped that digital connectivity through China would not impose on Pakistani nationals 'the many censorship mechanisms' Beijing imposed on Chinese netizens.[162]

Difficult fiscal challenges did force Islamabad to review major CPEC projects and their funding processes.[163] However, by August 2018, CPEC completed nine projects while 13 were under construction, employing 70,000 Pakistanis. It added 6910 MWe of a planned 11,110 MWe to Pakistan's national grid.[164] Khan's election also shone a light on CPEC's *systemic-subsystemic* context. While making a congratulatory call to Khan on his assumption of office, Secretary of State Mike Pompeo triggered a controversy. Washington reported that Pompeo 'raised the importance of Pakistan taking decisive action against all terrorists operating in Pakistan and its vital role in promoting the Afghan peace process'.[165] His counterpart Shah Mahmood Qureshi denied Pompeo had mentioned 'terrorists operating in Pakistan'.[166] Washington responded with a counter-statement that it 'stood by' its readout of the

---

[159] Report (2018f).
[160] Hyatt (2018).
[161] Editorial (2018b) and Report (2018g).
[162] Editorial (2018b).
[163] Anderlini et al. (2018).
[164] Awan (2018) and Yao (2018).
[165] Spokesperson (2018a).
[166] AP (2018).

call. It was against that backdrop of worsening US–Pakistani relations that the USA cancelled $300m in pledged military aid to Pakistan, shortly before Pompeo arrived on a visit.[167] This could only be indirectly attributed to the USA's CPEC phobia, as US views of Pakistan had long been on a sliding trajectory.[168]

However, just before embarking on his South Asian trip, Pompeo stressed US opposition to any IMF loans that could help ameliorate Pakistan's fiscal challenges if such a bailout enabled Islamabad to service Chinese debts, especially those relating to CPEC financing. 'Make no mistake, we will be watching what the IMF does', Pompeo warned, 'There's no rationale for IMF tax dollars—and associated with that, American dollars that are part of the IMF funding—for those to go to bail out Chinese bondholders or China itself'.[169] Imran Khan's long-standing critique of US combat operations in Afghanistan and drone attacks in Pakistan, support for Sino-Pakistani cooperation, and a deepening US–Indian 'strategic partnership' vis-à-vis China,[170] explained fraught polarisation between the Pakistan–China and USA–India dyads.

CPEC would likely remain a lightning rod for geopolitical tensions and could even sharpen these. Islamabad and Beijing acted to confront uncertainty and stabilise 'CPEC construction'. After Imran Khan assumed office, Chinese Foreign Minister Wang Yi visited Islamabad, noting that of the 22 CPEC projects worth $19bn in hand, nine had been completed and 13 were under construction. These had created 70,000 jobs for Pakistanis and raised the country's GDP growth by 1%–2% points. Far from aggravating Pakistan's debt burden, as alleged, Wang insisted 'the construction of CPEC' had delivered 'tangible outcomes'.[171] Western economic advisers noted, CPEC promised 'very crucial' gains to both China and Pakistan: it offered Beijing an 'alternate secure route to import Energy' and find new markets for Chinese goods and services; CPEC enabled Pakistan to 'counter Indian influence' by becoming a major hub connecting Eurasia, South Asia and Southeast Asia, and provided Pakistan with 'a much needed base to kick-start its economic growth'.[172] Partisan politicking did not alter that belief.

Shortly after a Pakistani minister hinted several CPEC projects could still be suspended for a year-long review, Pakistan's Chief of Army Staff, General Qamar Bajwa, arguably one of the two most powerful Pakistanis, arrived in Beijing to reassure his hosts. During his meeting with Xi Jinping, the latter noted, China placed a 'high premium' on Sino-Pakistani ties at a time when 'the international and regional situation is undergoing complex changes, and the two countries are supporting each other, helping each other, and enhancing mutually beneficial cooperation'. Xi urged the PLA and its Pakistani partners to cooperate 'in various fields and at all levels' to ensure CPEC's security and counter terrorism. 'As long as a high-degree of mutual

---

[167] Smith (2018).
[168] Tanzeem (2018).
[169] Pompeo and Turak (2018).
[170] Pompeo et al. (2018) and Spokesperson (2018b).
[171] Wang (2018).
[172] Report (2016a, p. 5).

trust and concrete measures are in place', Xi told Bajwa, 'the CPEC construction will succeed and deliver benefits to people of the two countries'.[173] So, continuity was assured.

Confirmation came during Imran Khan's November 2018 meetings in Beijing with Xi Jinping, Li Keqiang, and other Chinese leaders. The 'iron brothers' reaffirmed their 'All-Weather Strategic Cooperative Partnership', restated commitments made in the April 2005 Treaty of Friendship, Cooperation and Good-neighbourly Relations,[174] pledged to deepen collaboration and speed up CPEC implementation, notably, completion of Gwadar Port, 'the central pillar of CPEC', and socio-economic developmental projects. Khan pledged to ensure the security of Chinese personnel and institutions in Pakistan, and of CPEC itself.[175]

Progress was discussed and plans for CPEC's second phase construction were formalised during Khan's visit to China leading a high-powered delegation attending the April 2019 BRF. The latter focused, at Khan's request, on social sector programmes and agricultural development, with additional grant of $1bn financing early harvest and pilot projects advancing vocational training, medical facilities and agricultural extension. CPEC's second phase would also advance construction of the Western Routes across Khyber Pakhtunkhwa and Balochistan provinces, the ML-1 railway project between Peshawar and Karachi, and the various SEZs along the alignments. The two governments exchanged project details for the second phase before Khan arrived at the 2019 BRF. Khan had presided over the groundbreaking ceremony for the construction of Gwadar's new international airport before flying out to Beijing.[176] Six months later, on another visit to Beijing before Xi embarked on an 'informal summit' with Mr Modi in India, Khan and Xi lauded Sino-Pakistani ties, and vowed to accomplish the realisation of the CPEC blueprint.[177] At least at that point, CPEC appeared to be progressing according to plan.

References

Abdi S (2017) Time for India to avenge 1962 humiliation. New Indian Express, 24 July 2017
Achom D (2018) China taking Pak economic corridor all the way to Afghanistan. NDTV, Beijing
AFP (2017) Pakistan deploys force of 15,000 to protect Chinese nationals. SCMP
AFP (2019) Pakistan evacuates Chinese nationals as firing in Kashmir kills three. Straits Times, 1 Aug 2019
Afzal M (2017) Why the Trump administration's policy on Pakistan is likely to fail. Brookings, Washington
Ahmad A (2016) Status of Gilgit-Baltistan. Pakistan Today, 16 Jan 2016
Ahmed I (2018) China plans to build offshore naval base in Pakistan's Gwadar port, says report. Hindustan Times, 9 Jan 2018

---

[173] Li (2018) and Blanchard (2018).
[174] Wen and Aziz (2005), Hu and Zardari (2008).
[175] Hu (2018), Li and Khan (2018) and Xinhua (2018).
[176] Butt (2019) and Harris (2019).
[177] Report (2019b).

Akhund M (2018) The Gilgit-Baltistan question: a historical analysis. Times, 16 June 2018
Ali S (1999) Cold war in the high Himalayas: The USA, China and South Asia in the 1950s. St. Martin's Press, New York, pp 120–189
Ali S (2017a) Opportunities and challenges of China-Pakistan strategic collaboration: CPEC's systemic and sub-systemic context and implications. Seminar presentation, NDU, Islamabad, 24 Nov 2017
Ali S (2017b) US-Chinese strategic triangles. Springer, Heidelberg, pp 6–16
Ali S (2017c) US-Chinese strategic triangles: examining Indo-Pacific Insecurity. Springer, Heidelberg, pp 165–193
Amedeo J (2018) America needs a clear strategy to counter China's expansion in the SCS. NI, 1 Aug 2018
Anadolu (2017) 10 workers linked to Chinese project killed in Pakistan. Tolo News, Kabul, 13 May 2017
Anderlini J, Sender H, Bokhari F (2018) Pakistan rethinks its role in Xi's Belt and Road plan. FT, 10 Sept 2018
Anwar Z (2011) Gwadar deep sea port's emergence as regional trade and transportation hub: prospects and problems. J Polit Stud 1(2)
Anwar Z (2014) Gwadar deep Sea Port's emergence as regional trade and transportation hub: prospects and problems. J Polit Stud 1(2):97–112
AP (2018) Pompeo's Call to Pakistan's new premier stirs controversy. NYT, 24 Aug 2018
APP (2016) KKH reopens for traffic. Pakistan Observer, 16 Apr 2016
APP (2018) PM's Nepal visit concludes with CPEC, SAARC in focus. Nation, 6 Mar 2018
Asif M, Yang B (2019) International media reporting and legal validity of Gilgit-Baltistan. Asian Soc Sci 15(2):177–187
Awan Z (2018) BRI and CPEC, five years on. CD, 28 Aug 2018
Azam A (2012) The KKH: a friendship channel. Youlin, 28 July 2012
Aziz F (2007) Singapore's PSA takes over Pakistan's Gwadar port. Reuters, Gwadar, 6 Feb 2007
Babar M (2017) China, Pakistan reject US criticism of CPEC. News, 8 Oct 2017
Baker A (2007) Storming the red mosque. Time, 10 July 2007
Basit A (2018) Attacks on Chinese nationals and interests in Pakistan are likely to continue. Here's why. SCMP, 27 Nov 2018
Bennett J (2017) India fears Chinese encirclement, citing 'overwhelming' Sino presence in S. Asia. ABC, 5 June 2017
Bhatti J, Liu T (2018) Economic corridor changes Pakistan's business, economic landscape. Xinhua, Islamabad, 8 June 2018
Bilal M (2018) Belt and road opens new opportunities for Pakistan. CD, 21 Aug 2018
Blackwill R, Tellis A (2015) Revising US grand strategy toward China. CFR, New York, Mar 2015, pp 3–39
Blanchard B (2018) China's Xi says places 'high premium' on Pakistan ties, as army chief visits. Reuters, Beijing, 20 Sept 2018
Boao Forum for Asia (2018) Asian competitiveness annual report. Boao, 8 Apr 2018
Brown K, Hoffman F, Miller F (2018) Author's interviews. London, Apr 2018
Butt N (2019) Under second phase of CPEC: Pakistan, China may sign MoUs, agreements on projects. Business Recorder, 4 Apr 2019
Butt K, Butt A (2015) Impact of CPEC on regional and extra-regional actors: analysis of benefits and challenges. In: Paper presented at international conference on CPEC, GC University, Lahore, 9–10 Dec 2015
Chan M (2018) First Djibouti now Pakistan port earmarked for a Chinese overseas naval base, sources say. SCMP, 5 Jan 2018
Chaudhury D (2018) Pakistan rejection of China's dam aimed at showing OBOR in line with global rules. Economic Times, 12 July 2018
Chen X, Joseph S, Tariq H (2018) Betting Big on CPEC. Eur Financ Rev 2018:61–70

Conrad P (2017) China's access to Gwadar Port: strategic implications and options for India. Marit Aff J Nat Marit Found India 13(1):55-62
CPEC Coordinator (2017) The agreement on the long term plan for China-Pakistan economic corridor. CPEC Secretariat, Islamabad, 21 Nov 2017
CPEC Secretariat (2018) CPEC has brought many positive changes to Pakistan's economy: Imran Khan. MPDR, Islamabad, 24 July 2018
Crabtree J (2018) China is pumping cash into Pakistan, but that might not convince other foreign investors. CNBC, 20 Feb 2018
Dasgupta S (2016) Success of Gwadar Port is not a threat to India: Chinese experts. ToI, 13 Nov 2016
DeAeth D (2018) 'China Hawks' and 'Panda Huggers': China causing an uproar in Australian politics. Taiwan News, 28 May 2018
Deloitte (undated) How will CPEC boost Pakistan economy? https://www2.deloitte.com/content/dam/Deloitte/pk/Documents/risk/pak-china-eco-corridor-deloittepk-noexp.pdf Accessed 15 July 2019
Dorsey J (2018) Disappeared Chinese engineer holds ties with Pakistan hostage. SCMP, 27 Jan 2018
Dossani E, Erich N (2017) China's field of dreams in Pakistan. RAND, Santa Monica, 16 Oct 2017
Dumbaugh K (2010) Exploring the China-Pakistan relationship: roundtable report. CNA, Arlington
Dutt V, Ninan S (2015) Indian defence experts worried about KKH's strategic implications to security. India Today, 31 Jan 2015
Editorial (2017a) Foreign policy white paper commits us to containing China. Canberra Times, 23 Nov 2017
Editorial (2017b) CPEC security cost. Dawn, 22 Mar 2017
Editorial (2018a) CPEC: a win-win situation for Pakistan and China. Times, Islamabad, 31 Dec 2018
Editorial (2018b) CPEC achievement. Dawn, 16 July 2018
EIU (2019) What is the future of CPEC? Economist, London, 28 Mar 2019
Fallon J (2017) Checkmate: China's containment of US containment policy. Defence Viewpoints, UK Defence Forum, London, 9 Oct 2017
Fan Z (2019) In the dark: how much do power sector distortions cost South Asia?. WBG, Washington, p 207
Farr G (2017) Pakistan's role in China's OBOR initiative. International Relations, 10 July 2017
Fazil M (2016) Five Reasons Gwadar Port Trumps Chabahar. Diplomat, 9 June 2016
Gady F (2016) China confirms export of 8 submarines to Pakistan. Diplomat, 19 Oct 2016
Ganguly S, Pardesi M (2009) Explaining sixty years of India's foreign policy. India Rev 8(1):4–19
Gertz B (2018) China building military base in Pakistan. Washington Times, 3 Jan 2018
Gishkori Z (2016) Pakistan raises special security division successfully. Dawn, 3 Sept 2016
Goldstein L (2018) China's naval expansion is no threat. NI, 6 June 2018
Gul A (2017) China turning Pakistan port into regional giant. VoA, Gwadar, 24 Oct 2017
Gul A (2018) China tries to bring Pakistan, Afghanistan closer. VoA, Islamabad, 15 May 2018
Gul A (2019) China giving Pakistan $3.5bn in loans, grants. VoA, Islamabad, 15 Feb 2019
Gupta A (2015) KKH: a security challenge to India. Indian Defence Review, 2 Oct 2015
Haas L (2018) Passing the torch to China? US News, 6 Mar 2018
Habib M (2018) China's 'belt and road' plan in Pakistan takes a military turn. Economic Times, 21 Dec 2018
Hadid D, Sattar A (2018) Militants, police killed in failed attack on Chinese consulate in Karachi. NPR, 23 Nov 2018
Hansen G (1967) The impact of the border war on indian perceptions of China. Pac Aff 40(3/4):235–249. (Autumn 1967–Winter 1967–1968)
Harris M (2019) Imran Khan completes groundbreaking new Gwadar Airport ceremony. CPIC News, 3 Apr 2019

Hassan A (2005) MA thesis: Pakistan's Gwadar port—prospects of economic revival. Naval Postgraduate School, Monterey, June 2005, pp v, 3–4

Hays J (2009) Karakoram highway in China. Facts and details. http://factsanddetails.com/china/cat15/sub104/item441.html Accessed 7 July 2018

Hu Y (2018) Nations fast-track Pakistan port plan. CD, 5 Nov 2018

Hu J, Zardari A (2008) Joint Statement between the PRC and Islamic Republic of Pakistan. State Council, Beijing, 16 Oct 2008

Huaxia (2015) China-aided highway changes life of Pakistanis. Xinhua, Islamabad, 19 Apr 2015

Hussain Z (2017) The legacy of Lal Masjid. Dawn, 9 July 2017

Hyatt D (2018) Will Imran Khan pivot Pakistan from US to China? GT, 29 July 2018

Iqbal A (2017) The long term plan of the China-Pakistan economic corridor. News International, 20 Dec 2017

Jacob J (2017) China's BRI: perspectives from India. China World Econ 25(5):80

Jacobs A (2013) The high road on the Karakoram Highway. Bulletin-NYT, 17 Nov 2013

Jain P (2018) Was Mao's decision to attack India to 'humiliate Nehru'? Quora, 21 Mar 2018

Jia W (2018) CPEC is not a gift: Remarks at the CPEC 2018 Summit. Institute of New Structural Economics, Karachi, 2 June 2018

Jorgic D (2017a) Hoping to extend maritime reach, China lavishes aid on Pakistan town. Reuters, Gwadar, 17 Dec 2017

Jorgic D (2017b) China pours money into Pakistan port in suspected strategic power push. Sydney Morning Herald, 18 Dec 2017

Joshi P (2015) India's Karakoram conundrum: a legacy of the great game. Foreign Policy News, 18 Dec 2015

Kanwal G (2018) Pakistan's Gwadar port: a new naval base in China's string of pearls in the Indo-Pacific. CSIS, Washington, 2 Apr 2018

Kapila S (2013) India's Foreign Policy 2012: a critical review in relation to China And Pakistan military threats. Ni Hao-Salam, 12 Feb 2013

Kapur S (2018) Asia's nuclear nemesis. EAF, 19 July 2018

Kazim H (2012) The KKH: China's asphalt powerplay in Pakistan. Spiegel, 17 July 2012

Kemenade W (2009) Iran's relations with China and the West: cooperation and confrontation in Asia. Clingendael, Hague

Khan Z (2012) Singapore port operator on way out of Gwadar. Tribune, 9 Aug 2012

Khan F (2014) Attabad lake: teardrop miracle. Express Tribune, 25 May 2014

Khan R (2016) 15,000 troops of special security division to protect CPEC projects, Chinese nationals. Dawn, 12 Aug 2016

Khan I (2018a) Prime Minister-designate's address to the nation. PTI Information Office, Islamabad, 26 July 2018

Khan I (2018b) Prime Minister-elect's address to the nation. PTI Information Office, Islamabad, 26 July 2018

Khan Y (2019) Long-delayed Gwadar's master plan is finally ready. Geo News, 6 Aug 2019

Krishnamurthy R (2018) Why doesn't India destroy the KKH? It passes through India's POK. Quora, 3 Apr 2018

Kugelman M (2017) The CPEC: what it is, how it is perceived, and implications for energy geopolitics. In: Downs E, Herberg M, Kugelman M, Len C, Yu K (eds) Asia's energy security and China's BRI. NBR, Nov 2017

Kumar R, PTI (2019) Not for other countries to comment on India's internal affairs: MEA on Xi Jinping-Imran Khan Kashmir talks. Economic Times, 9 Oct 2019

Lai D (2012) The US and China in power transition. AWC, Carlisle, Mar 2012

Lai D (2016) The US-China power transition: stage II. Diplomat, 30 June 2016

Lambah S (2016) The tragic history of Gilgit-Baltistan since 1947. Indian Foreign Aff J 11(3)

Lampton D (2018) Author's interview, Kuala Lumpur, June 2018

Layne C (2018) The US-Chinese power shift and the end of the Pax Americana. Int Aff 94(1):89–111

Legarda H, Nouwens M (2018) Guardians of the belt and road. Mercator Institute for China Studies. Berlin, 16 Aug 2018, p 11

Leonardo R (2017) Assessment of the US-China power transition and the new world order. Real Clear Defence, 11 Sept 2017

Li G (2018) Xi meets Pakistani army chief. Xinhua, Beijing, 19 Sept 2018

Li K, Khan I (2018) Joint statement between the PRC and the IRP on strengthening China-Pakistan all-weather strategic cooperative partnership and building closer China-Pakistan community of shared future in the new era. State Council, Beijing, 4 Nov 2018

Macfie N, Zahra-Malik M, Blanchard B, Rajagopalan M (2015) Pakistan PM approves deal to buy eight Chinese submarines: official. Reuters, Islamabad/Beijing, 2 Apr 2015

Malik M (2019) CPEC not 'debt-trap' but development schema for Pakistan. Xinhua, Islamabad, 26 May 2019

Malik M, Jennings R (2017) Threats from America will move a wary Pakistan even closer to China. Forbes, 18 Sept 2017

Mangaldas L (2018) Trump's Twitter attack on Pakistan is met with both anger and support in South Asia. Forbes, 2 Jan 2018

Markey D, West J (2016) Expert brief: behind China's gambit in Pakistan. CFR, New York, 12 May 2016

Mattis J (2017) Political and security situation in Afghanistan: SASC testimony. DoD, Washington, 3 Oct 2017

Menon S (2016a) As China's Pakistan ties deepen, India needs a strategy to mitigate the fallout. Wire, 11 July 2016

Menon S (2016b) As China's Pakistan ties deepen, India needs a strategy to mitigate the fallout. Brookings, Washington, 12 July 2016

Miller A (2018) Answers to advance policy questions for nominee for commander, resolute support mission and commander, US Forces-Afghanistan. SASC, Washington, 19 June 2018

Ministry of Heritage and Culture (2010) Brief history of the relationship between Oman and Baluchistan. Omani Stud J (16). http://www.atheer.om/en/7162/brief-history-of-the-relationship-between-oman-and-baluchistan/ Accessed 13 July 2018; others maintain Gwadar was gifted to Oman in 1783

Mishra R (2001) Nuclear and missile threats to India: China-Pakistan nexus in South Asia. South Asia Analysis Group, Paper 296, 17 Aug 2001

Moody A (2016) Looking East: shift in the balance of power. Telegraph, 3 Oct 2016

MPDR (2014a) Pakistan 2025: one nation—one vision. GoP, Islamabad, 29 May 2014

MPDR (2014b) Vision 2025. GoP, Islamabad, July 2014

MPDR (2017a) Long-term plan (LTP) for CPEC. GoP, Islamabad, 20 Dec 2017, pp 4–5

MPDR (2017b) LTP for CPEC. GoP, Islamabad, 20 Dec 2017

Mu X (2019) Pakistan proposes some 1.3bln USD for CPEC in fresh budget. Xinhua, Islamabad, 11 June 2019

Musharraf P (2005) President's remarks at the ground-breaking ceremony of the Gwadar Port Project. Office of the President, Islamic Republic of Pakistan, Gwadar, 22 Mar 2005

Nasim H (2019) Pak-China fibre optic link activated for commercial use. Express Tribune, 2 Feb 2019

Nawaz S (2018) Trump's flawed Pakistan policy: why Islamabad is unlikely to change. Foreign Affairs, 10 Jan 2018

Niazi Z (2017) CPEC and Gwadar. Express Tribune, 11 Aug 2017

Nye J (2011) US-China relationship: a shift in perceptions of power. Belfer Centre, Harvard/Cambridge, 6 Apr 2011

Official Spokesperson (2018) Response to a query on possible cooperation with China on OBOR/BRI. MEA, Delhi, 5 Apr 2018

Orchard P (2017) Containing China on the open seas. Real Clear World, 2 Nov 2017

Orchard P (2018) One belt, many headaches. Geopolitical Futures, 21 Aug 2018

Panda A (2015) Xi Jinping on Pakistan: 'I feel as if I am going to visit the home of my own brother'. Diplomat, 20 Apr 2015

Pastricha A (2019) Indian pilot captured by Pakistan returns home. VoA, 1 Mar 2019

Peters C (2017) US policy in Afghanistan: Q &A session. SASC, Washington, 3 Oct 2017

Planning Commission (2013) 11th five year plan 2013–18. GoP, Islamabad, Apr 2013

Pompeo M, Turak N (2018) Pompeo spotlights Pakistan as latest tension point between Washington and Beijing. CNBC, 31 July 2018

Pompeo M, Mattis J, Swaraj S, Sitharamn N (2018) Closing remarks at the US-India 2+2 Dialogue. DoS, Delhi, 6 Sept 2018

Press Release (2018) CPEC project updates. Chinese Embassy, Islamabad, 25 Apr 2018

Project Director (Water) (2017) Request for proposal: necessary facilities of fresh water treatment, water supply and distribution Gwadar phase-1 (Water Transmission main from Swad Dam to Gwadar City). GDA, Gwadar, Feb 2017

PTI (2016) India's Chabahar port plan is to counter our Gwadar port plan: Chinese media. Hindu, 7 June 2016

PTI (2017a) Gwadar port to have implications for US, Gulf too: ex-Pakistan envoy. Economic Times, 9 May 2017

PTI (2017b) On OBOR, US backs India, says it crosses 'disputed' territory. ToI, 4 Oct 2017

PTI (2017c) Pakistan's Gwadar port leased to Chinese company for 40 years. Economic Times, 20 Apr 2017

PTI (2017d) India hatching conspiracies against CPEC: Pakistan. Indian Express, 29 Dec 2017

PTI (2017e) US conspiring against CPEC along with India, alleges Pak NSA. Business Standard, 19 Dec 2017

PTI (2018a) Gwadar port about China's interests, not Pakistan's: US think-tank. Economic Times, 9 Feb 2018

PTI (2018b) SCO could give platform to India, Pakistan to resolve disputes: Chinese media. Economic Times, 12 July 2018

PTI (2018c) CPEC being extended to Afghanistan: report. Economic Times, 8 Apr 2018

PTI (2019a) UNSC holds closed-door meeting on Kashmir. NDTV, 16 Aug 2019

PTI (2019b) Indian national arrested for spying, claims Pakistan police. India Today, 1 Aug 2019

PTI (2019c) Pakistan army plans new unit to protect CPEC projects. Gulf News, 19 May 2019

Qamar K (2017) One Belt One Road: CPEC–Pakistan. CD, 6 Feb 2017

Rafiq A (2017) China's $62bn bet on Pakistan: letter from Gwadar. Foreign Affairs, 24 Oct 2017

Rajagopalan R (2018) A China military base in Pakistan? Diplomat, 9 Feb 2018

Ram M (1978) India-Pakistan-China-Karakoram highway. Econ Polit Wkly 13(26)

Rana S (2017) Pakistan stops bid to include Diamer-Bhasha Dam in CPEC. Express Tribune, 15 Nov 2017

Ratcliffe R (2019) Pakistan warns India its actions in Kashmir could provoke war. Guardian, 27 Sept 2019

Raza S (2017) Three CPEC projects hit snags as China mulls new financing rules. Dawn, 5 Dec 2017

Report (2012) What they said: India Tests Agni V Nuclear Missile. NYT, 19 Apr 2012

Report (2014) India's top secret 1962 China war report leaked. Diplomat, 20 Mar 2014

Report (2015) China converts $230m loan for Gwadar airport into grant. News, 23 Sept 2015

Report (2016a) How will CPEC boost Pakistan economy? Deloitte, London

Report (2016b) Kashmir territories: full profile. BBC, London, 1 Mar 2016

Report (2017a) CPEC game changer for the region: Nawaz Sharif. Gulf News, 28 Mar 2017

Report (2017b) Five reasons why China is so scared of India. Economic Times, 7 Mar 2017

Report (2017c) China defends ally Pakistan after Trump criticism. Reuters, Beijing, 22 Aug 2017

Report (2017d) Tillerson says India crucial for US-Afghan strategy, slams Pak for sheltering terrorists. Business Standard, 25 Oct 2017

Report (2017e) China asks US to shed bias after Tillerson criticises Beijing's development. Business Standard, 20 Oct 2017

Report (2017f) Pakistan gives China a 40-year lease for Gwadar Port. Maritime Executive, 27 Apr 2017
Report (2017g) India alarmed by China's plan to deploy warships in Pakistan's Gwadar Port. Sputnik, Delhi, 12 Jan 2017
Report (2017h) Kulbhushan Jadhav, former Indian Navy Officer, sentenced to death by hanging in Pakistan. India Today, 10 Apr 2017
Report (2017i) Chinese embassy warns of threat to envoy. Nation, 22 Oct 2017
Report (2018a) BLF attacks military camp on CPEC route. Balochistan Post, 14 May 2018
Report (2018b) CPEC an engine of growth for Pakistan: Imran Khan. Times Monitor, 24 July 2018
Report (2018c) Enemies conspiring against CPEC, Pakistan's development: Ahsan. Pakistan Today, 15 Feb 2018
Report (2018d) Iran interested in CPEC partnership with Pakistan. Financial Tribune, 15 Apr 2018
Report (2018e) CPEC: opportunities and risks. ICG, Brussels, 29 June 2018, p.7
Report (2018f) CPEC an engine of growth for Pakistan: Imran Khan. Times, 24 July 2018
Report (2018g) Pakistan-China optic fibre project a first of its kind. CPEC Secretariat, Islamabad, 17 July 2018
Report (2019a) Pakistan FDI. Trading economics https://tradingeconomics.com/pakistan/foreign-direct-investment. Accessed 1 Sept 2019
Report (2019b) Xi meets Pakistani PM, calls for forging closer community of shared future. Xinhua, Beijing, 10 Oct 2019
Research Snipers (2018) Chinese investment in CPEC will cross $100bn. Pakistan Institute of Development Economics. Islamabad, 10 Feb 2018
Reuters (2013) Pakistan hands port operation from Singapore's PSA to Chinese. Straits Times, 31 Jan 2013
Reuters (2018) Chinese worker killed in 'targeted attack' after his car was riddled with bullets in Karachi, Pakistan. SCMP, 6 Feb 2018
Rizvi Z (2008) Gwadar port: history-making milestones. Dawn, 14 Apr 2008
Roggio B (2018) Trump blasts Pakistan for its 'lies and deceit'. Long War Journal. 2 Jan 2018
Rolland N, Boni F, Nouwens M, Samaranayake N, Khurana G, Tarapore A (2019) Where the Belt meets the Road: security in a contested South Asia. Asia Policy 14(2):1–41
Rudd K (2018) Understanding China's rise under Xi Jinping: Address to Cadets at US Military Academy. Asia Society Policy Institute, West Point, 5 Mar 2018
Safi M, Zahra-Malik M (2019) Pakistan returns Indian pilot shot down over Kashmir in 'peace gesture'. Guardian, 1 Mar 2019
Sajjad S (2017) SMEs, the engine of growth, ignored in Pakistan. Express Tribune, 14 Aug 2017
Sen R (2018) China's CPEC game plan threatens India. DNA India, Delhi, 16 Jan 2018
Sering S (2012) Expansion of the Karakoram corridor: implications and prospects. IDSA, Delhi, pp 3–6
Shakil F (2017) Bad terms: Pakistan's raw deal with China over Gwadar port. Asia Times, 29 Nov 2017
Sharma R (2013) How India is preparing to counter the China threat. First Post, 16 Feb 2013
Sheth N (2015) Sorry Modi—China still doesn't take India seriously. Diplomat, 17 July 2015
Siddiqa A (2017) Reforms in Gilgit-Baltistan. Strateg Stud 37(1)
Siddiqui S (2017) CPEC investment pushed from $55bn to $62bn. Express Tribune, 12 Apr 2017
Singh G (1981) The Karakoram Highway and its strategic implications for India. Indian J Polit Sci 42(1)
Singh P (2013) Gilgit-Baltistan between Hope and Despair. IDSA Monograph No. 14, Delhi, Mar 2013
Singh S (2016) Story of Gilgit-Baltistan: snatched by British, occupied by Pakistan. Indian Express, 18 Aug 2016
Singh P (2018) Pakistan's dam despair. IDSA Comment, Delhi, 23 Jan 2018
Sloan S (2018) Author's interview, London, Apr 2018
Small A (2015) The China-Pakistan axis: Asia's new geopolitics. Hurst, London, pp 2–5

Smith P (2013) The tilting triangle: geopolitics of the China-India-Pakistan relationship. Comp Strategy 32(4):313–330

Smith S (2018) Trump admin cancels $300m aid to Pakistan over terror record. NBC, 2 Sept 2018

Spokesperson (2018a) Secretary Pompeo's call with Pakistani Prime Minister Imran Khan. DoS, Washington, 23 Aug 2018

Spokesperson (2018b) Joint Statement on the Inaugural US-India 2+2 Ministerial Dialogue. DoS, Washington, 6 Sept 2018

Steinbock D (2017) The global economic balance of power is shifting. Global Agenda, World Economic Forum, Cologne, 20 Sept 2017

Tanzeem A (2018) US-Pakistan relations worsen as both sides dig in. VoA, Islamabad, 6 Feb 2018

Tarar M (2017) Vibrant security measures for CPEC. CPEC Secretariat, Islamabad, 1 Feb 2017

Thakur R (2018) The US-Pakistan-China nexus. Strategist, 9 Feb 2018

Tharoor S (2019) India's democratic dictatorship. PS, 13 Sept 2019

TNN (2012) India needs to revisit 1962 humiliation for catharsis. ToI, 10 Oct 2012

Topping S (1979) Karakoram. NYT, 2 Dec 1979

Trump D (2017) Remarks on the strategy in Afghanistan and South Asia. White House, Fort Myer, Arlington, 21 Aug 2017

Tunningley J (2017) Can China overcome the Malacca Dilemma through OBOR and CPEC? Global Risk Insights, 8 Mar 2017

Vajpayee A (1998) Vajpayee to Clinton. PMO, Delhi, 13 May 1998

Vickery R (2018) The United States and India need a maritime initiative. AMTI/CSIS, Washington, 16 May 2018

Walsh D (2007) Red mosque siege declared over. Guardian, 11 July 2007

Wang Y (2018) Far from aggravating the debt burden of Pakistan, the construction of CPEC has delivered tangible outcomes. MoFA, Beijing, 8 Sept 2018

Wang X, Iqbal A (2017) Memorandum of understanding: the agreement on the long-term plan for China-Pakistan economic corridor (2017–2030). NDRC/MPDR, Islamabad, 21 Nov 2017

Wen J, Aziz S (2005) Treaty of friendship, cooperation and good-neighbourly relations. Chinese Embassy/Xinhua, Islamabad, 5 Apr 2005

Xi J (2017) Official message to the people of Pakistan on the launch of the CPEC website by the GoP. Chinese Embassy, Islamabad, 6 Dec 2016

Xinhua (2017) CPEC flagship project of China's BRI. GT, 5 Nov 2017

Xinhua (2018) China eyes closer ties with Pakistan. CD, 3 Nov 2018

Yao J (2018) A community of shared future with Pakistan. CD, 28 Aug 2018

Yao J, Shah L (ed) (2019) CPEC quarterly. MPDR, Islamabad, Vol 3, Oct 2019, p ii

You J (2007) Dealing with the Malacca Dilemma: China's effort to protect its energy supply. Strateg Anal 31(3):467–489

Yousufzai G (2018) Five wounded in attack on bus ferrying Chinese workers in Pakistan. Reuters, Quetta, 11 Aug 2018

ZD (ed) (2018) BRI reshaping Asia's int'l relations. Xinhua, Boao, 8 Apr 2018

Zeb R (2018) Pakistan in the SCO: challenges and prospects. Central Asia-Caucasus Analyst, 8 Feb 2018

Zhao L (2018) in APP (2018) CPEC projects in all 4 provinces of Pakistan: Chinese embassy report. Times of Islamabad, 6 Oct 2016

Chapter 6
Case Study 2: The Twenty-First Century Maritime Silk Road

6.1 Maritime Silk Road Deepens Anxiety

In September 2018, leaders of 53 African countries and the African Union (AU), including 40 presidents, a vice president, 10 prime ministers and the AU Commission Chair, joined UN Secretary-General Antonio Guterres and delegates from 26 international organisations in Beijing as President Xi Jinping's guests at the Forum on China–Africa Cooperation (FOCAC).[1] The only absentee, King Mswati of Eswatini (ex-Swaziland), maintained diplomatic ties to Taipei. The summits' documents—'the Beijing Declaration: Towards an Even Stronger China–Africa Community with a Shared Future', and 'the FOCAC Beijing Action Plan: 2019–2021'—affirmed post-2000 Sino-African cooperation and outlined planned collaboration under the twenty-first century Maritime Silk Road (MSR) rubric,[2] until their 2021 summit.[3]

Western commentary described Sino-African engagement as efforts to counteract China 'growing old before growing rich'. Africa offered 'poor and young' workers willing to supplant labour shortages afflicting Chinese manufacturing.[4] Observers noted differences between Chinese financing announced at the 2015 FOCAC and that offered in 2018. Although both pledged $60bn, the 2018 package was supposedly more onerous.[5] Some saw the summit as an attempt to rebut allegations of 'neocolonialism' and 'debt-trap' diplomacy.[6] US officials urged African leaders 'to consider the implications of partnership with China and how it aligns with your own sustain-

---

[1] Mo (2018).
[2] For more information, you can find the 'Official BRI Map reproduced by CCTV' in 2014, here: http://files.chinagoabroad.com/Public/uploads/v2/uploaded/pictures/1504/new_0.jpg (last accessed October 16th, 2019).
[3] Shinn (2018), Xinhua (2018a).
[4] Johnson (2018).
[5] Shinn (2018), Moore (2018).
[6] Tiezzi (2018), Kuoppamaki and Bathka (2018).

ability and prosperity goals'.[7] This mirrored the debatable US belief that China posed 'the biggest long-term threat to the US geopolitical and geoeconomic interests'.[8]

UN Secretary-General, Antonio Guterres, in contrast, praised FOCAC: 'It's very important to support African countries to fully benefit from China's cooperation, and to fully integrate it into their own development programs'. Describing FOCAC as a vector of benefits flowing from China's progress, 'an important contribution to reducing poverty around the world, and to meeting other development goals', Guterres urged 'synergies' enabling 'the global economy to be more prosperous and to bring more benefits to the global population'.[9] The divergence in the US and UN perspectives looked diametrically opposed. Logically, both could not be correct.

China's trade, aid and investment ties with Africa, forged in the 1950s, grew rapidly in the early 2000s. Chinese FDI stocks there rose from $7bn in 2008 to $26bn in 2013. China overtook the USA as Africa's biggest trade partner in 2009. In 2016, China–Africa trade reached $200bn, more than twice Afro-US commerce.[10] In 2015–2018, Beijing honoured its 2015 FOCAC pledges of grants, loans and export credit, completing several railways, highways, ports, airports and other projects. Those under construction were carried over into the next three-yearly plan.[11] At the Beijing FOCAC, Xi proposed a 'five-nos' framework, founded on the principle of 'giving more and taking less, giving before taking and giving without asking for return'[12]:

- No interference in African countries' pursuit of development paths befitting their national conditions
- No interference in African countries' internal affairs
- No imposition of China's will on African countries
- No attachment of political strings to assistance, and
- No seeking of 'selfish political gains' in investment and financing cooperation

Xi's offer of $60bn in assistance included $15bn in grant, interest-free loans, and concessional credit, a $10bn development-financing fund, and $5bn for boosting African exports. China added 'major initiatives' to the 10 'cooperation plans' launched at the 2015 FOCAC[13]:

- Industrial promotion: a China–Africa economic and trade expo in China encouraging Chinese firms to increase investment in Africa
- Food security by 2030 via 50 agricultural assistance programmes: RMB 1bn in emergency food aid, 500 ago-experts to train/advise African specialists, researchers and entrepreneurs in advanced agro-business; aid African firms by

---

[7] Sullivan (2019).
[8] Johnston (2019), Blackwill (2019), Brands (2018a), Pillsbury (2015).
[9] Information Office (2018a.
[10] Biswas and Tortajada (2018), Solomon (2017).
[11] Xi (2018a), Tuegno (2018).
[12] Xi (2018a).
[13] Xi (2018a).

6.1 Maritime Silk Road Deepens Anxiety

using local currency settlement, expand credit via three Africa-focused financial institutions, and establish corporate social responsibility alliances
- Infrastructure connectivity: forging 'China–AU infrastructure connectivity' and building Sino-Africa corporate coherence via AIIB, NDB and SRF financing
- Trade facilitation: increasing non-resource imports from Africa, effecting 50 trade facilitation projects, supporting a pan-African FTA and promoting e-commerce
- Green development: 50 projects and a new China–Africa environmental cooperation centre
- Capacity building: 10 vocational training workshops for African youth, training for 1000 'high-calibre Africans', 50,000 official scholarships, shorter training sessions for another 50,000 youths and 2000 African exchange students invited to China
- Healthcare: upgrade 50 medical and health programmes, especially the African Centre for Disease Control and China–Africa Friendship Hospitals; data exchange and help with communicable diseases prevention and control; training more health workers, improving medical facilities, and targeted help for women and children
- People-to-people exchanges: an Institute of African Studies; the upgraded China–Africa Joint Research and Exchange Plan; new cultural centres, 50 cultural, sports and tourism events, and several exchange-boosting festivals
- Peace and security: a China–Africa Peace and Security Fund, continued military aid to the AU, 50 new security assistance programmes to protect MSR/BRI projects and boost state capacity for upholding security and fighting terrorism

Progress heightened Western concerns over 'China's "neo-colonialist" behaviour as it acquires raw materials', i.e. metals, minerals, hydrocarbons and farm produce from Africa. Empirical–rational reviews found little evidence supporting these anxieties.[14] Africa was central to the MSR's first of three 'Blue Economic Passages' highlighted in the 2017 NDRC White Paper: China–Indian Ocean–Africa–Mediterranean Sea, China–Oceania–South Pacific, and China–Arctic Ocean–Europe.[15] In June 2019, more than 10,000 officials and traders from 53 African countries attended the first China–Africa Economic and Trade Expo, as promised at the Beijing FOCAC, in Changsha, Hunan Province, to build on FOCAC proceedings. In Jan–May 2019, China–Africa trade grew by three per cent year-on-year to $84.8bn; Chinese investment increased by $1.5bn in that period.[16] Strategic angst triggered by Chinese policy and presence, which long preceded Xi's BRI proclamations, consequently deepened. With FOCAC processes formalised within MSR activities, Sino-African relations looked even more unsettling.[17]

The warmth evinced by African elites at these gatherings only partly explained the West's MSR-rooted anxiety catalysing even graver insecurity than SREB did. Assistant Secretary of Defence for Asia-Pacific Security Affairs, Randall Schriver, insisted MSR projects concealed Beijing's quest for securing PLAN access to overseas ports and foreign bases: 'The military is supportive of a comprehensive strategy

---

[14]Chen et al. (2016).
[15]Rolland(2019), NDRC, State Oceanic Administration (2017a),NDRC, MoFA, MOFCOM (2015).
[16]Mu (2019).
[17]Bloomberg (2018a, Pilling (2017), Solomon (2017); Dollar et al. (2015).

and in many ways the leading edge is predatory economics'. Shriver said, 'where China is using economic tools, they're often doing so in order to create access and potential bases'.[18] OPIC, channelling US private capital, credit and state-level insurance to developing economies, assumed salience in the US' counter-MSR drive. OPIC's then-CEO, Ray Washburne, saw malign motivations fuelling Beijing's MSR blueprint: China was 'not in it to help countries out, they're in it to grab their assets, their rare earths and minerals and things like that as collateral for their loans'.[19]

His successor, David Bohigian, rhetorically restrained, built counter-MSR coalitions with the European Development Finance Institutions (EDFI) and Development Finance Institute of Canada (DFIC). A fortnight before Xi's second BRF, OPIC fashioned a 'DFI (Development Finance Institution)-Alliance' fusing public–private funding to 'co-finance, share risks and other tangible initiatives' enabling the alliance to 'catalyse financing and investment in sustainable projects in emerging markets and countries of mutual interest'. The alliance pledged to be 'role models for other investors'. Counter-MSR coalitions promised to provide alternative financing to areas where MSR projects were under construction or consideration.[20]

Official outrage driving bloc-building targeted China's port-infrastructure projects along MSR's IOR axes. However, Beijing's articulation of its expanding maritime interests after China's foreign trade exploded following its WTO accession long preceded Xi's 2013 MSR speech in Jakarta. President Hu Jintao, on an official visit to Madrid in November 2005, when he and King Juan Carlos jointly presided over the first China–Spain business summit in the Spanish capital, harkened back to ancient trade links which provided a concrete foundation for rebuilding commercial ties: 'Although China and Spain are far from each other, businessmen of the two countries started trade as early as in the sixteenth century and opened the world famous maritime Silk Road, which helped promote the east and west exchanges and exerted great impact upon human civilisation and progress'.[21]

Think tanks, e.g. the CSIS in Washington, NBR in Seattle, CNA in Arlington and SIPRI in Stockholm, mirroring DoD views, shaped the trans-Atlantic policy discourse. They reinforced deepening concern over MSR's purported geopolitical threats to Western pre-eminence and values presented as 'global norms' since the Soviet collapse.[22] Policy-documents exposed the conceptual and normative gulf separating the Chinese and US MSR narratives.

Beijing's June 2017 MSR White Paper framed four defining principles: 'shelving differences and building consensus; openness, cooperation and inclusive development; market-based operation and multi-stakeholder participation; and joint development and benefits sharing'. MSR would fashion 'blue-partnerships' with other

---

[18] Shriver(2018), OSD (2018a).

[19] Churchill (2018).

[20] Bohigian et al. (2019).

[21] Hu (2005).

[22] OSD (2018b, Greenert (2018), Ghiasy et al. (2018), Szechenyi (2018), Krishnan (2017a), Clemens (2015a).

maritime states to generate 'green development', 'ocean-based prosperity', 'maritime security', 'innovative growth' and 'collaborative governance' of the seas. The NDRC stressed joint efforts in pursuing shared interests by engaging APEC, EAS, FOCAC, UNESCO and other multilateral organs.[23]

Washington's response analysed OBOR/BRI within assessments of China's 'military power', describing the initiatives as 'indicative of China's intention to use economic means to advance its interests and enhance its global role by integrating hard infrastructure development with trade and financial architecture'.[24] A more detailed DoD study, underscoring the USA's mainly military perspective on MSR/BRI, explained US fears. MSR's port-building efforts presented 'military force posture, access, training, and logistics implications' for the rivals. DoD expected Beijing to expand PLAN's access across MSR's alignment, increasing 'China's ability to deter use of conventional military force, sustain operations abroad and hold strategic economic corridors at risk. The PLA's expanding global capabilities provide military options to observe or complicate adversary activities' in war. Beijing could 'preposition the necessary logistics support to sustain naval deployments' to defend Chinese interests across 'the Indian Ocean, Mediterranean Sea and the Atlantic Ocean'.[25] This was intolerable.

6.2 Naval Dynamics

Few projects, some begun years earlier and later wrapped into MSR, caused greater concern than Beijing's construction of the Doraleh Multipurpose Port (DMP) in Djibouti on the Horn of Africa. Djibouti's location at the junction of Africa, Asia and the Middle East, and proximity to the Gulf of Aden's Bab el-Mandeb Strait, was its attraction. Its command of major SLoCs traversing the Red Sea, the Gulf and the Indian Ocean, its colonial history, and great-power interests had drawn in French, US, British and German military presence, but not investments, to the tiny, poverty-stricken, country. Djibouti quickly joined Beijing's BRI enterprise, attracting significant MSR finance. In May 2017, Djiboutian and Chinese officials inaugurated the DMP with hopes of the port generating accelerating revenue flows.

6.2.1 Djibouti Base Jitters

Then on 1 August, the PLA's ninetieth founding anniversary, Djibouti's defence minister and a PLAN deputy commander hoisted flags, opening China's first overseas 'logistics base' seven miles north-west from the US's Camp Lemonnier, horrifying

---

[23] NDRC, State Oceanic Administration (2017b).
[24] OSD (2018a).
[25] OSD (2018b, pp. ii–iv, 3–4, 12–16, 19–20, 23).

BRI/MSR critics.[26] PLA contractors had begun building their 36-hectare base close to the DMP in March 2016 after securing a 10-year lease. During negotiations, Beijing announced the facility would resupply PLAN and other Chinese vessels conducting anti-piracy patrols in the Gulf of Aden and off the Somali and Yemeni coasts, and aid Chinese peacekeeping missions and humanitarian and disaster relief (HADR) operations in the region. The base, mirroring Camp Lemonnier, would build dock facilities for repairing ships and helicopters, and *materiel* storage. The PLA could station up to 10,000 service personnel to man such operations until the lease expired in 2026.[27]

US observers, discerning threats to the USA's largest African military base, key to operational efficacy and confidence, questioned Djibouti's goals. Washington assessed that even in peacetime, with this first overseas PLA facility on the Horn of Africa, Beijing expanded its political-economic influence across the region, creating a more geopolitical and strategic, than operational or tactical, challenge.[28] US concerns spiked in early 2018 when Washington complained that 'military grade' laser beams, allegedly fired from the Chinese base, had hit two pilots flying a USAF C-130 Hercules aircraft, presumably overflying approaches to the PLA base. The pilots suffered no long-term harm but the DoD insisted, 'This activity poses a true threat to our airmen. It's a serious matter, so we're taking it very seriously'.[29]

Accusing PLA personnel of conducting several purported attacks, with details classified, the USA urged China to investigate the 'incident'. China's Defence and Foreign Ministries rebutted the allegation, with MoFA responding, 'You can remind the relevant US person to keep in mind the truthfulness of what they say, and to not swiftly speculate or make accusations'.[30] The exchange, in the period following Washington's National Security Strategy and National Defence Strategy documents identifying China as a rival power, underscored the adversarial dynamic reinforced by the PLA's first overseas base.

In July 2018, Djibouti's President Ismail Omar Guelleh hosted the presidents of Somalia, Sudan and Rwanda, the prime minister of Ethiopia, and the AU Chairperson, at the launch of the Chinese-built Djibouti International Free Trade Zone (DIFTZ). The $370m project, built on a 600-ha barren plateau comprising four industrial clusters specialising in trade and logistics, export processing, and business support, would emerge as a hub for Africa's commerce with Asia and the wider world. At the launch, work on only 240 ha was completed, but over the next decade, it would cover 40 km^2, and an investment of $3.5bn would transform DIFTZ into Africa's largest FTZ.[31]

That aspiration, ignoring the presence of naval facilities operated by the USA, Japan, France and Italy in Djibouti, was clouded by accusations that China was

---

[26] Egozi (2018), Brands (2018b), Krishnan (2017b).
[27] Fei (2017), Blanchard (2017).
[28] Osborn (2018).
[29] Mehta (2018).
[30] Ali et al. (2018).
[31] Mutethiya (2018).

exploiting Djibouti's poverty and naivety to take 'control' over a strategic MSR nodal point, and challenge the US-led order. Senators representing a bipartisan consensus warned Secretaries Pompeo and Mattis against 'major strategic benefits to China', and risks of 'undermining the balance of power in East Africa and around the Bab al-Mandeb strait, a major artery of maritime trade'. They feared, 'China's control of Doraleh could allow it to impede US military operations…as well as those of US allies'.[32] Djibouti's leaders perceived no threats but Western commentary articulated profound anxiety, although research-based analyses offered reasoned views.[33] Washington's Quad-partners, too, expressed concerns, but US anxiety attracted the most attention.

Former CNO, Admiral Jonathan Greenert, noted that China's ability to accumulate power and Xi Jinping's 'somewhat revisionist "China Dream" goals' made China the USA's 'greatest long-term security challenge'. Greenert posited, China could 'fundamentally challenge US interests' across the Indo-Pacific. Washington's response to the 'China challenge' could 'be the most consequential strategic decision of the first half of the twenty-first century'.[34] He insisted Beijing sought 'the capability to deny freedom of manoeuver to the US and allied forces', while China's SCS island-building challenged US maritime presence across the Western Pacific. Beijing thus 'calls into question' US commitment to allies, and to a 'free and open Indo-Pacific'.[35] MSR purportedly backstopped that perceived grand-strategic threat.

DoD, ignoring the 100-plus 'allies and strategic partners' housing around 800 US bases,[36] insisted BRI/MSR was 'intended to develop strong economic ties' with clients, 'shape their interests with China's, and deter confrontation or criticism of China's approach to sensitive issues'. Partners would 'develop economic dependence on Chinese capital, which China could leverage to achieve its interests'. Former acting-DCI Michael Morell described China as 'the most formidable challenge, perhaps the most daunting the United States has faced since World War II'. China wanted to 'restore its place as Asia's dominant nation', and 'become the most powerful and influential country in the world'. Seeking military power 'stronger than any Washington encountered' since the Cold War, Beijing mixed sophisticated naval, aerial and space capabilities with 'legitimate economic and diplomatic tactics', and 'illegitimate' ones, e.g. 'intellectual property theft', and 'economic coercion and the seizure' of contested SCS reefs.[37] The consensus? This growing threat to US *systemic primacy* must be thwarted.

MSR reinforced that view. A former practitioner, introducing the CSIS study, 'China's MSR: Strategic and Economic Implications for the Indo-Pacific Region', conceded there was 'a shortage of infrastructure investment to meet the needs of developing nations across the Indo-Asia Pacific', but raised 'growing questions about

---

[32] Coons and Rubio (2018).
[33] Economy (2018), Manek (2018), Ali and Stewart (2018), Downs et al. (2017).
[34] Greenert (2018, pp. 1–2).
[35] Greenert (2018, pp. 7–8).
[36] Dufour (2018).
[37] Morell (2018).

the economic viability and the geopolitical intentions behind China's proposals'. On port-building projects, he asked 'whether these investments are economic or military in nature'.[38] Contributors examined three MSR cases: Kyaukpyu in Myanmar, Hambantota in Sri Lanka, and Gwadar in Pakistan, and Delhi's Iranian counterpoint—Chabahar. They enumerated geopolitical gains Beijing allegedly sought, and the pushback from Delhi, which, with Washington, Tokyo and Canberra, forged the Quadrilateral Security Dialogue—the Quad. The authors concluded that a competitive dialectic dynamic was apparent. Still, given all recipient states' deep interest in China-funded infrastructure projects, and apparently mutual economic gains, conclusions were ambiguous.

6.2.2 Commercial Rationales and Strategic Mistrust

An examination of MSR projects' economic drivers found, 'Many of the same attributes that make a port commercially competitive can also increase its strategic utility'. Notably, 'deep-water ports can accommodate larger commercial vessels as well as larger military ships'. Specifically, 'more than half of the 7.6m barrels of crude oil that China imports each day comes from countries along the Persian Gulf'.[39] So, China's 'increased military presence in the Indian Ocean should not come as a surprise'. Beijing was only 'following in the traditional path of other rising powers' in 'expanding its military operations to match its interests abroad'. While its engagements were 'changing regional security dynamics', Chinese influence would likely trigger 'greater concern, particularly from Indian strategists'. Happily, in this view, Chinese 'forces and facilities would be highly vulnerable if a major conflict were to break out'.[40]

CSIS analysts posited that, against the backdrop of China's expansive maritime-territorial claims, reclamation of disputed reefs, and installation of military assets there, MSR posed 'unique' challenges to Quad-members. These dynamics flowed from intricate interactions of geoeconomics and geopolitics across the region and beyond[41]:

- Washington sought to 'counteract Chinese influence' in 'the vacuum' created by the US withdrawal from the Trans-Pacific Partnership
- Delhi aimed at negating Beijing's 'encroachment on its zone of strategic interest', and 'encirclement from Chinese projects in Pakistan'
- Tokyo sought to mitigate 'China's ability to influence the energy supply chains' on which its, and East Asia's, economic well-being depends
- Canberra targeted fears that Beijing's project aid 'could render fragile states more vulnerable to coercion'

---

[38] Green (2018).
[39] Funaiole and Hillman (2018).
[40] Cooper (2018).
[41] Gale and Shearer (2018).

6.2 Naval Dynamics

This analysis averred that shared anxiety over 'major geopolitical shifts in the Asia-Pacific', reinforced by Beijing's 'expanding maritime strategy and increasing assertiveness', triggered a 'revitalisation' of the 2007-vintage Quad as a counter-China strategic pushback. Although Quad-members differed in the details, they were determined to secure FoN, maritime security and 'respect for international law'. The Quad reflected and reinforced an 'underlying structural dynamic', providing a 'strong foundation' for collaboration against the 'Chinese challenge'.[42] In 2018, Quad-members secured British and French support.[43]

Across the Atlantic, SIPRI analysts concluded MSR aimed to 'diversify and secure' China's SLoCs as part of its 'maritime renaissance', further develop its $1.2tn 'blue economy', improve food and energy security, defend territorial sovereignty and enhance Beijing's 'international discourse power'. China's marine-economy, around 10% of its GDP, grew annually by 7.5% during 2012–2017; Beijing planned to boost its growth to 15% of GDP by 2035.[44] SIPRI acknowledged that the $3.4tn, or 21% of global trade which crossed the SCS in 2016, included 64% of China's maritime trade, making the SCS-SLoC 'pivotal' to Chinese, as well as other economies.[45]

SIPRI posited that MSR's scale and speed, complemented by SREB projects, could fill 'a substantial maritime-terrestrial connectivity gap' and instil 'a spirit of cooperation', but also trigger 'scepticism over possible security implications'. MSR-SREB connectivity could generate economic security by fashioning a 'mega production and trade network', eroding the impact of external disruption to key supply chains. However, it added to Western 'security complexity' across the contested SCS-IOR space.[46] Given the ambivalence among China's critics, their views of MSR would likely be mixed. Beijing nonetheless saw the necessity of and benefits from MSR outstrip risks. This explained the growing portfolio of equity purchased, investment made, leases secured and development begun or expanded on foreign ports and coastal/insular sites along the MSR's China–SCS–Australasia–IOR–Gulf–Africa–Europe alignments, as shown in Table 6.1[47]:

Although several EU-member-states remained welcoming to Chinese maritime investments, the USA pushed back. This led, in April 2019, to the cancellation of a 40-year lease on the Long Beach Container Terminal in California acquired by the Hong Kong-based China Ocean Shipping Company (COSCO) in 2012. In 2018, the US Departments of Homeland Security and Justice, urged by the Trump Administration, ordered COSCO to sell the terminal to a 'suitable third party', citing a potential risk

---

[42] Gale and Shearer (2018, p. 33).
[43] Kelly and Perry (2019), AP (2019a), Panda (2019), Akita (2018), Report (2018a), Bajwa (2018), Jennings (2018), Singh (2018a).
[44] Xinhua (2018b).
[45] Ghiasy et al. (2018, p. 12).
[46] Ghiasy et al. (2018, p. 1).
[47] Ghiasy et al. (2018, p. 6), GNA (2018), Port of Zeebrugge (2018), Horowitz and Alderman (2017) Chastised by EU, a Resentful Greece Embraces China's Cash and Interests. NYT, 26 Aug 2017, Mathews (2017), Kwok (2017), Zhen (2016).

Table 6.1 MSR port/terminal acquisitions

Year	Port/country	Lease/equity	Duration
2015	Gwadar	Lease	40 years
2015	Kyaukpyu	Lease	50 years
2015	Kuantan, Malaysia	Lease	60 years
2016	Obok/Doraleh, Djibouti	Lease	10 years
2016	Melaka Gateway, Malaysia	Lease	99 years
2016	Piraeus, Greece	67% equity	35 years
2016	Darwin, Australia	80% equity	99 years
2017	Hambantota, Sri Lanka	Lease	99 years
2017	Muara, Brunei	Lease	60 years
2017	Feydhoo Finolhu, Maldives	Lease	50 years
2017	Valencia, Bilbao, Spain	51% equity in Noatum Holding Company	
2018	Zeebrugge, Belgium	Lease—'concession agreement' on a terminal	
2018	Jamestown, Ghana	$50m in construction aid, $16m in grant[a]	
2019	Genoa, Italy	49.9% equity, 2xterminals to COSCO/Qingdao[b]	

[a]Report (2018b)
[b]Report (2019a)

to US national security.[48] This forced sale of Chinese assets acquired in the West, especially in the USA, could set an anti-MSR precedent.

Chinese officials, SOEs and private firms, separately or jointly, negotiated with owners of ports across Asia, Africa, the Middle East, Europe and Central America in 2018–19. Agreement proved slow and occasionally, abortive. Talks among the governments of China and Italy, and the Northern Adriatic Port Association, over investment in, development of and privileged access to Venice, Trieste and Ravenna in Italy, Capodistria in Slovenia and Fiume in Croatia, began in 2017. After an October 2016 agreement between COSCO Shipping and Italy's Vado Holdings, COSCO began building Italy's first automated container port at Vado Ligure near Milan with an annual capacity of 860,000 TEUs.

By the time Italy joined BRI during Xi's March 2019 visit, raising EU and US hackles, 80% of work was completed. MoUs exchanged then included management accords between Trieste and Genoa ports and China Communications Construction Company (CCCC) boosting Europe–China links.[49] Venice, Tianjin and Ningbo port authorities had already signed a MoU on deepening such ties. Resonating with the EU's TEN-T blueprint to fashion a Venice–Ravenna–Capodistria–Trieste cluster, this

---

[48]DeAeth (2019).
[49]Ching (2019), Ye and Han (2019), NSC (2019), MoFA (2019).

would be MSR's 'privileged link' to south-eastern Europe. Chinese cargo arriving on 18,000-plus TEU container ships would be distributed from Northern Italy across Southern Germany, Switzerland, Austria, Hungary and the Balkans.[50]

Notably, Chinese port-building overseas began long before MSR was proclaimed. One of the biggest, Kribi in Cameroon in West Africa, has been under construction by CHEC in phases since June 2011. The first phase, with Beijing funding 85% of its $568m budget with a concessional loan, was completed in three years. CHEC began building the second phase directly, with China's Exim Bank furnishing 85% of the $1.3bn budget for the two phases. China and Cameroon planned to install 24 berths by 2040, securing an annual capacity of 100m tons.[51]

6.2.3 Duqm Dramatics

Negotiations over developing a port-cum-industrial and commercial complex around the fishing harbour of Duqm, 280 miles south of the Omani capital, Muscat, offered an atypical instance of Chinese investors acquiring commercial control. The Chinese party comprised private Muslim investors from an autonomous region, rather than SOEs or the Chinese government. The Duqm project, progressing gradually, garnered praise from local beneficiaries partnering in the joint venture, and critique from China's detractors based elsewhere.[52]

Hurt by low energy prices and determined to diversify away from dependence on hydrocarbon exports, Oman initially sought but failed to obtain foreign capital, technology and management skills from Western partners. When much looked lost, a private consortium from China's Hui Muslim-majority Ningxia region responded. Supported by the two governments, it established the China–Arab Wangfang Investment Management Company in 2015 to build projects in Middle Eastern countries. This firm's Omani subsidiary, Oman Wangfang, implemented projects agreed with Muscat's agency responsible for developing Duqm, a harbour with a dry dock, on the Western Arabian Seashores, which Omani leaders had already decided to develop.

An agreement to build a 1172-ha industrial park was signed in May 2016. Muscat projected it would attract $10.7bn in Chinese FDI by 2022, with the first $370m devoted to infrastructure. Chinese banks financed Oman Wangfang-managed projects; on completion, the land would be leased to Chinese investors. Ali Shah, Oman Wangfang's chairman, planned to complete the park's initial units—a $2.8bn methanol refinery, a cement plant, oil-industry pipe fabrication facilities, an $84m automobile assembly plant, a 1GWe solar power station, a $203m five-star hotel, a $138m building materials storage-and-distribution complex, a $100m hospital, and a school, by 2022. In April 2017, Ningxia's deputy governor visited Duqm to lay the

---

[50]Escobar (2018).
[51]Report (2018c), Huang (2015).
[52]Nikkei (2018), Iyer-Mitra (2018), Shepard (2017), Jabarkhyl (2017), Reuters (2016).

park's foundation stone. Sponsors hoped Duqm would erode Dubai's dominance as the region's trans-shipment hub within a decade.

The Quad was troubled by other concerns. The Indian Prime Minister, Narendra Modi, one of the four leaders forging the Quad-coalition, played a key role in building consensus on resisting Beijing's MSR endeavours. Visiting the White House in June 2017, he and President Trump announced that as 'responsible stewards in the Indo-Pacific region', they reiterated 'the importance of respecting freedom of navigation, overflight, and commerce throughout the (Indian Ocean) region'. They then called upon 'all nations to resolve territorial and maritime disputes peacefully and in accordance with international law'.[53] Although they did not name China, their joint statement clearly targeted Chinese maritime activities and recorded their determination to take appropriate countervailing measures.

Modi was the first to initiate concrete action. During a February 2018 visit to Oman, he secured access rights for the Indian Navy to Duqm's port and logistical facilities, enabling Delhi to sustain extended presence across Western Indian Ocean.[54] Muscat allowed India to service IN warships at Duqm's dry dock as part of Delhi's 'maritime strategy to counter Chinese influence and activities'.[55] The Indo-Omani MoU formalised Delhi's rotational deployments of surface vessels, submarines and ISR aircraft at Duqm. Bracketing the Indian Ocean, Modi paid equal attention to securing the IOR's eastern margins. He urged ASEAN leaders to jointly fashion a tacit coalition to defend shared interests across the SCS/IOR: 'India shares ASEAN's vision for peace and prosperity through a rules-based order for the oceans and seas. Respect for international law, notably UNCLOS, is critical for this. We remain committed to work with ASEAN to enhance practical cooperation and collaboration in our shared maritime domain'.[56]

ASEAN's response to Quad's counter-China efforts remained consistently mixed.[57] In the IOR's western reaches, Delhi received more active support in resisting apparent Chinese gains. Britain and Japan, too, acquired similar rights to China and India for rotating, servicing, repairing and maintaining advanced naval combatants, escorts and aircraft at Duqm, instancing another counter-MSR coalition coalescing.[58]

The growing challenges confronting China's maritime initiatives did not go unnoticed. SIPRI analysts noted ASEAN's dilemma in reconciling MSR's economic, commercial and infrastructure-related attractions, and extant maritime-territorial, jurisdictional disputes. Anxious to see ASEAN sustain autonomous 'centrality' within the region's security architecture, and aware that 90% of China's trade by volume and 60% by value was maritime, they urged ASEAN-members to maintain a balance of interests between China and the West.[59] They noted the 'critical' economic and

---

[53] Trump and Modi (2017).
[54] Singh (2019).
[55] Singh (2018b), Roy (2018).
[56] Modi (2018).
[57] Wong (2018), Simon (2017), Mollman (2016).
[58] Nikkei (2018), Iyer-Mitra (2018), Panda (2018a).
[59] Ghiasy et al. (2018, pp. 17–22), Report (2017a).

6.2 Naval Dynamics

security interests regional and extra-regional actors perceived in trans-IOR SLoCs: all major trading powers—China, Japan, South Korea, ASEAN-states, the USA and EU-members—had an interest in regional peace, security and stability.

Additionally, the Indian Ocean generated substantial marine resources, e.g. 40% of global offshore oil output, and 15% of the world's fish-catch, in 2016.[60] SIPRI analysts underscored potential challenges to the EU's SLoCs and FoN, and intensifying competition between Beijing and Quad-members. They urged the EU to engage with China, ASEAN and other stakeholders to secure its SLoCs, a 'red line' for EU-members, while exploring benefits of the 'emerging interplay' of maritime and terrestrial security spaces.[61] In 2018, reinforcing China's Quad-linked competitors, France and Britain began deploying warships to Indo-Pacific waters.[62]

Tacitly acknowledging the need to allay Western anxiety, Xi Jinping returned to the theme of the need to build benign and mutually beneficial commercial connectivity, described in the Chinese narrative as the MSR's core objective, upon historical foundations. In his message to a gathering of ministers and officials with maritime responsibilities from China, and Latin American and Caribbean states in Beijing, Xi said, 'In the distant past, the people of China and Latin America overcame great difficulty in crossing vast seas and jointly created the maritime Silk Road spanning the Pacific'.[63] That effort, underscoring a desire to revive pacific commerce, rather than fashion naval control over distant waters, did not, however, reassure MSR's critics.

In March 2019, DoD secured access rights to both Duqm and Salalah, 300 miles to the southwest, for both naval and aerial assets. Overt focus on Iran overlay unstated targeting of China.[64] MSR-reinforced relational dynamics focused simultaneous expectations of benign, threatening and uncertain outcomes, demanding imaginative combinations of diplomacy and deterrence. While Beijing stressed MSR/BRI's shared benefits, and partner-states responded positively, critics saw a gathering geopolitical threat to the US-led order masquerading as commercial and connectivity projects. They acted to counteract MSR's allegedly devious conspiracy to suborn simple-minded elites from poorer states with a view to securing expanding dominance.

6.3 Clashing Perspectives, Purposes and Narratives

Western concern over MSR's purported strategic intent, specifically its naval aspects, long- predated Xi Jinping's proclamations. US DoD-funded think tank analysts

---

[60] Jaishankar (2016).
[61] Ghiasy et al. (2018, p. 33).
[62] Ghiasy et al. (2018, p. 23).
[63] Xi (2018b), Zhang (2018).
[64] US Embassy (2019), Stewart (2019).

from CNA, CSBA and RAND, for instance, wrote reports and spoke at conferences focusing on threats posed to US *primacy* by Beijing's military revitalisation.[65] That tradition continued in examining MSR's likely maritime-threats to the US, and US-aligned, interests. For some, the 'vital issue' was the degree to which 'China's increased economic activity' would generate 'increased military presence', i.e. the number, location, purpose and capabilities of any 'permanent installations and support bases'.[66] Beijing's future ability to leverage commercial linkages into naval access was the key issue driving the US-led counter-China coalition's trans-IOR military diplomacy.

One analyst described MSR as China's 'most vital SLoC', giving Beijing access to markets and resource sources in Southeast Asia, South Asia and the Middle East, enabling the shipment of vital hydrocarbons, minerals and other commodities. That, and Beijing's anxiety to circumvent negative effects of the Obama-era 'Asian Pivot', indicated a Chinese quest for 'control' over the IOR. He, however, noted that 'sea control' required a navy 'several times the size of the current PLAN', an unrealistic ambition.[67] He emphasised Beijing's interest in protecting its growing workforce and FDI-funded projects overseas, if necessary, militarily, as in PLAN's March 2015 evacuation of Chinese and other nationals from Yemen. This analysis illuminating defensive and deterrent motivations driving MSR was, however, a minority view.

6.3.1 The Chinese Framework

The author quoted a PLA Academy of Military Science (AMS) fellow, saying, 'China has only two purposes in the Indian Ocean: economic gains and the security of SLoC'.[68] Few Western observers agreed. The MSR generated two divergent narratives, one highlighting the geopolitical challenges it posed to non-Chinese interests, and the Chinese one highlighting prospects of 'a shared destiny for mankind', i.e. universal peace through shared prosperity. The 'truth' resided in the middle. Profound anxiety, triggered by the geopolitical backwash from China's geoeconomic activism, coloured the West's China discourse which itself was conflicted between mutually opposed views propounded by 'strategists' and 'bankers'. Contradictory perspectives nourishing a dialogue of the deaf capable of triggering avoidable missteps that could spiral uncontrollably demanded rigorous analyses of the MSR blueprint.

The NDRC, MoFA and MOFCOM issued the first Silk Road white paper in March 2015, setting out China's 'vision and actions on jointly building' SREB and MSR, with key objectives[69]:

---

[65] van Tol et al. (2010), Medeiros et al. (2008), Cliff et al. (2007).
[66] Clemens (2015b).
[67] Clemens (2015b, p. 3).
[68] Zhou (2014).
[69] NDRC (2015).

6.3 Clashing Perspectives, Purposes and Narratives   245

- Embracing the trend towards a multipolar world, economic globalisation, cultural diversity and greater IT application
- Upholding the global free-trade regime and the open world economy in the spirit of open regional cooperation
- Promoting orderly and free flow of economic factors, 'highly efficient' resource allocation and deep integration of markets
- Encouraging policy-coordination and implementation of broader, 'more in-depth regional cooperation of higher standards'
- Jointly creating an 'open, inclusive and balanced regional economic cooperation architecture that benefits all'
- Promoting connectivity of Asian, European and African continents and their adjacent seas, establishing/strengthening partnerships; setting up 'all-dimensional, multi-tiered and composite' connectivity networks, and realising 'diversified, independent, balanced and sustainable' development in BRI economies
- Aligning and coordinating development strategies, tapping regional market potential, promoting investment and consumption, creating demand and job opportunities, enhancing people-to-people and cultural exchanges and mutual learning

With their very first objective challenging the US-led *primacy*-centred global hierarchy, NDRC authors noted China's key interests: BRI would enable Beijing to 'further expand and deepen its opening-up, and to strengthen its mutually beneficial cooperation with countries in Asia, Europe and Africa and the rest of the world'. China was committed to 'shouldering more responsibilities and obligations within its capabilities, and making greater contributions' to peace and development. Given the import of trading within China's economy, and of seaborne commerce within its international trade, MSR acquired heightened salience. To forge a 'blue partnership' and a 'blue engine' for sustainable growth, MSR advanced four principles[70]:

- Shelving differences and building consensus
- Openness, cooperation and inclusive development
- Market-based operations and multi-stakeholder participation
- Joint development and benefit-sharing

These principles would inform cooperation between China and MSR-partners, building the China–Indian Ocean–Africa–Mediterranean Sea 'Blue Economic Passage' linking the China–Indochina Peninsula Economic Corridor from the SCS to the Indian Ocean, and connecting the CPEC and BCIM 'economic corridors'. A second 'Blue Economic Passage' followed the China–Oceania–South Pacific alignment, while a third connected China to Europe via the Arctic Ocean. The NDRC's 2017 white paper stressed multilateral collaboration via APEC, the East Asian Summit, FOCAC, UNESCO, the Indian Ocean Rim Association and others. It listed complementary institutions which China would establish with partner-states to advance a 'joint', mutually beneficial, and sustainable MSR.[71]

---

[70] NDRC (2017).
[71] NDRC (2017).

Even earlier, the NDRC designated Fujian Province, with Fuzhou, Xiamen and Quanzhou ports deeply engaged in international trade, MSR's 'core area'. Once the stepping stone on the ancient MSR, Fujian was helped to expand port facilities and leverage 'the unique role of overseas Chinese and the Hong Kong and Macau Special Administrative Regions' to advance MSR's construction.[72] Fusing MSR and SREB termini, as the starting point for many trains and ships carrying goods from coastal 'factory-cities' to European markets, Fujian acquired a special status in China's MSR/BRI framework.[73] BRI/MSR's economic/commercial focus dominated the discussions among central- and provincial-level Chinese officials as the 2015 white paper was drafted, but US anxiety was already apparent. And that had a history.

Following up on the DoD's February–April 1992 Defence Policy Guidance instructing US forces to permanently secure the USA's post-Soviet *primacy*, a seminal preview of prospects came in the 1999 horizon-scanning Summer Study on Asia's 2025 geopolitical landscape. Sponsored by the DoD's Office of Net Assessment (ONA), it identified China as an emerging 'constant competitor', and India, a likely 'swing-state'.[74] Later, a trio of civilian DoD-contractors postulated the 'String of Pearls' rubric to describe Beijing's port-building activities along its IOR SLoC years before Xi Jinping's Jakarta address.[75] That threatening portrayal of Beijing's alleged efforts to 'control' SLoCs linking its production hubs to the sources of energy imports powering China's manufacture-and-export economic model proved catchy.

Presented to then-SoD Donald Rumsfeld, it made a lasting impression on the USA's national security establishment.[76] The phrase's evocative imagery, and implications for US force projection undergirding *primacy*, stuck, colouring subsequent discourse. Zhou Bo, a fellow at the PLA's AMS, one of two premier military theoretical research establishments,[77] pointed out that those DoD-commissioned analysts had insisted Beijing was building naval bases and/or electronic listening posts in Myanmar, Bangladesh, Sri Lanka and Pakistan but, nine years on, 'these "bases" are found nowhere in the Indian Ocean'.[78] Counterfactuality changed few minds.

In December 2011, Seychelles Foreign Minister, Jean-Paul Adam, invited Beijing to build a PLAN base on his island-state's territory, but China would only 'consider' replenishment visits and port-calls. Emphasising that Beijing's only 'purposes' in the Indian Ocean were, 'economic gains and the security of SLoCs', Zhou noted China's deepening reliance on SLoCs for sustained development, and pointed to the USA and India as the two 'most important' countries for its 'freedom of navigation in the Indian Ocean'. China's counter-piracy operations in the Gulf of Aden, begun in late-2008 on the basis of a UNSC resolution, manifested this perspective. 'Access,

---

[72] Wing (2015).
[73] Xinhua (2017).
[74] Under Secretary of Defence (Policy) (1999).
[75] MacDonald et al. (2004).
[76] Conrad (2012), Gassaway (2011), Spinetta (2006).
[77] The other being the National Defence University (NDU), Beijing.
[78] Zhou (2014).

6.3.2 Maritime History as a Mirror

Countering DoD's 'String of Pearls' imagery, Beijing underscored the benign aspects of the Ming Dynasty Admiral Zheng He's seven voyages of discovery, the first preceding Vasco da Gama's arrival in Western India by 80 years, Christopher Columbus's discovery of the New World by 67 years, and Ferdinand de Magellan's planetary circumnavigation by 114 years. Chinese analysts stressed the commercial drivers behind the Tang, Song and Yuan dynasty-era MSR, beginning in the ports of Quanzhou, Guangzhou, Ningbo and others. After warfare blocked the Tang Dynasty terrestrial Silk Road, trade shifted south, and during the Song dynasty, the MSR's ancient precursor became the main trade channel.[80] Zheng He's voyages epitomised the peak of China's maritime activism along the ancient MSR.

Zheng He (1371–1435), commanding the largest fleet the world had seen until then, sailed from coastal China through the SCS, via Southeast Asia to the Gulf and East Africa, from 1405 to 1433. His fleets typically comprised 300-plus vessels, with 'treasure ships' more than 400-foot long, and supply vessels, water tankers, combat-escorts and patrol-ships. The fleet carried around 28,000 personnel, calling at ports in modern-day Indonesia, Malaysia, Thailand, India, Sri Lanka, Yemen, Saudi Arabia, Somalia and Kenya.[81] Zheng He explored trade routes, negotiated trade pacts, fought pirates, brought back tributes and specimens of littoral flora and fauna, engaging diplomatically and commercially.[82] He did, on occasion, involve himself in coastal politics, installing local princes on disputed thrones, as in Melaka, but took no land.[83]

Chinese specialists, examining modern-day MSR's scientific–technical features, questioned Western postulates, asserting, MSR was a 'powerful means to create a cooperative, peaceful, and harmonious environment' for interstate partnership offering 'a good opportunity and external environment for China's comprehensive deepening reform'. MSR was a template for boosting China–ASEAN collaboration that could 'build a community with a common destiny'. It would 'tighten mutual interests and strengthen sea lane interconnection', advancing 'common prosperity and progress' in shipping, marine-energy, economy and trade, scientific and technological innovation, environmental protection, and 'human communication'. These objectives reflected an existing need.[84]

---

[79]Zhou (2014).

[80]Chongwei et al. (2018, pp. 1–2).

[81]Johnson (undated).

[82]Musgrave and Nexon (2017), Murphy (2010), Xinhua (2005a, b).

[83]Menzies (2003).

[84]Chongwei et al. (2018).

Western critics, challenging that beneficent imagery, insisted Zheng He's voyages were designed to exercise power, influence local opinion, expand Ming China's tributary system and establish the Emperor's suzerainty wherever the visually and power-potentially overwhelming fleet docked.[85] They implied, Beijing's portrayal of Zheng He's voyages was selective, even inaccurate; that his sorties were more political-economic than commercial, and the new MSR followed that practice of using material incentives to enhance China's image, expand its influence, and advance Beijing's covertly revisionist goals. Indian analysts concurred.

They quoted India's first Prime Minister, Jawaharlal Nehru, in distilling modern lessons from maritime history. 'We cannot afford to be weak at sea', Nehru warned, 'History has shown that whoever controls the Indian Ocean has, in the first instance, India's seaborne trade at her mercy and, in the second, India's very independence itself'.[86] India's pre-eminent maritime power proponent, K.M. Panikkar, noted Delhi's geography-rooted geopolitical compulsions: 'the peninsular character' of India and the 'essential dependence of its trade on maritime traffic give the sea a preponderant influence on its destiny'. Panikkar reinforced Nehru's belief that 'the economic life of India will be completely at the mercy of the power which controls the seas'.[87] These dicta, defining Delhi's maritime calculus, identified China and the MSR—ironically premised on mirror-image assumptions—as the principal threat to Indian interests.[88]

A well-regarded naval analyst insisted, for instance, that the first MSR blueprint presented by Beijing indicated it was an 'ASEAN-centred project' designed to 'enhance connectivity and cultural links in China's strategic backyard—the SCS'. China, in his view, 'later expanded the scope of the project to include the Indian Ocean'. This implied subterfuge suggested to him that Beijing concealed malign intent until the MSR blueprint gained regional acceptance as part of the ideational landscape. The analyst stressed China's 'opacity' in avoiding details reinforcing suspicions, failing to 'dispel any impressions of it being a cover for maritime military bases'.[89] So, in this view, MSR's 'essential rationale' was the 'leveraging of Chinese soft-power' to 'shore-up China's image as a benevolent state'.[90]

Delhi feared China's infrastructure investments securing its SLoC could 'pose a challenge to India's stature as a "security provider" in the region', thus 'adversely affecting New Delhi's strategic purchase in its primary area of interest'.[91] The NDRC's 2015 BRI white paper and 2017 MSR-focused document did not allay suspicions. One analyst noted, the development of 'more projects such as Gwadar

---

[85] Musgrave and Nexon (2017), Lin (2015), Yang (2014), Zhou (2014), Branigan (2010), Johnson (undated).
[86] Nehru (2012).
[87] Panikkar (2012).
[88] Gurung (2018), Alam (2018), Bhattacharjee (2017a), Srinivasan (2014), Raina (2014), Prakash (2011).
[89] Singh (2014).
[90] Singh (2014).
[91] Singh (2014).

6.3.3 Hambantota Histrionics

Physical proximity to MSR projects in Sri Lanka, specifically, a 99-year lease on the China-built Hambantota port in a debt-swap, triggered Indian alarm. Delhi was 'concerned as its neighbours, including Sri Lanka, join the BRI and welcome Chinese projects on their territories'.[94] Hambantota, a fishing harbour in rural south-western Sri Lankan constituency of then-President Mahinda Rajapaksa, was, at his behest, developed into a major port. Initial work was concluded in 2011, and the project completed in 2015, by which time Rajapaksa had lost office. Colombo's inability to repay loans taken for the port led on 29 July 2017 the Sri Lanka Ports Authority (SLPA) and the China Merchant Port Holdings (CMPH) to exchange $1.1bn of the $1.3bn loan for an 80% equity on a 99-year lease.[95] Prime Minister Ranil Wickremesinghe, Rajapaksa's political nemesis, announced, 'We are giving the country a better deal without debt'.[96] That view, resonating with Rajapaksa's, indicated elite-consensus on Sri Lanka's relations with China generally, and Colombo's MSR engagement in particular.

The Hambantota project sharply focused Indian and allied critique of the MSR's purported goal of securing 'control' over China's SLoCs via 'debt-trap imperialism'.[97] Few recalled that Sino-Sri Lankan ties began in the 1950s when both led the post-colonial non-aligned campaign. In 1970–71, when the Sinhala-nationalist 'Janatha Vimukthi Peramuna' (people's liberation army) majoritarian campaign ignited a civil war, China was among the few countries offering *materiel* support. Later, Colombo's operations against 'Liberation Tigers of Tamil Eelam' (LTTE) rebels, especially its bloody 2009 victory over the insurgents, triggered Western sanctions. Colombo then turned to Beijing for reconstruction-aid.

China funded several infrastructure projects, including modernising the Colombo Port. Sri Lanka included Hambantota port in its 2002 national development plans, assigning French consultants to conduct a feasibility study in 2003. The study was eventually rejected because at the time, Colombo handled 95% of Sri Lanka's international trade, and a nearby alternative could erode the capital's maritime attraction. Calculations changed in 2004. When the Boxing-Day Tsunami killed over

---

[92] Taneja (2016).
[93] Miglani (2017).
[94] Patranobis (2018).
[95] Stacey (2017).
[96] Wickremesinghe (2017).
[97] Marlow (2018), Hillman (2018), Stacey (2017), Dutta (2017).

31,000 people, displaced 440,000, destroyed 100,000 homes and damaged another 40,000, affecting 1m–2m of 19m Sri Lankans,[98] the Central Bank and the ADB estimated asset losses at $1bn, and output losses, at $300m.[99] Hambantota was 'badly affected'.[100]

After being elected president in November 2005, Rajapaksa focused on reconstruction in his constituency, which included Hambantota. Anxious to build a deep-sea port and the country's 2nd city in his hometown, physically closer to major Indian Ocean shipping lanes than Colombo, Rajapaksa ordered another feasibility study. The Danish consultancy appointed to conduct it reported in 2006 that with the economy growing at over 6%, overflow from Colombo, Galle and Trincomalee ports would find Hambantota attractive. Dry and break-bulk cargo would be the main traffic until 2030, when containers would take over. Consultants surmised, by 2040, Hambantota could handle almost 20m TEUs, making it one of the world's busiest harbours.[101] Rajapaksa endorsed SLPA's plans to build a port, an international airport, a conference centre, a cricket stadium, multi-lane highways, and production-warehousing-and-distribution complexes in and around Hambantota.

Other investors proved indifferent to Rajapaksa's vision while Chinese banks and SOEs were already engaged in Sri Lankan projects. A $307m Chinese loan at 6.3% (LIBOR+) interest financed Hambantota's first phase. The project area of 6070 ha was expected to attract total investments of $1.4bn.[102] CHEC began building a roll-on-roll-off terminal, two oil terminals, a multipurpose terminal and a container terminal, all with a draught of 17 m, and protected by two 312m and 988m-long breakwaters. A 210m-wide access channel provided a 600m-diameter turning circle, enabling the mooring of vessels of up to 100,000 DWT.

On completion of the $808m second phase, begun in November 2012, Hambantota boasted a main container berth (838.5m/17m), a feeder container berth (460m/12m), a multipurpose berth (838.5m/17m), and a transition berth (208m), with some facilities built on 42.6 ha of reclaimed land. To support bunkering and aircraft refuelling operations, the project included a tank farm built 1.2 km inland. The farm's 14 tanks, connected to the harbour with pipelines, stored 80,000 m^3 of bunkering fuel, aviation fuel and LPG.[103] The project's 800-ha industrial zone sought investment in petroleum-based industry, plants for processing food, fertiliser, aluminium and glass, and timber, and refining sugar and waste-fuel, and assembly and fabrication, logistics and warehousing, and power-generation.[104] The lease allowed CMPH to operate Hambantota Port for 99 years, reimburse SLPA's creditors using port-revenue, and secure its share of profits.

---

[98] Independent Evaluation Department (2014), Mulligan and Shaw (2007).
[99] Mulligan and Shaw (2007, p. 70).
[100] Mulligan and Shaw (2007, p. 66).
[101] Hillman (2018).
[102] Ibid., Stacey (2017), Report (undated).
[103] SLPA (2018).
[104] SLPA (2018).

6.3 Clashing Perspectives, Purposes and Narratives

Since 2014, after a PLAN submarine docked in Colombo, Delhi expressed concerns that control over Hambantota could allow Chinese naval operations unacceptably close to Indian waters.[105] Colombo assured Delhi of Sri Lankan authority over the Hambantota port, and that 'the Agreement clearly says no military ships will be allowed in the port'. To exercise sovereign control, Colombo moved its Galle naval base, located 125 km away, to Hambantota.[106] CMPH, with an 85% stake in the Hambantota International Port Group (HIPG), the operator, played no role in port security, which was managed by HIPG Service Co, with Colombo holding 50.7% share. CMPH also agreed to reduce its stake in HIPG to 65% after 10 years.

Disturbed by the PLAN submarine docking in Colombo in 2014, India sought to prevent repetitions. A retired Indian General articulated fears that Beijing was untrustworthy, that instead of berthing at the Colombo naval base, the sub moored at the Colombo South Container Terminal (CSCT), a PRC-financed deep-water anchorage, effectively a 'Chinese enclave', in 'clear violation of the China-Sri Lanka protocol'. Also, having insisted Gwadar was built as a commercial port, Beijing stationed a marine unit and PLAN vessels alongside Pakistan Navy assets there. He accused China of building a second naval base at Jiwani, 35 km east of Iran's Indian-built Chabahar port. Delhi was outraged that having stymied its attempt to build a base, Seychelles invited Beijing to construct a 'strategic supply base'. With Djibouti operational, India saw Hambantota as evidence of PLAN's growing IOR footprint.[107]

Delhi signed 'logistical' agreements with Oman and France, also boosting 'maritime ties with Mauritius, Seychelles, Mozambique and Madagascar'.[108] Colombo informed Delhi that citing Indian objections, another PLAN sub had been refused permission to dock in 2017.[109] Still, to pre-empt Beijing's leverage in pressing Colombo into amending the 'no military activities' clause, Delhi took over Hambantota's $210m Chinese-built Mattala Rajapaksa Airport. Denying reports that the $300m, 40-year lease was to allay Indian fears of potential risks posed by China's control over Hambantota port, Sri Lanka insisted that with little interest from domestic or overseas carriers to service the airport, and with operating losses mounting, Colombo sought bids from foreign operators in 2016; only Delhi applied. But many, including the Sri Lankan opposition, perceived a linkage.[110]

After President Sirisena Maithripala accused Indian intelligence of plotting his assassination, Indian leaders lionised Prime Minister Ranil Wickremesinghe during an October 2018 visit to Delhi.[111] On his return, Wickremesinghe was dismissed by Sirisena who appointed Rajapakse the new premier. When Wickremesinghe rejected his dismissal, Sirisena prorogued the legislature, plunging Sri Lanka into a political

---

[105] Parashar (2014).
[106] Reuters (2017), Aneez and Sirilal (2017a).
[107] Katoch (2018a), PTI (2018a), Singh (2018b), Pant (2018), Sen (2017), Dasgupta (2015).
[108] Singh (2018b).
[109] Aneez and Sirilal (2017b).
[110] Brewster (2018a), Chan (2017).
[111] Srinivasan (2018a).

crisis, before being forced by the judiciary to reinstate Wickremesinghe.[112] Colombo-watchers blamed the turbulence on Sino-Indian rivalry.[113] Wickremesinghe then negotiated a $989m loan from China's Exim Bank to fund 85% of a $1.16bn highway linking the tea-growing Kandy highlands to Hambantota port, before polls returned the Rajapaksas to power.[114]

6.3.4 Maldives Malarkey

A low-lying archipelago of 26 atolls comprising 1192 islands and reefs—188 inhabited—with an area of 298 km^2 spread between the Laccadive Sea to the east and the Arabian Sea to the west, 300 miles southwest from peninsular India, Maldives, with 400,000 inhabitants,[115] occupies a strategic location athwart Indian Ocean shipping lanes. That, and upscale tourism, domestic strife and external competition for influence, earned Maldives international attention. The transfer of power from one elected president to another in September 2018 after a year of repressed dissent, viewed as mirroring Sino-Indian proxy-rivalry, betrayed a volatile mix of local politics, economic drivers drawing elites to China-financed MSR projects, and geopolitical backwash from the US–China–India *strategic triangle*.[116]

Having repeatedly 'expressed concern' over the freshly ousted President Abdulla Yameen's policies, India was quick to congratulate the president-elect, Ibrahim Solih, assuring him, 'In keeping with our "Neighbourhood First" Policy, India looks forward to working closely with the Maldives in further deepening our partnership'.[117] India's official news-agency summarised results of the unexpectedly peaceful polls: 'Pro-China–Maldives president Yameen loses election, India welcomes result'.[118] The mood in Delhi and Beijing could not be more divergent. Facing yet another electoral surprise after those in Sri Lanka, Malaysia and Pakistan, China evinced sober reflection on the complexity of implementing the MSR blueprint.[119]

South Asia's pre-eminent actor, India, helped maintain stability in Maldives during Maumoon Abdul Gayoom's three-decade long rule. In November 1988, when a Sri Lankan separatist group, the People's Liberation Organisation of Tamil Eelam (PLOTE), engaged by a Maldivian faction under Abdullah Luthufi, landed 60-plus mercenaries to capture Maldives' defenceless capital, Male, Gayoom requested help. Delhi deployed a 1600-strong airborne force, rescued and restored Gayoom to office, defeated the plotters, and handed surviving fighters to Gayoom's government. Three

---

[112] Report (2018d, e, f).
[113] AFP (2018a), Shams (2018a), Miglani and Aneez (2018), Gopalaswamy (2018).
[114] AP (2019b).
[115] DCI (2018), UNDP (2018), Subramanian (2018).
[116] Shams (2018b), Report (2018g).
[117] MEA (2018a).
[118] PTI (2014).
[119] Reuters (2018a).

6.3 Clashing Perspectives, Purposes and Narratives   253

decades later, following a fresh crisis in Male, Delhi again placed its forces on standby for 'deployment at short notice' in case of an 'eventuality'.[120]

At the end of Gayoom's presidency, Maldives held its first free elections in October 2008 when Mohamed Nasheed, a democracy activist and victim of imprisonment and torture, won the presidency. He sought to improve and expand the tourism industry. The largest projects aimed at modernising international air-travel. With WBG support, Male launched a bid to upgrade and operate the country's main airport located on a neighbouring island, build a new terminal, and link Male to it. An Indian-Malaysian joint venture, winning the $511m contract in 2010, began work. Nasheed's opponents filed legal action protesting a new $25 surcharge levied without legal authority on every outbound passenger. In February 2012, after Nasheed ordered the chief justice be detained for allegedly obstructing justice by hindering trial of officials from the previous regime accused of corruption and abuse of power, mutinous policemen forced him to resign; Vice President Mohamed Waheed Hassan Manik took over.

Disturbed by developments, India, the United States and Britain urged an independent inquiry, which later reported that the president's ouster was 'legal and constitutional'.[121] Most powers accepted this conclusion, but Maldives' relations with India froze. In November 2012, the president, having failed to renegotiate the airport-contract, cancelled it. The consortium's Indian leader, GMR Corporation, was given seven days to leave. Delhi announced it would 'expect that Maldives would fulfil all legal processes and requirements in accordance with the relevant contracts and agreement it has concluded' with GMR.[122] Indian observers saw Beijing's influence in Male working against Delhi's interests, allegations Waheed rebutted.[123] In June 2014, a Singapore-based arbitral tribunal ruled against Male, and in October 2016, awarded the contractors compensation worth $270m.[124]

Although Maldives' leadership had in the meanwhile changed hands, its relations with India stayed cool. In the first round of September 2013 polls, Nasheed won 45% of the votes, with his closest rivals securing 25% and 24% respectively, but after a contestant alleged irregularities, the Supreme Court annulled the results.[125] In November, UNDP-assisted voting produced two front-runners, Nasheed, and Abdulla Yameen Abdul Gayoom, the long-running former-ruler's half-brother. In mid-November, with voter turnout at more than 91%, Yameen won the run-off vote 'by a tight margin', and was sworn in the following day,[126] but strains persisted. In 2015, convicted of having ordered as president the arrest of a sitting judge, Nasheed was jailed for 13 years. Released in 2016 on medical parole, he received asylum in Britain and led

---

[120] Report (2018h).

[121] Tisdall (2012).

[122] Robinson (2012).

[123] PTI (2012), Radhakrishnan (2012).

[124] Usmani A (2016).

[125] Report (2013a).

[126] Report (2013b).

a campaign from exile questioning Yameen's legitimacy.[127] Tensions rose as the end of Yameen's five-year term approached in 2018.

A crisis ensued when the Supreme Court annulled Nasheed's trial, ordering the release and reinstatement of imprisoned legislators even if they changed parties. Facing judicial challenges, Yameen declared a state of emergency and arrested two judges, and former president Gayoom, now a critic. The Court's remaining judges then overturned the original ruling, legitimising Yameen's actions.[128] With a crackdown on Nasheed's supporters in progress, Prime Minister Modi and President Trump 'expressed concern', while the Delhi cognoscenti, endorsing Nasheed's plea, urged Indian intervention.[129] In Beijing, MoFA Spokesman Geng Shuang noted: 'The international community should play a constructive role on the basis of respecting the Maldives' sovereignty instead of taking measures that could complicate the current situation'.[130] With a flotilla of 11 PLAN vessels sailing around Eastern Indian Ocean, China urged Maldivians to resolve their domestic disputes without external interference.[131]

Delhi, calling for the withdrawal of emergency regulations, restoration of democracy, release of the detained judges and Mr Gayoom, desisted from physical intervention. Indian and Chinese media highlighted contradictory perspectives. The impression of Beijing deterring Delhi from intervening in Maldives flowed from Sino-Maldivian convergence on promoting Male as a key waypoint along MSR's alignment. Maldives had joined BRI early on and, during a 2014 state visit to China, Abdulla Yameen evinced 'great interest in China's initiative to build the twenty-first century MSR'.[132] On a return visit to Male the following month, Xi Jinping pledged significant investment in Maldives' desired upgrading of and diversification from fishery and tourism, building infrastructure enhancing intra-archipelagic connectivity, and responding to national anxiety to develop defences from rising sea-levels.

Xi assured, China respected 'the independence, sovereignty, and territorial integrity of Maldives, and respects the political systems and development paths chosen by the Maldivian people'.[133] The two sides signed MoUs on diplomatic cooperation, boosting trade and investment, building infrastructure and urban housing, and promoting science and technology. The leaders 'agreed to engage ourselves in the building of twenty-first century MSR'. Xi offered help in constructing the 'China–Maldives Friendship Bridge' linking Male to the main airport located on a neighbouring island.[134] Two months later, the two parties authorised Chinese consultants to conduct a pre-feasibility study of the proposed bridge. This followed a

---

[127] Bengali (2018).
[128] AFP (AFP 2018b).
[129] Web Desk (2018), Srinivasan (2018b), Junayed and Aneez (2018).
[130] Geng (2018).
[131] Blanchard (2018), Editorial (2018a).
[132] Xinhua (2014).
[133] Xi (2014).
[134] Xi (2014), Yameen (2014).

6.3 Clashing Perspectives, Purposes and Narratives

Chinese SOE, Beijing Urban Construction Group, being given the $511m contract to expand the Male Airport, two years after Maldives had cancelled the work order issued to India's GMR.[135]

In December 2016, Male leased Feydhoo Finolhu, an uninhabited atoll near the capital and the airport, to a Chinese hotel-development firm. The $4m, 50-year lease deeply troubled Delhi.[136] Key projects were formalised during Yameen's December 2017 state visit to China. He reiterated 'support for and participation in the twenty-first century MSR proposed by China'. A bilateral FTA signed during the visit promised to 'energise' trade as both leaders pledged to encourage Chinese investment in Maldives. The partners vowed to 'deepen pragmatic cooperation' under the BRI/MSR rubric, and 'enhance connectivity'. Agreements on economic and technological cooperation, human resource development, seawater desalination, solid waste treatment and recycling, finance, health and meteorology, too, were signed.[137]

Xi underscored Beijing's commitment to the completion of the 'Friendship Bridge' and the 'Expansion and Upgrading of Velana International Airport', 'the two most flagship projects' for Yameen's government.[138] Officials would finalise Chinese aid to these and other projects helping communications, housing and urbanisation, tourism, women, youth and sports.[139] Growing warmth towards China paralleled a cooling of relations with India. In June 2018, Male reduced the number of visas and work permits issued to Indian workers, mostly from the state of Kerala, whose chief minister complained to Delhi. India deported an MP from Yameen's party, visiting Chennai for treatment, from Chennai airport, raising tensions. Male asked Delhi to take back within the month two Indian Navy helicopters and their 50-man crew Delhi had stationed in Maldives to support disaster management efforts.[140]

Demonstrating leverage, Delhi slashed the export of essential foodstuffs, e.g. rice, flour, potatoes, pulses, eggs, onions and sugar, to Maldives. Indian media, accusing Male of boosting ties to China, Pakistan and Saudi Arabia at India's cost, described Maldives as a 'British colony becoming a Chinese satellite'.[141] Indian analysts viewed Delhi-Male-Beijing dynamics in zero-sum terms.[142] Six weeks before presidential elections, Yameen inaugurated the 'China–Maldives Friendship Bridge', described by Chinese media as a BRI/MSR 'hallmark project'. China provided 91.8% of the $184.44m spent building the bridge in grants and concessional credit; Male funded the remaining 8.2%. Ground was broken in March 2016, and 2nd Harbour Engineering Company, a CCCC subsidiary, completed work in July 2018.[143]

---

[135] TBP (2014).
[136] Chaudhury (2016).
[137] MoFA (2017).
[138] MoFA (2017).
[139] MoFA (2017).
[140] Miglani (2018).
[141] Dutta (2018).
[142] Reuters (2018b), Report (2018i).
[143] Wang (2018a, b).

This first major MSR project seemingly cemented Maldives' BRI role whatever the local political dynamics. Presidential elections in September 2018 did, however, indicate popular discontent with policy-priorities pursued by Yameen, who sought re-election to a second five-year term. His defeat at the hand of opposition leader Ibrahim Solih, who won 58.3% of the votes, was seen as a victory for India and a defeat for China.[144] Having expressed scepticism over how free or fair the polls would be, Delhi enthused, 'This election marks not only the triumph of democratic forces in the Maldives, but also reflects the firm commitment to the values of democracy and the rule of law'. In contrast, Chinese reports, quoting Maldives' election commission, simply provided factual details, including 89.22% participation, vote distribution and the candidates' respective shares.[145]

Indian Prime Minister Narendra Modi, warmly felicitating Solih on his victory, symbolically visited Male to attend Solih's swearing-in ceremony.[146] Solih returned the gesture in December. Delhi extended a $1.4bn aid package, notably including $800m in infrastructure-building line-of-credit. Indian aid to Male rose from $18.1m in 2018–19 to $83.3m in 2019–20. A new visa facilitation programme eased people-to-people contacts while the second Defence Cooperation Dialogue revived India's oceanic security arrangements eroded under Yameen. Maldive's $3.4bn debt to China and Delhi's fear of the PLAN's future use of Maldive's MSR projects drove India's reinforced engagement.[147] Five months after his electoral defeat, Yameen was arrested on charges of money-laundering related to the Feydhoo Finolhu lease. A month on, the High Court, citing a lack of evidence, freed him, but his party's defeat in parliamentary polls ensured his marginalisation.[148]

Having seen electoral changes oust 'China-friendly' elites in Sri Lanka, Malaysia and Pakistan, China confronted yet another upset potentially disruptive of MSR's implementation. Questioning the zero-sum narrative advanced by detractors, Beijing proposed that Delhi aid Maldives' developmental drive alongside China, cooperating rather than competing with each other.[149] Prospects looked mixed. While a few planned MSR projects faced post-election uncertainty, most new governments adjusted policy to accommodate long-standing economic and diplomatic cooperation in the 'new era'. Maldives proved more ambivalent. Nasheed, now speaker of the Maldivian parliament, accused China of exorbitant and unfair economic practices: 'They over-invoiced us and charged us for that and now we have to repay the interest rate and the principal amount'. He complained that from 2020, Male would have to devote 15% of its annual budget to repay Chinese debt.[150] Foreign Minister Abdulla Shahid more diplomatically noted, 'The heightened cooperation between our two countries has indeed been mutually beneficial to our peoples. China will remain

---

[144] Riedel (2018), Withnall (2018), PTI (2018b).
[145] Report (2018j).
[146] AFP (2018c).
[147] Report (2019b), MoD (2019), MEA (2018b).
[148] Junayd (2019).
[149] Chaudhury (2018a).
[150] IANS (2019).

as an important economic and bilateral development partner for the Maldives'.[151] Ambivalence typified much MSR-building efforts.

Volatility marked Sino-Indian competition for insular support across the IOR. Disturbed by Beijing's apparent success in securing access to various ports, Delhi pursued counter-arrangements with neighbours. In 2015, Prime Minister Modi signed 'confidential' agreements with Seychelles and Mauritius permitting Delhi to build naval, air and electronic-operational bases on both archipelagos. The parties agreed that Delhi would construct major facilities on the Seychelles' Assumption Island and Agalega in Mauritius. When India's Assumption Island plans for basing military personnel, conducting hydrographic surveys, naval aviation, communications and surveillance operations, training Seychelles' defence forces, and funding the entire project, were leaked, Seychelles' opposition-coalition refused to ratify the accord.

In 2018, India and Seychelles signed a modified agreement but apart from the coastal surveillance radar Modi inaugurated in 2015, little new Indian *materiel* was evident.[152] Beijing had demurred after Seychelles offered base-facilities to visiting Defence Minister Liang Guanglie in 2011, only considering 'seeking supply facilities at appropriate harbours in the Seychelles and other countries'.[153] Angered by China's alleged plans to build 'strategic support' facilities in the IOR, analysts urged Delhi to act.[154] After Seychelles purportedly turned to China, India drew Mauritius closer, expanding access to its Agalega Islands south of Seychelles and 600 miles north of Mauritius. India's Afcons won an $87m, Delhi-financed, contract to double its runway-length to 3000m, to handle large ASW aircraft, build a jetty for IN vessels, and install an extensive Identification Friend or Foe transponder system.[155]

6.3.5 Maritime Silk Road's Malaysian Misadventure

In May 2018, a disparate coalition of political factions, forged into an opposition front by the 93-year old former prime minister, Mahathir Mohamad, and his once-estranged, imprisoned former deputy, Anwar Ibrahim, unseated the long-ruling *Barisan Nasional* coalition led by Prime Minister Najib Razak in Malaysia's fourteenth general elections (GE-14), and took office. Accusing Najib of corruption, Mahathir cancelled or suspended several large-scale infrastructure projects. Three of these were part of Malaysia's Najib-negotiated BRI/MSR-funded plans: the $13bn East Coast Rail Link (ECRL) connecting Tumpat near the Thai border and a deepened Kuantan Port on the east coast, to Port Klang on the west, and the 2bn Trans-Sabah Gas-, and Multi-Product pipelines in Sabah.[156] That Malaysia paid 88% of the

---

[151] Zhu and Tang (2019).
[152] Katoch (2018b).
[153] Office of the Spokesman (2011).
[154] Katoch (2018c), Singh (2015), Chankaiyee2 (2014).
[155] Pilling (2018).
[156] Nambiar (2019).

pipelines' costs to Chinese contractor when only 13% of the work was done proved troubling.[157]

Najib had projected that by 2030, carrying 5.4m passengers and 53m tons of cargo annually, ECRL would present an effective, if pricey, alternative to Singapore.[158] Mahathir criticised other Chinese-financed undertakings, e.g. Forest City real estate development in Johor and the multi-aspect Melaka Gateway projects, but allowed these private-invested enterprises to continue.[159] His campaign rhetoric had painted China as protecting Najib's corrupt practices,[160] but in office, especially during a visit to China, he persuaded Beijing to stop the three projects without damaging ties. In fact, Mahathir burnished Malaysia's BRI/MSR credentials.[161] Still, as an occasionally acerbic nationalist, he appeared to question Najib's MSR-financed blueprint. Chinese SOEs, investment funds and private firms partnered with Malaysian counterparts in myriad joint ventures, but only the largest infrastructure/construction projects generated some controversy. Table 6.2 lists the ones that attracted most critical public attention[162]:

Initial unease notwithstanding, Mahathir's cabinet colleagues reassured Chinese investors of their interest in continuing collaboration. The Malaysia–China Kuantan

Table 6.2 Major Chinese projects in Malaysia

Project	Principal developer	Cost RM	Cost US$
Forest City real estate, Johor	Country Garden	105bn	25.30bn
Forest City allied services, Johor	Country Garden	26bn	6.26bn
Melaka Gateway, Melaka	Power China International	15bn	3.60bn
Trans-Sabah, Multi-product pipelines	China Petroleum Pipelines	10.4bn	2.49bn
50 MWe solar plant, Kedah	China General Nuclear Power	10bn	2.40bn
Kuantan Port and MCKIP	Guangxi Beibu Gulf Intl. Ports	8bn	1.92bn
Bakun Dam, Sarawak	Power China International	7.5bn	1.80bn
East Coast Rail Link, Phase-I	CCCC	5.5bn	1.32bn
Penang-mainland 2nd Bridge	CHEC	4.5bn	1.08bn
Penang undersea tunnel	CRCC	3.7bn	891.56m
TRX Signature Tower, KL	China State Constructn. Engng.	3.5bn	843.37m
Four Seasons Hotel, KL	CRCC	2.5bn	602.40m
Xiamen University campus, Selangor	Sino Hydro	1.3bn	313.25m

---

[157] Saieed (2018), Lee et al. (2018).
[158] Teoh (2017).
[159] Report (2018k).
[160] Wey (2018).
[161] Xinhua (2018c, d).
[162] Shankar (2019), Todd and Slattery (2018), Huaxia (2018), median $-RM exchange rate as in Oct 2018.

Industrial Park (MCKIP) offered an instance. Mahathir urged local partner Alliance Steel to tear down the high fencing around its site, raising anxiety levels. But his Deputy International Trade and Investment Minister, Ong Kian Ming, during a visit to MCKIP, assured Chinese investors that Malaysia welcomed FDI that generated benefits. By then, MCKIP-linked approved, committed and potential FDI totalled $10.25bn (RM42.56bn).[163] MCKIP's Malaysian co-owners, encouraged by the USA–China trade war, were already negotiating further investment with other Chinese firms. With calm restored, private Chinese investment in Malaysia picked up. Weeks after Mahathir's first Beijing visit, China's Pacific Construction Group announced plans to invest $2.3bn (RM 10bn) in Malaysian infrastructure over the next decade. Other potential investors, too, saw Malaysia as a 'business-friendly and most competitive' FDI destination.[164]

After Mahathir's anointed successor, Anwar Ibrahim, visited Beijing in October 2018, mutual trust appeared restored.[165] Days before Mahathir arrived in Beijing to attend the second BRF, agreement between his government and CCCC, ECRL's principal contractors, was announced. ECRL would be 40 km shorter, at 648 km; local participation in civil works rose from 30 to 40%; CCCC accepted some operating and maintenance risks. ECRL's first two phases now cost ringgit 44bn ($10.7bn) rather than ringgit 65.5bn.[166] Resumption of work would 'surely benefit Malaysia and lighten the burden on the country's financial position'.[167] Post-GE14 volatility over, the one uncertainty in KL-Beijing ties resided in Malaysia's reassertion of its SCS maritime-territorial claims and demand for sustained FoN. With Sino-US rivalry intensifying across the Western Pacific, this could challenge MSR's implementation.[168]

6.4 The Quad's 'Free and Open' Countermoves

The USA and its regional partners, discerning long-term 'threats' to the post-Soviet order sustained with US strategic pre-eminence, acted against purported MSR/BRI challenges. Quad-members relaunched their proto-alliance to neutralise China's competing vision, but they and ASEAN-states placed diverse emphases on security-versus-economic goals.[169] The EU's views, too, varied. Brussel's 'Global Strategy on Foreign and Security Policy' urged EU-members to strengthen ties to a 'connected Asia' with a 'coherent approach'.[170] As noted, 27 of 28 EU envoys in Beijing

---

[163] Ho (2018a).
[164] Ho (2018a).
[165] Information Office (2018b), Ho (2018b).
[166] Editorial (2019).
[167] Jaipragas (2019).
[168] Yong (2018).
[169] Hussain (2019), Tarapore (2018).
[170] EU External Action (2016).

presented a rare, candid, joint-critique of BRI. In early 2018, the European Commission (EC) began consultations on 'its own comprehensive vision' of a 'Euro-Asian connectivity strategy', addressing transport, energy, digital economy and people-to-people contacts 'in line with its values and interests'.[171] Intra-EU consultations on proposed EU–Asia connectivity stressed principles: 'respect for international standards and good practices', and governance from planning through lending to project-implementation. The unstated offer was a normative counterpoint to BRI.

6.4.1 Geoeconomics Reinforce Geopolitics

The EC stressed 'sustainable connectivity and state resilience' to spur trans-Eurasian development 'providing economic opportunities for European and Asian businesses alike' and creating 'incentives for countries to cooperate and maintain peaceful relations between them'.[172] The EU's initiative was openly geoeconomic. After lengthy consultations, in September 2018, Brussels launched 'Connecting Europe and Asia—Building blocks for an EU Strategy', to be approved by the Council and the European Parliament before the October 2018 Asia–Europe Meeting (ASEM).[173] Highlighting the 'well-developed Trans-European Network for Transport (TEN-T) framework', the initiative would connect it with BRI/SREB's cross-border, multimodal Asian networks, without naming the latter.[174]

EU plans covered sea-transport, notably, working with the International Maritime Organisation (IMO), promoting non-hydrocarbon fuel-use in European and Asian ports, simplifying customs procedures, digitalising documentation, and facilitating maritime traffic.[175] Brussels would invest $70bn in guaranteeing infrastructure projects, and using the resulting confidence, attract investments worth $330bn in 2021–2027 to implement its blueprint while reinforcing regulatory standards along the SREB. High Commissioner Federica Mogherini denied any links between this vision and China, BRI, SREB or MSR, but her emphasis on 'values' and 'principles' indicated a normatively framed geoeconomic response to BRI.[176]

Donald Tusk, EU Council President, hosted 51 Asian and European leaders and the ASEAN Secretary-General, at the ASEM-12 summit in Brussels. The EC-drafted Chair's Statement vowed consensus on ASEM being '**a building-block for effective multilateralism and the rules-based international order** anchored in international law and with the UN at its core'. The leaders 'underlined shared interest' in boosting 'Europe–Asia sustainable connectivity' upholding 'market principles and agreed

---

[171] EEAS-ASIAPAC (2018).
[172] EEAS-ASIAPAC (2018).
[173] Mogherini (2018).
[174] Mogherini (2018, pp. 2–3).
[175] Mogherini (2018, p. 4).
[176] Emmott (2018), AFP (2018d).

international rules, norms and standards'. They noted their 'commitment to maintain peace and stability and to ensure **maritime security and safety**, freedom of navigation and overflight and to combat piracy in full compliance with international law'. They stressed 'the critical importance of peaceful settlement of disputes in accordance with international law, in particular the UN Charter and the UNCLOS'.[177]

The Chair's Statement, repeating phraseology from the more general Western critique of China's handling of its maritime-territorial disputes and its implementation of MSR/SREB projects, indicated a consensus critical of Chinese conduct and Beijing's partial acquiescence, manifest in Premier Li Keqiang's presence. Li pointed to a 'complex, fast-changing world with many uncertainties' confronting ASEM with 'more worries and doubts about the future'. He reiterated Xi's 'vision of building a community with a shared future for mankind' as a collective endeavour requiring states to 'strengthen dialogue and cooperation, respect the right to choose one's own path of development, seek political solutions to disputes through negotiation and consultation, and jointly safeguard overall peace and stability'.[178]

Urging ASEM members to 'fulfil our important responsibilities for global peace and prosperity...embrace openness and inclusiveness, and coordinate our actions to promote open and interconnected progress', Li sought action against 'the serious impact of unilateralism' with multilateralism firming up 'an open world economy'. He described BRI and its components as critical to 'interconnected development' and people-to-people and cultural ties cooperatively binding societies in improving the state of public health, education, ageing, tourism, disability, and women- and youth affairs. His stress on being 'steadfast in upholding the rules-based international order, the authority of the UN, and the purposes and principles of the UN Charter'[179] resonated with but also questioned Brussels's perspective. Mirroring US interests, the EU forged security-partnerships with Japan, South Korea and Indonesia, supported Korean denuclearisation, Afghan state-building, and FoN in maritime Asia. It endorsed 'respect for international law, including the Law of the Sea and its arbitration procedures', and the 'peaceful settlement of maritime disputes'.[180] This was aimed at China.

The thrust of Brussels' action was economic diplomacy. In contrast, US efforts targeted the MSR's alleged *primacy*-eroding geopolitical endeavour. Introducing Washington's 'economic vision' for the new 'Indo-Pacific' regional construct to US and foreign policy-practitioners, business executives and strategic analysts in July 2018, Mike Pompeo noted that the region was 'of great importance to American foreign policy'. 'Free and open Indo-Pacific' meant 'good governance and the assurance that citizens can enjoy their fundamental rights and liberties', with all states enjoying 'open access to seas and airways' and 'peaceful resolution of territorial and maritime disputes'. He underscored 'fair and reciprocal trade, open investment

---

[177] Tusk (2018).
[178] Li (2018).
[179] Li (2018).
[180] EU External Action (2016, p. 38).

environments, transparent agreements' and 'improved connectivity to drive regional ties'.[181]

Pompeo outlined plans for infrastructure-building, driving growth, and spreading prosperity. The US OPIC, USAID and Millennium Challenge Corporation, tacitly counteracting MSR, invested $113m in seed-money to build connectivity, create jobs and improve lives. OPIC coordinated with Quad-counterparts on offering alternative projects to BRI-partners. After talks between OPIC, Japanese and Australian counterparts, the three allies initiated joint-investment in trans-regional infrastructure and connectivity. The enterprise, designed to attract private capital, 'would build infrastructure, address key development challenges, increase connectivity and promote economic growth'.[182] Officials joined the US Chamber of Commerce in Washington in July 2018, briefing potential investors on 'economic opportunities in the Indo-Pacific region'.

At this gathering, OPIC's CEO, Ray Washburne, directly slammed MSR projects. Washburne insisted China was 'not in it to help countries out, they're in it to grab their assets', that Beijing used MSR, 'purposefully plunging recipient countries into debt' and then going after 'their rare earths and minerals and things like that as collateral for their loans'.[183] OPIC worked with Quad-allies to build a quadrilateral 'development partnership countering China's belt and road plan'. Washburne's agreement with India looked 'very much like the ones we have with Japan and Australia'.[184] The enterprise was eased by the recently forged Indo-Japanese–Australian geoeconomic coalition against BRI.[185]

An early counter-MSR exercise targeted Papua New Guinea (PNG), where Chinese firms would build communications networks leapfrogging PNG to fifth-generation (5G) ICT status. PNG received $5bn of the $6bn in project-funding Beijing pledged to South Pacific island-states. In 2013, PNG Telikom, with China's Exim Bank funding, contracted China's Huawei to design and install a high-speed broadband network linking Port Moresby, the capital, to four other cities via a submarine cable. Having stopped Huawei from building a similar cable linking Sydney, the Solomon Islands and PNG by offering a $98m Australian alternative, and banned Huawei from providing 5G services to Australia on national security grounds, Canberra joined Washington and Tokyo to present Port Moresby with a non-Chinese 5G system.[186] Boosting counter-MSR action, the USA enacted the 2018 BUILD Act, creating the International Development Finance Corp (USIDFC), subsuming OPIC and other agencies, mounting a direct challenge to MSR with its $60bn budget.[187]

---

[181] Pompeo (2018).

[182] Media office (2018).

[183] Washburne (2018a).

[184] Washburne (2018b).

[185] Chaudhury (2018b), Wardell and Packham (2018), Chaudhury (2017), Panda (2017).

[186] Smyth and White (2018).

[187] Guensburg and Widakuswara (2018), Washburne (2018c), Moss and Collinson (2018).

6.4 The Quad's 'Free and Open' Countermoves

Long before these joint responses to BRI/MSR, however, Japan had already been active.[188] Tokyo's low-profile investments challenged MSR projects with emphases on transparency, sustainability and rules-based norms. In 2015, consolidating and expanding existing programmes, Japan launched its trans-Asian 'Partnership for Quality Infrastructure' initiative, pledging to spend $110bn in five years, channelled mainly via the Japan International Cooperation Agency (JICA) and the Japan Bank for International Cooperation (JBIC).[189] Citing projects, e.g. Delhi Metro in India, 'Sun Bridge' Ulan Bator railway flyover in Mongolia, and the 'Vietnam–Japan Friendship Bridge' at Nhat Tan, as examples, Tokyo highlighted 'four pillars' of 'quality infrastructure':

- Expansion and acceleration of assistance via 'full mobilisation' of resources and tools
- Collaboration between JICA, JBIC and the Japanese-led ADB
- Boosting JBIC capacity to double funding for 'high risk' projects and
- Promoting 'Quality Infrastructure Investment' as an international standard.

A year later, Japan incorporated most of the MSR's South Pacific-to-Africa catchment area with funding raised to $200bn. In Africa, with the JICA-aided Infrastructure Consortium for Africa (ICA) leading, Tokyo stressed building 'transport corridors to maximise inclusiveness of development', enhancing power-generation using 'small-scale renewables', e.g. solar panels, mini-grids, geo-thermal, wind and 'run-of-the-river hydro', improving 'quality of internet connectivity' where low-bandwidth, erratic provision and high-costs prevented optimal use and expanding access to improved water supply and better sanitation.[190] Later, expanded to cover the Caribbean, and Central and South America, the initiative was rolled into Tokyo's 'Free and Open Indo-Pacific' geoeconomic strategy, resonating with the US geopolitical formulation, designed to counteract MSR.[191]

Nonetheless, during Abe's October 2018 visit to Beijing, the neighbours signed 500 deals worth $18bn, pledged to boost economic ties, and agreed to explore 'third-party cooperation' including in building infrastructure and connectivity. Abe's hope that the neighbours would usher in 'a new era when competition is transformed into coordination' suggested changing *systemic* dynamics encouraging Sino-Japanese cooperation, including vis-à-vis BRI/MSR.[192] Positive trends strengthened during Foreign Minister Taro Kono's April 2019 visit to Beijing, in preparation for Xi's June visit to Japan for the G20 Osaka summit. However, complex balancing of China–Japan competitive-cooperative imperatives vis-à-vis the USA persisted.[193]

---

[188] Usui (2005).
[189] MoFA (2015).
[190] JICA, ICA (2016).
[191] Maslow (2018), Akimoto (2018), Berkofsky (2018), International Cooperation Bureau (2017).
[192] Xinhua (2018e), An (2018), Editorial (2018b), PTI (2018c).
[193] Liang and Bianji (2019), Nakamaru (2019).

6.4.2 A Quadrilateral Maritime Focus

The Quad's one member directly located at the heart of the Indian Ocean Region (IOR)—India's response to BRI/MSR reflected Delhi's discomfiture with Beijing's regional activities. Notwithstanding China's profession of pacific beneficence, profound *Sinophobia* drove Indian policy.[194] After MSR's launch, India boosted or initiated naval cooperation with Singapore, the USA, France, Seychelles, Maldives, Mauritius, Mozambique and Oman. In 2017, the Indian Navy began regular patrols from the Malacca Strait to Mozambique's coasts. A US vessel refuelling an Indian frigate in the Sea of Japan during trilateral drills highlighted growing cooperation resulting from China-rooted anxiety.[195] Japan's post-2016 large-scale trans-IOR port-building and infrastructure projects, some managed jointly with India, as listed in Table 6.3, indicated similar China-focused concerns and reactive counter-action[196]:

How effective Quad-members' offers were to MSR-beneficiaries in weaning them off was unclear, as objects of Chinese and Japanese affection showed equal interest in the assistance extended by both. In a major counter-MSR coup, Quad-members persuaded BRI-partner Bangladesh to transfer its south-eastern deep-sea port project from China to Japan, but Dhaka then distributed elements of its south-western deep-sea port project among Chinese, Indian, Japanese and other benefactors, exemplifying an acute geopolitical–geoeconomic nexus.

Table 6.3 Japan's major IOR infrastructure projects since 2016

Location/country	Project summary	Cost in US$
Matarbari, Bangladesh	Deep-sea port and coal-fired power stations	3.7bn
Mumbai, India	Trans-harbour transport link	2.2bn
Dawei, Myanmar	Port and special economic zone	800m
Yangon, Myanmar	Container terminal and ancillary facilities	200m
Toamasina, Madagascar	Port and related ancillary facilities	400m
Nacala, Mozambique	Port and related ancillary facilities	320m
Mombasa, Kenya	Port and related ancillary facilities	300m

---

[194] Kazmin (2019), Yang (2018), Jaishankar (2018).
[195] Pant (2018), Jaishankar (2018).
[196] Brewster (2018b).

6.4.3 Quad-China Competition in Bangladesh

Bengal featured in China's maritime history. On Zheng He's fourth voyage, in 1414, his envoys called on Bengal's *Sultan*, receiving a giraffe brought as a gift from the ruler of Malindi in Kenya. Considered an auspicious *qi-lin*, this was the first of many African animals Zheng He brought back to the Ming court.[197] Nearly 600 years on, in 2009, in an effort to become a regional maritime hub, Bangladesh granted India, Nepal and Bhutan access to its two sea ports, Chittagong and Mongla. Prime Minister Sheikh Hasina made a similar offer during a visit to Beijing in March 2010.

Then-Foreign Minister Dipu Moni explained Dhaka's hopes for expanding and deepening Bangladesh's main port: 'It will be a great achievement if China agrees to use our Chittagong port, which we want to develop into a regional commercial hub by building a deep sea port in the Bay of Bengal'.[198] Chittagong handled 30.5m tons of bulk cargo and 1.1m TEUs annually; Dhaka hoped to build a $8.7bn deep-sea port, raising its capacity to 100m tons and 3m TEUs by 2055.[199] Hasina, considering China central to this vision, broached several proposals, expanding the portfolio of major projects China was already funding and building in Bangladesh. Ending her trip in Kunming, where the BCIM alignment had germinated and was being nursed to life, she discussed the port project with Yunnan's provincial leaders.

In Beijing, Hasina reiterated Dhaka's strong support for the China-proposed Kunming–Chittagong highway: 'Bangladesh is ready to put in its best efforts to increase regional connectivity and fight against poverty and other regional problems', and 'We planned to develop the deep-sea port which would benefit all the countries in the region, including China'. Admitting resource constraints, she sought Beijing's help; Wen Jiabao agreed. Dipu Moni insisted that a deep-sea port in Chittagong would help trade not only for Bangladesh, south-western China, Myanmar, Nepal and Bhutan, but also north-eastern India: 'It is not true that if we have good relations with India, we cannot build up a relationship with China'.[200] Vice President Xi Jinping, visiting Dhaka a few months later, pledged to build the deep-sea port.[201]

Delhi feared the worst: 'We can look at it as China assisting Bangladesh where it needs help, but from a geo-strategic point of view, we are certainly seeing an expanding footprint that undermines India's salience in the South Asian region'.[202] Anxiety was deepened in 2011 when, while commissioning British-built-and-Chinese-armed corvettes, Hasina disclosed, 'Last year, during my visit to China, I requested the Chinese Government to present two frigates with helicopters to the Bangladesh Navy. The Government of China agreed… Meanwhile, two ultra-modern missile-armed large patrol craft are being built in China'.[203] The intimacy manifest in Bangladesh's

---

[197] Zielinski (2013).
[198] Moni (2010a).
[199] Report (2010).
[200] Moni (2010b), BSS (2010).
[201] UNB (2010).
[202] Sahgal (2010).
[203] Hasina (2011).

leader asking Chinese counterparts for naval gifts troubled Delhi but, given how Dhaka-Beijing ties had evolved, appeared 'normal' to Bangladeshis.

Hasina and her predecessors built on foundations laid by President Ziaur Rahman during his 1977 visit, abandoning an alignment with the Indo-Soviet axis. Ties were cemented in 1978 when Vice-Premier Li Xiannian and Foreign Minister Huang Hua visited Dhaka, offering $58.3m in aid and a five-year trade agreement.[204] By 1980, when Zia made his third trip to China, it had become the principal source of *materiel*.[205] President HM Ershad's repeated visits, Beijing's infrastructure-assistance, and military-provision, deepened ties. Relations improved further after Prime Minister Khaleda Zia's May 2004 visit to Beijing. In April 2005, Premier Wen Jiabao announced an 'all-round cooperative partnership' in Dhaka.[206] Later, reopening the Nathu-La trade route with India, China hoped Bangladesh would benefit by using its Burimari land-port.[207] In 2006, China expanded its Asia–Pacific Trade Agreement to include Bangladesh,[208] which began exporting ready-made garments to China, its prime import-source.

Bangladesh received counter-terrorism assistance, satellite-imagery receivers, agro-advisers and ordnance. As SAARC's rotating-Chair, Dhaka welcomed Beijing as an observer. China offered post-graduate scholarships, built a $400m digital telecom network requested in 2004 and enabled tripartite accord on Bangladesh–Myanmar–China road-links.[209] Security ties, too, evolved. In 2007–08, facing maritime disputes with Yangon, Dhaka sought Beijing's intercession, restoring calm.[210] Hasina's 2010 visit to China, following one to India, stressed defence cooperation and deep-sea infrastructure. In 2012, Bangladesh was awarded an 111,631 km^2 EEZ in Law-of-the Sea Tribunal's arbitration over 38-year old maritime disputes with Myanmar.[211] In 2014, the Permanent Court of Arbitration awarded Bangladesh 19,467 km^2 out of 25,602 km^2 maritime area disputed with India for decades, extending Bangladesh's EEZ to 118,813 km^2.[212] In 2013, Hasina announced procurement of new naval-vessels, including two submarines, from 'a friendly country', to defend an expanded EEZ.[213]

Delhi was outraged. A former Naval Chief blasted the decision: 'Given Bangladesh's economic situation and the fact that it is surrounded on three sides by India, the acquisition of submarines is not only illogical but actually an act of provocation as far as India is concerned. Obviously, this transfer is a step further in

---

[204] Halim and Kamal (1996), Mohammad (2008).
[205] Ali (2010).
[206] Luan (2006a).
[207] Zhu (2005).
[208] Luan (2006b).
[209] Luan (2006c).
[210] Ali (2010, pp. 266–267).
[211] Jesus and Gautier (2012), Panday (2012).
[212] Wolfrum et al. (2014), Paul (2014), Bhattacharjee (2014a).
[213] Bhuiyan et al. (2015), Keck (2013).

6.4 The Quad's 'Free and Open' Countermoves

China's strategy of encircling India with its client-states'.[214] A serving IN commander was equally blunt: 'Why would Bangladesh need submarines? This decision by the government there is a matter of concern for us. We also suspect that Chinese submarines are sneaking into Indian territorial waters in the Bay of Bengal region, though none has been detected yet. This is reason enough for greater naval presence in the region'.[215] Dhaka's efforts to boost maritime security confronted Sino-Indian tensions.

That was the backdrop for the deep-sea port project. Bangladesh's $60bn foreign trade, fuelled by ready-made garment exports, suffered from shallow-draft Chittagong and Mongla harbours requiring slow and expensive cargo-lighterage. Surveys identified two areas near the ports where deep-sea anchorages could be built. After a 2006–2009 feasibility study by Japan's Pacific Consultant International, in 2013, Chinese engineers surveyed the Sonadia Island south of Chittagong. Bangladesh accepted the bid from CHEC to build the deep-sea port. Hasina was due to visit Beijing in June 2014 to issue the work order to CHEC[216] but, while the joint statement issued during her trip included five projects and Beijing's offer to build 'a Chinese Economic and Industrial Zone in Chittagong', the port was not mentioned.[217] Indian analysts posited Dhaka rejected CHEC's plans to 'design, execute and operate' the $14bn port.[218]

Hasina's prior trip to Tokyo proved decisive. After Abe briefed her on 'the situation in the SCS', they 'underscored the importance of the FoN and agreed that international disputes and issues should be resolved peacefully and all relevant countries should adhere to relevant international law' and 'globally agreed norms and practices'. They 'underscored the importance of the freedom of overflight over the high seas'. Japan offered help with 'creating sustainable employment' from Bangladesh's marine-economy, i.e. fishery and other marine resources, 'renewable energy, oceanographic research, shipping and tourism'. That and Hasina's support for Japan's inclusion as a UNSC permanent member, forged the Japan–Bangladesh Comprehensive Partnership, securing 600bn yen in aid. Hasina asked for specific project-finance; Abe proffered help including the transfer of 'clean coal technology'.[219]

This Hasina-Abe accord reflected 'an apparent drive by Abe to keep China in check as Beijing tries to extend its military clout in the East and South China Seas'.[220] In China just days later, Hasina sidestepped the deep-sea port issue. This reversal was rewarded during Abe's September 2014 visit to Dhaka with his 'Bay of Bengal Industrial Growth Belt (BIG-B)' initiative, pledging help with 'developing infrastructure and improving energy supply and investment environment'.[221] To support Japan's

---

[214] Prakash (2016).

[215] Gupta (2013).

[216] Report (2014) Chinese firm to construct Sonadia sea port. Prothom Alo, 4 June 2014.

[217] MoFA (2014).

[218] Bhattacharjee (2014b).

[219] Abe and Hasina (2014), Yoshida (2014).

[220] Yoshida (2014).

[221] Japanese Embassy (2014).

UNSC candidacy, Hasina withdrew Dhaka's own. In May 2015, anxious to 'tamp down China's influence in Bangladesh' with *materiel* and economic/infrastructural aid, Narendra Modi visited Dhaka.[222] China had long been Bangladesh's key source of military hardware, had built large infrastructure projects since the 1980s, and although Beijing was upgrading parts of the Chittagong port—allegedly a 'pearl' in China's 'string',[223] PLAN warships had only made brief port-calls and attended offshore drills with BN vessels. Bangladeshi officials saw Beijing as 'the best option' for building the deep-sea port near Chittagong, but Hasina proved sensitive to Indian and Japanese concerns.[224]

She endorsed Tokyo's offer to build a deep-sea port at Matarbari, 25 km from Sonadia, the site Dhaka had proposed to Beijing. JICA extended 80% financing for a $4.6bn port-complex and four 600 MWe coal-fired power-plants with $3.7bn in 30-year loans at 0.1% interest. Japan's Sumitomo, IHI and Toshiba corporations built the project.[225] Planning Minister Mustafa Kamal explained, 'Matarbari is designed in such a way that it will be comprehensive, with power-plants, an LNG terminal and a port. Matarbari is sufficient; we may have to give up the other port project'.[226] Hasina's aides conceded that since June 2014, the Sonadia deep-sea project had 'faced uncertainty as major regional and global powers joined hands together to restrict China's involvement with the project'. These powers felt 'threatened' by 'Chinese dominance in the pan-Pacific territory in terms of port-related investment'.[227] Prime Minister Modi proved especially persuasive against Dhaka accepting Beijing's help.[228] The shift, triggered by Bangladesh's vulnerability to great-power rivalry[229] mirroring *systemic transitional fluidity*, was balanced with a second, $3bn, deep-sea port at Payra in the southwest of the country, and a 750-acre industrial park in Chittagong with CHEC holding 70% equity.[230]

Xi Jinping's October 2016 visit elevated China–Bangladesh ties to 'a strategic partnership of cooperation'. Twenty-seven projects worth $24bn were agreed, pushing Chinese investment towards $40bn.[231] Dhaka endorsed the BRI, receiving strong support for its development plans. Chinese funding towered over Modi's 2015 offer of a $2bn line of credit.[232] Xi's gifts built on a record of major infrastructure projects, especially 'friendship bridges' spanning the delta's myriad rivers. The biggest, the 6 km, $3.7bn, Padma Bridge, on which Beijing spent over $3bn, exemplified collaboration. By late-2018, Beijing's state-funded offers to Dhaka hit $31bn, with

---

[222] Reuters (2015).
[223] Editorial (2016a), Hasan (2016), Mukherjee (2014), Marshall (2012).
[224] Reuters (2015).
[225] JICA (2018), BETS Consulting (2018), Miglani and Paul (2015).
[226] Kamal (2015).
[227] Stacey (2018a, b), Rahman (2015).
[228] Bloomberg (2018b), Mazumdar (2017), Panda (2015).
[229] Cookson and Joehnk (2018), Editorial (2016b), Hasan (2016).
[230] Quadir (2018), Islam (2017), Mamun (2016).
[231] Parmar (2016).
[232] Paul (2016), Kabir (2016), Report (2016).

6.4 The Quad's 'Free and Open' Countermoves 269

another $11bn in private investment, troubling Delhi.[233] Major projects included the $2.86bn coastal marine-drive motorway, $1.6bn Dhaka Airport-Ashulia motorway, and $753m Joydebpur–Ishwardi rail-track.[234]

Questioned by Indian journalists, Hasina insisted, 'We want investment and cooperation from whoever offers it. We want development of the country. We have to think about our people as they are the beneficiaries of development. India has nothing to be worried about it'.[235] Competition in Bangladesh notwithstanding, Japan's position vis-à-vis China's BRI/MSR blueprint moderated as US–PRC–Japan dynamics betrayed fluidity. Shinzo Abe, returning Li Keqiang's earlier trip to Tokyo, visited Beijing with a 500-strong business entourage in late-2018, pledging a 'new era' of cooperation.[236] Shortly after Abe's return from China, Indian Prime Minister Narendra Modi arrived in Tokyo to boost Indo-Japanese economic-strategic coordination. Western and Indian analysts discerned in the summit an urgency to jointly challenge China's influence flowing from BRI/MSR.[237]

Indian anxiety coloured implementation of Bangladesh's Payra Port project, designed to relieve congestion at Mongla and Chittagong, support transit trade, and help develop a neglected zone. It needed a 2000-m jetty terminal and backup facilities, dredging for channel navigability, river training, two 660 MWe coal-powered stations, a coal terminal, and road-links to Khulna and Dhaka.[238] Four Chinese SoEs bid for two of the port's 19 components; Dhaka chose CHEC and China State Construction Engineering (CSCEC). Under a $600m contract, CHEC built the main port-infrastructure worth $140m, while CSCEC, responsible for the channel, also constructed housing, medical and educational facilities. Payra began partial operations in August 2016 although major construction would take until 2023.[239] Firms from India, the UK, Belgium, the Netherlands and Denmark, too, sought work on 13 FDI-financed components of the 7000 acre, $11bn–$15bn port. Government-to-government contracts built the rest.[240]

India, anxious to counteract Chinese-funded projects close to its Kolkata and Haldia ports, urged Dhaka to use these as transhipment points for European trade. Having begun building Payra and expanding Chittagong ports, Dhaka declined.[241] Delhi moved directly after Dhaka established the Payra Port Authority in 2014 with $143.37m allocated for immediate construction. Delhi offered $750m out of its cumulative $4.5bn line-of-credit extended to Dhaka, to build a multipurpose terminal at Payra. India Ports Global, a public–private joint venture, offered to 'design, fund and build' Payra by itself. Indian analysts noted, 'since the Payra sea port is

---

[233] Stacey (2018b).

[234] Bhaumik (2018).

[235] PTI (2018d).

[236] An (2018), Kajimoto (2018), Bodeen and Wang (2018), Report (2018l).

[237] Panda (2018b), Hurst (2018), Singh (2018c).

[238] Public-Private Partnership Authority (2017).

[239] Hazarika (2016), Mamun (2016).

[240] Report (2017b).

[241] Bose (2018), Special Correspondent (2018).

a strategically important project, India does not want its adversary China to build maritime infrastructure just next to the country's coastline'. Delhi offered to 'take up the project though it may not be financially viable'.[242] This view resonated with Bangladesh's interest in maximising FDI-funded infrastructure development in a competitive milieu.[243] To maintain peace in its dealings with two antagonistic major powers, Dhaka had to pay a price. During her October 2019 visit to Delhi, Hasina accepted India's demand, among others, to establish a coastal surveillance radar network in Bangladesh as part of their Maritime Security Partnership.[244] China was not named but the partnership's objective could not be clearer.

6.4.4 The Indian Ocean Region's Big Guns

Determined to drive a military-led campaign parallel to Quad-members' joint geoeconomic exertions targeting China, specifically MSR, they revamped their naval activities, inviting British and French navies to join them across Indo-Pacific waters.[245] The two Quad-members 'defending and securing' a 'Free and Open Indo-Pacific', i.e. India and the USA, formalised strategic collaboration at their first-ever foreign and defence ministers' meeting at the '2+2' conclave, in 2018. Indo-US geopolitical cooperation accreted in a palimpsest of agreements which, 'founded on the values of freedom, justice, and commitment to the rule of law', boosted their 'strategic partnership' while promoting 'synergy in their diplomatic and security efforts'.

The partners agreed to 'work together on regional and global issues' in 'bilateral, trilateral and quadrilateral formats'. The USA had designated India a 'Major Defence Partner' (MDP) and exchanged reciprocal basing rights under a Logistics Exchange Memorandum of Agreement (LEMOA); Washington now pledged swift transfer of *materiel* and technology to accelerate Delhi's military rise by including India among allies 'entitled to license-free exports, reexports, and transfers under License Exception Strategic Trade Authorisation (STA-1)'.

To protect classified information exchanges, permitting transfer of advanced US hardware, and to 'enable India to optimally utilise its existing US-origin platforms', ministers signed a 'Communications Compatibility and Security Agreement' (COMCASA). Enabling closer weapons-production collaboration, they negotiated an Industrial Security Annex (ISA), prioritising co-development and co-production through the Defence Technology and Trade Initiative (DTTI), forging defence innovation collaboration.[246] After decades of working with the Hawaii-based US IndoPacom on joint operations, Delhi acceded to a long-standing US request, linking the

---

[242] Bhattacharjee (2017b).
[243] Kabir (2017).
[244] Modi and Hasina (2019).
[245] Joshi (2017), Press Office (2017), Media Centre (2017).
[246] Spokesperson (2018).

Indian Navy (IN) and the Naval Forces Central Command (NAVCENT).[247] Synergising Indian action with US strategic presence from the Pacific to the Middle East, the IN collaboratively responded to China's maritime activities, including MSR.

6.5 Maritime Silk Road's Insecure Origins

Divergent narratives and contradictory perspectives on the purposes of China's MSR blueprint demand an empirical examination of MSR's evolution.[248] While China and its critics question the merit of each other's assertions and the real goals behind their mutually targeted action, data available in the public domain offer an opportunity for undertaking forensic scrutiny. China's external economic/commercial engagement, i.e. growing merchandise trade, provided a key measure. That growth, resultant imbalances, their political fallout, and policy-anxiety triggered the Trump Administration's 2018–2019 tariff war. Even before Beijing's WTO accession in 2001, reforms negotiated by Zhu Rongji with Washington and its OECD allies rapidly boosted China's trade.[249] Post-accession, growth accelerated, although much Chinese exports, assembled/fabricated by FDI-ventures, earned Beijing only modest labour- and value-addition revenues. China's imports, fuelled by demand for raw materials, chemicals, energy, commodities, consumer products, machine-tools, and high-technology components, similarly exploded. Merchandise trade data, as shown in Table 6.4, indicated China's dependency:[250]

Merchant shipping's rising salience within China's fast-growing economy, and an anxiety to secure shipping capacity to sustain growth, shifted policy and resource allocation towards building, modernising and expanding China's ship-building and ocean-freight capacity. In May 2017, Chinese shippers, with 1996 vessels with an aggregate capacity of 85,347,681 TEUs, topped the global order. For South Korea, second in the list, the figures were 1017 and 40,924,768 TEUs. third-ranking Malaysia had 36,663,697 TEUs; 4th-ranking USA had 36,154,504 TEUs.[251]

Growing trade volumes, the role of maritime commerce in economic well-being, and a perceptibly malign landscape, drove Beijing towards securing its SLoCs from Korea to Eastern Africa from the *systemic primate* which, since the 1995–1996 Taiwan Strait 'missile crisis', challenged Chinese 'core interests'. China laid a network of buoys, surface vessels, underwater gliders and satellites, collating salinity and temperature data across MSR alignments. Processed at centres in Guangdong Province, the Paracels and 'a joint facility in South Asia', the intelligence was disseminated to

---

[247] Spokesperson (2018).
[248] Rolland et al. (2019).
[249] Ianchovichina and Martin (2004).
[250] General Administration of Customs PRC (2019), UNCTAD STAT (2018), WBG Data (2019), Report (2019c).
[251] Hoffmann and Sirimanne (2017).

Table 6.4 China's international trade growth: 1996–2018 (mainland data)

Year	Exports $bn (current)	Imports $bn	Total trade $bn	GDP $bn	Trade/GDP %
1996	151.048	138.943	289.989	863.747	33.573
1997	182.792	142.189	324.981	961.604	33.795
1998	183.712	140.305	324.017	1029.000	31.488
1999	194.931	165.788	360.719	1094.000	32.972
2000	249.203	225.024	474.227	1211.000	39.159
2001	266.098	243.553	509.651	1339.000	38.062
2002	325.596	295.170	620.766	1471.000	42.200
2003	438.228	412.760	850.988	1660.000	51.264
2004	593.326	561.229	1154.555	1955.000	59.056
2005	761.953	659.953	1421.906	2286.000	62.200
2006	968.978	791.461	1760.439	2752.000	63.969
2007	1220.456	956.116	2176.439	3552.000	61.273
2008	1430.693	1132.567	2563.326	4598.000	55.748
2009	1201.612	1005.923	2207.535	5.110.000	43.200
2010	1577.754	1396.247	2974.001	6101.000	48.746
2011	1898.381	1743.484	3641.865	7573.000	48.090
2012	2048.714	1818.405	3867.119	8561.000	45.171
2013	2209.005	1949.990	4158.995	9607.000	43.291
2014	2342.293	1959.233	4301.526	10,482.000	41.037
2015	2273.468	1679.566	3953.034	11,065.000	35.725
2016	2097.632	1587.925	3685.557	11,191.000	32.933
2017	2263.329	1841.889	4105.218	12,238.000	33.544
2018	2480.000	2140.000	4620.000	13,608.150	33.950

PLAN subs patrolling MSR routes. The US and allied ASW sensor-networks were larger but the PLAN chain honed its submarine-targeting capabilities.[252]

That defensive development evolved over a decade while Beijing struggled with its 'Malacca Dilemma'. Much Chinese commerce traversed the Malacca and Lombok/Makassar straits, the latter by Very Large Crude Carriers (VLCC). Fears of rival navies choking these points led Chinese analysts to complain in 2004: 'It is no exaggeration to say that whoever controls the Strait of Malacca will also have a stranglehold on the energy route to China'.[253] Beijing lacked resources to secure the choke points but recognised that if it did, hostile response could render its action

---

[252] Chen (2018).
[253] China Youth Daily (2006).

6.5 Maritime Silk Road's Insecure Origins

counterproductive.[254] Chinese fears had substance. Prospects of blockades were serious enough for US and Japanese analysts to conduct risk-analyses.[255] US, Indian and allied strategists examined optimal paths to reinforcing China's 'Malacca Dilemma' after Beijing popularised the refrain.[256] The IN's doctrinal document noted, 'Control of the choke points could be useful as a bargaining chip in the international power game'.[257]

Jiang Zemin's 1993 'Military Guidelines for the New Period' laid the foundations for China's naval modernisation, promising PLAN fleets an ability to conduct sea-denial operations around and beyond Taiwan, protect China's coastal economic heartland, and deter or delay US intervention in a Taiwan-focused conflict. Jiang's 'strategic guidelines of the active defence', underscoring the need to wage a 'local war under high-technology conditions', drove PLAN towards conducting limited offensives to 'enforce sovereignty and territorial claims' in the ECS and SCS.[258] These moves dialectically deepened US anxieties, catalysing a three-year review of its global military posture to rebalance the deployment of US expeditionary forces.

The turn of the century proved pivotal. Beijing may have been unaware of the 1999 DoD/ONA Summer Study designating China an emergent 'constant competitor' confronting the 'sole superpower' determined to perpetuate *primacy*. Still, the destruction of Beijing's Belgrade chancery by US stealth bombers during NATO's Kosovo campaign, igniting a wildfire of China-wide protests, NDAA2000's China-targeting provisions, and President Clinton forcing Israel to cancel contracts for two PLA *Phalcon* AWACS systems, deepened foreboding. PACOM's intensified close-in aerial, surface and sub-surface ISR missions precipitated defensive insecurity. The April 2001 collision between a US ISR aircraft and a Chinese fighter jet, killing the PLA pilot and forcing the damaged EP-III to land at Lingshui in Hainan, confirmed Chinese fears, triggered a 12-day crisis, and further deepened mutual distrust.

The US 'posture' review, designed to 'maintain its current position as a nation of global influence through leadership and the efficient and effective application of informational, military, economic, and diplomatic power', pursued 'a global strategy, a cornerstone of which is increased access and forward presence in key areas'.[259] In the Asia–Pacific, the review built upon the US's 'current ground, air, and naval access to overcome vast distances, while bringing additional naval and air capabilities forward into the region'.[260] PACOM updated operational plans, strengthened command and control, added Stryker brigades and C-17 airlifters, deployed bombers and submarines to Guam, and proposed homeporting another CSG in the Pacific. It boosted pre-positioned logistics and deepened defence relations with 'good friends',

---

[254] Caesar-Gordon (2016), Yu (2016), Lanteigne (2008), You (2007), Storey (2006).
[255] Holmes (2019), Shibasaki and Watanabe (2011).
[256] Mirski (2018), Reddy (2016), Conole (2015), Joshi (2013), Spinetta (2006).
[257] Holmes et al. (2009).
[258] Jiang (1993), Cooper (2009).
[259] Jones (2004).
[260] Rumsfeld (2004a, b).

e.g. Singapore, Malaysia, India, Indonesia, Mongolia, 'and many others'.[261] Assuming CMC Chairmanship months later, Hu Jintao unveiled the PLA's 'New Historic Missions'.

Framed within Hu's 'scientific development' theory, an 'important guiding strategy for national defence construction and army building', these missions were 'logical manifestations' of the application of 'scientific development' as a lens to comprehend the dynamic international military-security milieu.[262] Hu derived his postulate from China's domestic and international imperatives – including concerns about political stability and territorial integrity, especially Taiwan, and new challenges flowing from China's growing global interests and engagements. He laid down the PLA's 'new historic missions' as the 'three provides, and one role'[263]:

- Providing an important guarantee of strength for the Party to consolidate its ruling position
- Providing a strong security guarantee for safeguarding the period of important strategic opportunity for national development
- Providing a powerful strategic support for safeguarding national interests and
- Playing an important role in safeguarding world peace and promoting common development.

DoD, driving RMA-based networking of sensors, data-fusion-and-analyses, targeting and 'shooting' platforms into 'kill-chains' to dominate battle-spaces, deployed new units across China's periphery to implement the 2004 Posture Review. Using Japan's 1940s use of land-based air power against hostile shipping as a template, Major Lawrence Spinetta, USAF, underscored Beijing's 'String of Pearls strategy' as an 'aggressive' quest 'to secure the raw materials necessary to fuel its economy'. PLAN's growth would 'deny the US access to the region, negate US influence, and intimidate neighbours into political accommodation'. China's aim was 'to challenge US maritime hegemony in the region'. Spinetta advised, 'Land-based air power, in conjunction with naval assets', promised the USA an effective method 'to counter China's "String of Pearls" and maintain an advantage in the maritime domain'.

Noting that 'Land-based air power can help control the littorals', he urged DoD to 'designate maritime interdiction as a primary Air Force mission'. The USAF 'must embrace, train, and fund maritime operations'. The USA must 'strengthen strategic partnerships in the region to ensure access and basing'.[264] Commander William Pharis, USN, averred that China's growing naval capabilities reflected Beijing's 'desire to be seen as a global superpower on par with the USA, to counter US influence and regain prominence within maritime Asia'. China's modern navy 'will seek to maximise enemy force attrition, deny access to critical areas, and protect vital shipping and SLoCs'.[265] The USA faced a 'significant quandary' in 'balancing the

---

[261] Fargo (2004).
[262] Editorial (2007a, b).
[263] Jia et al. (2007).
[264] Spinetta (2006, pp. 1–2).
[265] Pharis (2009).

opportunity for engagement with regional fear of a China with regional hegemonic aspirations'. Washington should not 'overestimate the Chinese naval threat, but, at the same time, history is full of examples of how dangerous it can be to underestimate an emerging power'.[266]

Other analysts reinforced the belief that PLAN's focus on deterring or delaying US intervention in Taiwan's support eroded the US power-projection capacity to deploy CSGs asserting US dominance along China's periphery. They urged 'developing and procuring highly capable ships, aircraft, and weapons for defeating Chinese anti-access systems', assigning 'a greater percentage' of USN assets to the US Pacific Fleet, 'homeporting more of the Pacific Fleet's ships' in Hawaii, Guam and Japan, increasing ASW operations, improved monitoring of PLAN activities, and enhancing understanding of Western Pacific 'operating conditions'.[267]

It was against this backdrop that Hu articulated China's 'Malacca Dilemma'. Dialectic dynamics ensured a Chinese response. In May 2007, PLAN Commander, Admiral Wu Shengli, proposed to Admiral Timothy Keating, Commander PACOM, a division of labour: US Pacific Fleet protect the Pacific between the US West Coast and Hawaii while PLAN defended the waters west of Hawaii. Keating declined.[268]

Beijing's sixth defence white paper illuminated linkages between China's growing economic success, and the PLA's responsibility to defend it. The document named the USA as the principal external threat, explaining the challenges posed by the United States' regional influence, China's economic, military and scientific disadvantages, and its expanding economic footprint demanding enhanced military capabilities.[269] The CMC ordered the PLA to protect China's broader interests beyond immediate territorial sovereignty concerns and conduct appropriate threat assessments to guide the creation of necessary capabilities.

Ordered to accomplish 'diversified military tasks', the PLA, especially PLAN, reformed organisational, procurement and *materiel* parameters. Beijing's power-projection capabilities could not threaten US dominance, but *primacy*'s absolutist premises rendered China a growing threat. PLAN's 2009–2020 modernisation aimed at advancing its CMC-proclaimed goals[270]:

- Becoming a viable 'strategic arm'
- Developing strike packages for sustained 'green water offensive operations' out to the first Island Chain and across the SCS
- Providing combat and support assets for limited force projection beyond peripheral waters
- Providing 'leadership, doctrine, tactics and training' for integration into joint and multinational naval operations

---

[266] Pharis (2009).
[267] O'Rourke (2010).
[268] Keating to author at the Shangri-La Dialogue, Singapore, 31 May 2007.
[269] CMC (2009).
[270] Cooper (2009).

Although deterring and preventing 'Taiwan independence' drove PLA modernisation, that 'core' interest, and 'new missions' propelled the PLAN towards 'bluewater' potential. Maritime-territorial disputes between China and US-aligned neighbours across the ECS and SCS, fusing *sub-systemic* and *systemic* flashpoints, assumed salience. US strategists noted, 'China's navy is viewed as posing a major challenge to the US Navy's ability to achieve and maintain wartime control of blue-water ocean areas in the Western Pacific—the first such challenge the US Navy has faced since the end of the Cold War'. This was 'a key element of a Chinese challenge to the long-standing status of the USA as the leading military power in the Western Pacific'.[271] The PLAN and MSR together posed an intolerable threat. As Quad-members aided China's rivals while Beijing sought to insulate 'regional' dynamics from 'extra-regional interference', the maritime domain emerged as the 'New Cold War's key fault-line.

With Beijing defending its growing interests along MSR's alignments with power and presence, the US-led coalition discerned threats to their collective *systemic* interests. Trump's Africa Strategy targeted MSR/BRI: 'China uses bribes, opaque agreements, and the strategic use of debt to hold states in Africa captive to Beijing's wishes and demands'. China's BRI/MSR links to Africa were not only challenging US interests, they were also, according to the US narrative, harming African economies. 'Its investment ventures are riddled with corruption and do not meet the same environmental or ethical standards as US development programs. Such predatory actions are sub-components of broader Chinese strategic initiatives, including OBOR'.[272]

China's SCS island-bastions and the Quad's naval and aerial operations in proximate waters symbolised the emergent bipolarity. Conflicting trajectories coloured by mutually defensive drivers ensured MSR triggered deepening anxiety among actors defending the status quo, precipitating a dialectic dynamic. MSR thus reflected and reinforced *systemic transitional fluidity* flowing from intensifying USA–Chinese structural and strategic competition, arguably the defining feature of the early decades of the twenty-first century.

References

Abe S, Hasina S (2014) Joint statement: Japan-Bangladesh comprehensive partnership. MoFA, Tokyo, 26 May 2014
AFP (2018a) Sri Lanka reinstates ousted PM Wickremesinghe. Straits Times, 17 Dec 2018
AFP (2018b) Maldives top court reverses decision to free jailed politicians citing president's 'concerns'. SCMP, 7 Feb 2018
AFP (2018c) New Maldives president signals end to pro-China stance as he embraces India's Narendra Modi at swearing-in ceremony. SCMP, 18 Nov 2018
AFP (2018d) EU launches Asia strategy to rival China's 'new Silk Road'. New Straits Times, 26 Sept 2018
Akimoto S (2018) How Japan can save the Indo-Pacific strategy. JT, 24 July 2018

---

[271] O'Rourke (2019).
[272] Bolton (2018).

Akita H (2018) Can Japan and China move beyond a tactical détente? EAF, 19 Sept 2018

Alam B (2018) In the shadow of maritime security: India, Japan and the China factor. Indian Def Rev, 8 Jan 2018

Ali S (2010) Understanding Bangladesh. Columbia University Press, New York, p 133

Ali I, Stewart P (2018) 'Significant' consequences if China takes key port in Djibouti: US general. Reuters, Washington, 7 Mar 2018

Ali I, Gao L, Wen P (2018) China denies US accusation of lasers pointed at planes in Djibouti. Reuters, Washington/Beijing, 4 May 2018

An B (2018) China-Japan ties at 'historic' point. CD, 27 Oct 2018

Aneez S, Sirilal R (2017a) Sri Lanka to shift naval base to China-controlled port city. Reuters, Colombo, 3 July 2018

Aneez S, Sirilal R (2017b) Sri Lanka rejects Chinese request for submarine visit: sources. Reuters, Colombo, 11 May 2017

AP (2019a) France leads naval exercise with US, UK and Japan in American territory of Guam in the Pacific. SCMP, 12 May 2019

AP (2019b) China to give US$989m loan to Sri Lanka for major new motorway project. SCMP, 23 Mar 2019

Bajwa R (2018) With eye on China, Japan to send warship to SCS and Indian Ocean. International Business Times, 6 July 2018

Bengali S (2018) He brought world attention to his sinking island nation. Now he wonders if he can go back. LA Times, 17 Apr 2018

Berkofsky A (2018) 'Free and open Indo-Pacific': Tokyo's plans and priorities. Italian Institute for International Political Studies, Milan, 4 June 2018

BETS Consulting (2018) Matarbari Port development project. Ministry of Shipping, Dhaka, Nov 2018

Bhattacharjee R (2014a) Delimitation of Indo-Bangladesh Maritime Boundary. IDSA, Delhi, 19 Aug 2014

Bhattacharjee J (2014b) Sheikh Hasina's China visit: an assessment. Observer Research Foundation, Delhi, 19 June 2014

Bhattacharjee K (2017a) We are aware of China's maritime ambitions: India. Hindu, 27 Dec 2017

Bhattacharjee R (2017b) Payra port development will enhance India-Bangladesh ties. South Asia Monitor, 20 Jan 2017

Bhaumik S (2018) Bangladesh election: will Sheikh Hasina's China-India Balancing Act be enough to keep power? SCMP, 26 Nov 2018

Bhuiyan M, Ali M, Rahman M, Selim S (2015) Maritime boundary confirmation of Bangladesh. Cost Manag 43(5):18–4

Biswas A, Tortajada C (2018) A new Silk Road connects two old continents in a modern age. Channel News Asia, 7 Sept 2018

Blackwill R (2019) Trump's foreign policies are better than they seem. CFR, New York, Apr 2019, pp 8–17

Blanchard B (2017) China opens first overseas military base. Reuters, Beijing, 1 Aug 2017

Blanchard B (2018) Chinese warships enter East Indian Ocean amid Maldives constitutional crisis. Independent, 20 Feb 2018

Bloomberg (2018a) China's investment in West Africa challenges France for business in its former colonies. SCMP, 18 July 2018

Bloomberg (2018b) A new direction for China's Belt and Road. Star, 3 Nov 2018

Bodeen C, Wang E (2018) China-Japan drawing closer amid trade pressure from US. AP, Beijing, 26 Oct 2018

Bohigian D, Kleiterp N, Lamontagne P (2019) MoU on cooperation between OPIC/DFC, EDFI, and DFIC. OPIC, Washington, 11 Apr 2019

Bolton J (2018) Remarks on the Trump Administration's new Africa strategy. NSC/White House, Washington, 13 Dec 2018

Bose P (2018) India asks Bangladesh to use Kolkata, Haldia as transhipment ports. Hindu Businessline, 11 Oct 2018

Brands H (2018a) How China went from a business opportunity to enemy No. 1. Bloomberg, 6 Sept 2018

Brands H (2018b) China's master plan: a global military threat. Bloomberg, 10 June 2018

Branigan T (2010) Zheng He: messenger of peace, or of power? Guardian, 18 July 2010

Brewster D (2018a) Why India is buying the world's emptiest airport. Interpreter, 14 July 2018

Brewster D (2018b) Japan's plans to build a 'Free and Open' Indian Ocean. Interpreter, 29 May 2018

BSS (2010) Deep-sea port to benefit all neighbours: Hasina. Star, 22 Mar 2010

Caesar-Gordon M (2016) Securing the energy supply: China's 'Malacca Dilemma'. E-Int Relat, 26 Feb 2016

Chan T (2017) India is buying the world's emptiest airport in its battle for territorial dominance with China. Bus Insider, 13 Dec 2017

Chankaiyee2 (2014) China to build 18 naval bases in Indian Ocean. China Daily Mail, 22 Nov 2014

Chaudhury D (2016) Chinese company bags Maldivian Island on 50-year lease. Economic Times, 30 Dec 2016

Chaudhury D (2017) Japan teams up with India for Northeast, to extend Rs. 2,239-crore loans. Economic Times, 18 Sept 2017

Chaudhury D (2018a) China suggests to work with India in Maldives after poll shocker. Economic Times, 29 Sept 2018

Chaudhury D (2018b) India, Japan & Australia firm up partnership for free and open Indo-Pacific region. Economic Times, 12 July 2018

Chen S (2018) China's underwater surveillance network puts targets in focus along MSR. SCMP, 1 Jan 2018

Chen W, Hruby A, Laryea T (2016) China's Investments in Africa: what's the real story? Wharton School, Philadelphia, 19 Jan 2016

China Youth Daily (2006) In: Storey I (ed) China's 'Malacca Dilemma'. China Brief 6(8), 12 Apr 2006

Ching N (2019) Italy joining China's new silk road troubles US and EU. VoA, 23 Mar 2019

Chongwei Z, Ziniu X, Wen Z, Xiaobin C, Xuan C (2018) 21st century MSR: a peaceful way forward. Springer, Singapore, p 2

Churchill O (2018) China hasn't changed belt and road's 'predatory overseas investment model'. US official says. SCMP, 13 Sept 2018

Clemens M (2015a) The MSR and the PLA: maritime power conference paper. CNA, Arlington, 28 July 2015

Clemens M (2015b) The MSR and the PLA: maritime power conference paper. CNA, Arlington, 28–29 July 2015, p 1

Cliff R, Burles M, Chase M, Eaton D, Pollpeter K (2007) Entering the Dragon's Lair: Chinese Antiaccess strategies and their implications for the United States. RAND, Santa Monica

CMC (2009) China's National Defence in 2008. State Council Information Office, Beijing, p 2009

Conole R (2015) Maritime trade warfare: a challenge to the Chinese A2/AD System. NWC, Newport, 15 May 2014

Cooper Z (2018) In: Szechenyi N (ed) China's MSR: strategic and economic implications for the Indo-Pacific Region. CSIS, Washington, pp 26–29

Conrad M (2012) Does China need a 'string of pearls'?. Naval Postgraduate School, Monterey, p 2012

Cookson F, Joehnk T (2018) China and geopolitical tug of war for Bangladesh. EAF, 11 Apr 2018

Coons C, Rubio M (2018) Letter to Pompeo M, Mattis J. US Senate, Washington, 7 Nov 2018

Cooper C (2009) The PLA Navy's 'new historic missions': expanding capabilities for a re-emergent maritime power. RAND, Washington, 11 June 2009, p 2

Dasgupta S (2015) China says India must not think of Indian Ocean as its backyard. ToI, 1 July 2015

DCI (2018) The world factbook: Maldives. CIA, Langley

DeAeth D (2019) US forces China's COSCO to relinquish ownership of California port. Taiwan News, 10 May 2019

Dollar D, Tang H, Chen W (2015) Why is China investing in Africa? Evidence from the firm level. Brookings, Washington, 12 Aug 2015

Downs E, Becker J, deGategno P (2017) China's military support facility in Djibouti: the economic and security dimensions of China's first overseas base. CNA, Arlington, July 2017

Dufour J (2018) The worldwide network of US military bases. Global Research, Montreal, 15 Apr 2018

Dutta P (2017) Sri Lanka leases Hambantota port to China for non-military use. Should India worry? India Today, 29 July 2017

Dutta P (2018) India turns heat on Maldives as Chinese dragon spews fire in Indian Ocean. India Today, 25 June 2018

Economy E (2018) China's strategy in Djibouti. CFR, New York, 13 Apr 2018

Editorial (2007a) Opening up new prospects for national defence, army building: 6th commentary on conscientiously studying and implementing the spirit of 17th CPC National Congress. PLA Daily, 14 Nov 2007

Editorial (2007b) In-depth implementation of scientific development concept: 4th commentary on conscientiously studying and implementing the spirit of 17th CPC National Congress. PLA Daily, 8 Nov 2007

Editorial (2016a) China, India and Japan eye Bangladesh's project to build a deep water port. News in Asia, 23 Dec 2016

Editorial (2016b) China, India and Japan eye Bangladesh's project to build a deep-water port. News In Asia, 23 Dec 2016

Editorial (2018a) India must stop intervening in Male'. GT, 6 Feb 2018

Editorial (2018b) Neighbors show their shared resolve to keep ties on track. CD, 26 Oct 2018

Editorial (2019) Malaysian rail deal shows belt and road plan is still on track. SCMP, 17 Apr 2019

EEAS-ASIAPAC (2018) Elements for an EU strategy on connecting Europe and Asia. EC, Brussels, 6 Feb 2018

Egozi A (2018) Israelis to US: take on China around Djibouti. Breaking Defense, 28 Nov 2018

Emmott R (2018) EU unveils Asia infrastructure plan, denies rivalry with China. Reuters, Brussels, 19 Sept 2018

Escobar P (2018) Marco Polo in reverse: how Italy fits in the New Silk Roads. Asia Times, 12 Mar 2018

EU External Action (2016) Shared vision, common action: a stronger Europe—a global strategy for the EU's foreign and security policy. European Commission, Brussels, June 2016, pp 37–38

Fargo T (2004) The global posture review of US military forces stationed overseas: SASC testimony. DoD, Washington, 23 Sept 2004

Fei J (2017) China's overseas military base in Djibouti: features, motivations, and policy implications. China Brief 17(17), 22 Dec 2017

Funaiole M, Hillman J (2018) In: Szechenyi N (ed) China's MSR: strategic and economic implications for the Indo-Pacific Region. CSIS, Washington, pp 21–23

Gale J, Shearer A (2018) In: Szechenyi N (ed) China's MSR: strategic and economic implications for the Indo-Pacific Region. CSIS, Washington, pp 30–33

Gassaway C (2011) The strategic importance of Sri Lanka for IOR stability. NWC, Newport, 5 Apr 2011

General Administration of Customs PRC (2019) Review of China's Foreign Trade in 2018. State Council, Beijing, 14 Jan 2019

Geng S (2018) In Chaudhury D (ed) Maldives crisis: China comes to Yameen's rescue as India calls for restoration of democratic values. Economic Times, 7 Feb 2018

Ghiasy, Su F, Saalman L (2018) The 21st century MSR: security implications and ways forward for the EU. SIPRI, Stockholm, June 2018

GNA (2018) Ghana and China sign deal for Jamestown Fishing Port Complex. Accra, 6 Apr 2018

Gopalaswamy B (2018) Sri Lanka's political shake-up is a win for China. FP, 29 Oct 2018
Green M (2018) In: Szechenyi N (ed) China's MSR: strategic and economic implications for the Indo-Pacific Region. CSIS, Washington, p 1
Greenert J (2018) Tenets of a regional defense strategy: considerations for the Indo-Pacific. NBR, Seattle, Aug 2018
Guensburg C, Widakuswara P (2018) Trump administration rethinks foreign aid with eye toward China. VoA, 19 Oct 2018
Gupta J (2013) Indian Navy concerned over Bangladesh's decision to buy two submarines from China. ToI, 2 Dec 2013
Gurung S (2018) New threat in Indian Ocean: China to build at least six aircraft carriers. Economic Times, 14 July 2018
Halim M, Kamal A (1996) In Zafarullah H (ed) The Zia episode in Bangladesh politics. South Asian Publishers, New Delhi, p 135
Hasan Z (2016) Deep politics of deep-sea ports. Dhaka Tribune, 16 June 2016
Hasina S (2011) Address on the occasion of the commissioning of BNS Dhaleswari and BNS Bijoy. PMO, Khulna, 5 Mar 2011
Hazarika M (2016) China's CHEC and CSCEC to develop Payra deep-sea port in Bangladesh. Ship Technol, 11 Dec 2016
Hillman J (2018) Game of loans: how China bought Hambantota. AMTI/CSIS, Washington, 4 Apr 2018
Ho F (2018a) Trade war adds FDI lure to Kuantan. Star, 30 Sept 2018
Ho F (2018b) Beijing-KL ties brighten with Anwar's visit. Star, 28 Oct 2018
Hoffmann J, Sirimanne S (2017) Review of maritime transport-2017. UNCTAD, Geneva, p 104
Holmes J (2019) US-China war scenario: how would China's military attack a 'great wall in reverse'? NI, 30 Mar 2019
Holmes J, Winner A, Yoshihara Y (2009) Indian naval strategy in the twenty-first century. Routledge, Abingdon, p 37
Horowitz J, Alderman L (2017) Chastised by EU, a resentful Greece embraces China's cash and interests. NYT, 26 Aug 2017
Hu J (2005) Address at the First China-Spain Business Summit. FMPRC, Madrid, 15 Nov 2005
Huang Y (2015) Chinese-built deep sea port to boost Cameroon's economic growth: official. Xinhua, Yaounde, 16 Feb 2015
Huaxia (2018) Chinese nuclear power giant builds solar plant in Malaysia. Xinhua, Guangzhou, 9 Apr 2018; median $-RM exchange rate as in Oct 2018
Hurst D (2018) Abe, Modi Herald 'unparalleled potential' in Japan-India relations. Diplomat, 30 Oct 2018
Hussain N (2019) Regional consensus needed for a 'free and open Indo-Pacific'. EAF, 9 Feb 2019
Ianchovichina E, Martin W (2004) Impacts of China's accession to the WTO. World Bank Econ Rev 18(1):3–27
IANS (2019) China, Maldives clash over mounting Chinese debt as India warms up to Male. Economic Times, 8 July 2019
Independent Evaluation Department (2014) Sri Lanka: Tsunami-affected areas rebuilding project. ADB, Manila, Oct 2014, pp 1–2
Information Office (2018a) Guterres: FOCAC Beijing Summit helps African countries reduce poverty. State Council, Beijing, 5 Sept 2018
Information Office (2018b) Wang Yi meets with Former Deputy Prime Minister Anwar Ibrahim of Malaysia. MoFA, Beijing, 24 Oct 2018
International Cooperation Bureau (2017) Priority policy for development cooperation FY2017: 'free and open Indo-Pacific strategy'. MoFA, Tokyo, Apr 2017, pp 2–13
Islam S (2017) Payra to offer Bangladesh shippers new port option. IHS J Comm, 26 June 2017
Iyer-Mitra A (2018) Why India getting access to a strategic port in Oman may not amount to much. Business Standard, 23 Feb 2018

Jabarkhyl N (2017) Oman counts on Chinese billions to build desert boomtown. Reuters, Duqm, 5 Sept 2017

Jaipragas B (2019) Malaysia to go ahead with China-backed East Coast Rail Link. SCMP, 12 Apr 2019

Jaishankar D (2016) Indian Ocean region: a pivot for India's growth. Brookings, Washington, 12 Sept 2016

Jaishankar D (2018) A giant slowly rises. ISIS Focus, Issue 8, Feb 2018, pp 22–23

Japanese Embassy (2014) Japan-Bangladesh Summit Meeting. MoFA, Dhaka, 6 Sept 2014

Jennings R (2018) French, British Ships to sail disputed Asian Sea, Rile China. VoA, 15 June 2018

Jesus J, Gautier P (2012) Dispute concerning delimitation of the maritime boundary between Bangladesh and Myanmar in the Bay of Bengal. ITLOS, Hamburg, 14 Mar 2012, pp 90–115, 131–133, 141–148

Jia Y, Cao Z, Li X (2007) Advancing in big strides from a new historical starting point: record of events on how the Party Central Committee and the CMC promote Scientific Development in national defence and army building. Xinhua, Beijing, 7 Aug 2007

Jiang Z (1993) The international situation and military strategic guidelines. CMC, Beijing, 13 Jan 1993

JICA (2018) Preparatory survey on the Matarbari Port development: final report. CPA, Ministry of Shipping, Dhaka, Dec 2018

JICA, ICA (2016) Building quality infrastructure for Africa's development. African Development Bank, Abidjan, 22 Nov 2016, pp viii–ix

Johnson L (2018) Harvesting from 'poor old' China to harness 'poor young' Africa's demographic dividend? Bridges Afr 7(5), 5 July 2018

Johnson J (undated) Chinese Trade in the Indian Ocean. Asia Society. https://asiasociety.org/education/chinese-trade-indian-ocean Accessed 18 Sept 2018

Johnston A (2019) Shaky Foundations: the 'intellectual architecture' of Trump's China Policy. Survival 61(2):189–202

Jones J (2004) The global posture review of US military forces stationed overseas: SASC testimony. DoD, Washington, 23 Sept 2004

Joshi S (2013) Can India blockade China? Diplomat, 12 Aug 2013

Joshi P (2017) Quad 2.0's challenges for India: a Delicate Balancing Act. WotR, 7 Dec 2017

Junayd M (2019) Maldives high court orders police to release former president Yameen. Reuters, Male, 28 Mar 2019

Junayed M, Aneez S (2018) Maldives opposition calls for Indian intervention as president imposes emergency rule. Reuters, Male/Colombo, 6 Feb 2018

Kabir A (2016) Chinese President Xi Jinping's visit to Bangladesh gives boost to bilateral relations. Forbes, 14 Oct 2016

Kabir F (2017) BD to construct multipurpose terminal under 3rd Indian LoC. Financial Express. 14 Dec 2017

Kajimoto T (2018) China-Japan sign three-year FX swap deal to sign strengthen financial stability, business activity. Reuters, Tokyo, 26 Oct 2018

Kamal M (2015) In: Miglani S, Paul R (eds) Bangladesh favors Japan for port and power-plant, in blow to China. Reuters, Delhi/Dhaka, 10 Sept 2015

Katoch P (2018a) China's base in Sri Lanka part of its dominant Indian Ocean presence. Asia Times, 6 Aug 2018

Katoch P (2018b) After Seychelles faux pas, India must learn from China. Asia Times, 13 July 2018

Katoch P (2018c) China—consolidating in IOR at incredible pace. United Service Institution of India, Delhi, 29 June 2018

Kazmin A (2019) Kashmir terror attack revives old India-China tensions. FT, 18 Feb 2019

Keck Z (2013) China to sell Bangladesh 2 submarines. Diplomat, 22 Dec 2013

Kelly T, Perry M (2019) US, France, Japan and Australia hold first combined naval drill in Asia. Reuters, Tokyo, 16 May 2019

Krishnan A (2017a) China's great push forward: PLA's foreign military outposts a threat to India? India Today, 22 Dec 2017

Krishnan A (2017b) China's great push forward: PLA's foreign military outposts a threat to India? India Today, 23 Dec 2017

Kuoppamaki A, Bathka B (2018) Africa jumps aboard China's express train. DW, Beijing, 6 Sept 2018

Kwok D (2017) China's COSCO shipping buys $228m stake in Spain's Noatum Port. Reuters, Hong Kong, 13 June 2017

Lanteigne M (2008) China's maritime security and the 'Malacca Dilemma'. Asian Secur 4(2):143–161

Lee L, Latiff R, Blanchard B (2018) Malaysia's Mahathir cancels China-backed rail, pipeline projects. Reuters, Kuala Lumpur/Beijing, 21 Aug 2018

Li K (2018) Shouldering common responsibilities for meeting common challenges: speech at the 12th ASEM Summit. State Council, Brussels, 19 Oct 2018

Liang J, Bianji (2019) Abe to send envoy to BRF. PD, 17 Apr 2019

Lin C (2015) Admiral Zheng He and China's MSR. Times of Israel, 29 Sept 2015

Luan S (2006a) Chinese, Bangladeshi FMs hold talks. Xinhua, Beijing, 6 June 2006

Luan S (2006b) Tariff cuts on imports from 5 Asian nations. Xinhua, Beijing, 18 Aug 2006

Luan S (2006c) China, Bangladesh, Myanmar agree on road-connectivity. Xinhua, Beijing, 31 July 2006

MacDonald J, Donahue A, Danyluk B (2004) Energy futures Asia. Booz Allen Hamilton, McLean, Nov 2004

Mamun S (2016) Bangladesh signs MoU with China on Payra deep-sea port construction. Dhaka Tribune, 9 Dec 2016

Manek N (2018) Djibouti sees China involvement in port as no threat to US. Bloomberg, 15 Mar 2018

Marlow I (2018) Inside China's $1bn port in Lanka where ships don't want to stop. Bloomberg, 18 Apr 2018

Marshall D (2012) The string of pearls: Chinese maritime presence in the Indian Ocean and its effect on indian naval doctrine. Naval Postgraduate School, Monterey, Dec 2012, p 46

Maslow S (2018) Japan's 'Pivot to Asia': Tokyo discovers the Indo-Pacific. Asia & the Pacific Policy Society, 1 Aug 2018

Mathews J (2017) China's takeover of the Port of Piraeus in Greece: blowback for Europe. Asia-Pac J 15(13), No. 3, 1 July 2017

Mazumdar S (2017) India aims to woo Bangladesh away from China. DW, Delhi, 7 Apr 2017

MEA (2018a) Press release on presidential elections in Maldives. GoI, Delhi, 24 Sept 2018

MEA (2018b) Joint statement on the occasion of state visit of the President of the Republic of Maldives to India. GoI, New Delhi, 17 Dec 2018

Medeiros E, Crane K, Heginbotham E, Levin N, Lowell J, Rabasa A, Seong S (2008) Pacific currents: the responses of US allies and security partners in East Asia to China's rise. RAND, Santa Monica

Media Centre (2017) India-Australia-Japan-US Consultations on Indo-Pacific. MEA, Delhi, 12 Nov 2017

Media office (2018) OPIC convenes first trilateral meeting to promote infrastructure investment collaboration during visit to Southeast Asia. OPIC, Singapore, 17 July 2018

Mehta A (2018) Two US airmen injured by Chinese lasers in Djibouti. DoD says. Defense News, 3 May 2018

Menzies G (2003) In: Hua C (ed) Malacca most important port for Zheng He. Star, 1 Nov 2003

Miglani S (2017) India's 'new Silk Road' snub highlights gulf with China. Reuters, New Delhi, 21 May 2017

Miglani S (2018) Maldives seeks scaling back of Indian presence as it woos China. Reuters, Delhi, 10 Aug 2018

Miglani S, Aneez S (2018) Behind Sri Lanka's turmoil, a China-India struggle for investments and influence. Reuters, Colombo, 8 Nov 2018

Miglani S, Paul R (2015) Bangladesh favors Japan for port and power-plant, in blow to China. Reuters, Delhi/Dhaka, 10 Sept 2015

Mirski S (2018) Blockade: the ultimate way to win a war against China? NI, 30 July 2018

Mo J (2018) Visiting leaders hail success of FOCAC summit. CD, 7 Sept 2018

MoD (2019) India and Maldives to continue close cooperation on maritime security and counter-terrorism. GoI, New Delhi, 24 Jan 2019

Modi N (2018) Opening remarks by the PM at the Plenary Session of the India-ASEAN Commemorative Summit. PMO, Delhi, 25 Jan 2018

Modi N, Hasina S (2019) India-Bangladesh joint statement during official visit of Prime Minister of Bangladesh to India. MEA, Delhi, 5 Oct 2019, paragraphs 29–31

MoFA (2014) Joint Statement between the PRC and the PRB on Deepening the Closer Comprehensive Partnership of Cooperation. State Council, Beijing, 11 June 2014

MoFA (2015) Partnership for quality infrastructure: investment for Asia's future. GoJ, Tokyo, 21 May 2015

MoFA (2017) Joint Press Communique between the PRC and the Republic of Maldives. Beijing, 8 Dec 2017

MoFA (2019) Joint Statement on the 9th meeting of the China-Italy Joint Government Committee. GoI, Rome, 25 Jan 2019

Mogherini F (2018) Connecting Europe and Asia: building blocks for an EU strategy. High representative for Foreign Affairs and Security Policy, EC, Brussels, 19 Sept 2018

Mohammad H (2008) Foreign Policy under Ziaur Rahman. Daily Star, 31 May 2008

Mollman S (2016) ASEAN's statement about the SCS that was reportedly issued then retracted. Quartz, 15 June 2016

Moni D (2010a) In: Report: Bangladesh to persuade China to use Chittagong port. Reuters, Dhaka, 14 Mar 2010

Moni D (2010b) In: Krishnan A (ed) China offers to develop Chittagong port: Dhaka allays New Delhi's concerns over ties with Beijing. Hindu, 20 Sept 2010

Moore W (2018) 2018 FOCAC: Africa in the new reality of reduced Chinese lending. Centre for Global Development, Washington, 31 Aug 2018

Morell M (2018) We responded with urgency to 9/11. Now we need to respond as urgently to China. WP, 10 Sept 2018

Moss T, Collinson E (2018) Boom: a new US Development Finance Corporation! Centre of Global Development, Washington, 3 Oct 2018

Mu X (2019) China-Africa trade expo to forge closer economic partnership. Xinhua, Changsha, 28 June 2019

Mukherjee A (2014) Is Bangladesh the newest addition to China's 'string of pearls'? Geopolit Monitor, 28 July 2014

Mulligan M, Shaw J (2007) What the world can learn from Sri Lanka's post-tsunami experiences. Int J Asia-Pac Stud 3(2):65

Murphy Z (2010) Zheng He: symbol of China's 'peaceful rise'. BBC, London, 28 July 2010

Musgrave P, Nexon D (2017) Zheng He's voyages and the symbolism behind Xi Jinping's BRI. Diplomat, 22 Dec 2017

Mutethiya E (2018) Djibouti launches China-built free trade zone. CD, 6 July 2018

Nakamaru R (2019) Abe eyes diplomatic win with Xi Jinping visit but faces balancing act between China and US. Kyodo, Beijing, 16 Apr 2019

Nambiar S (2019) Buoying Malaysia's debt-heavy economy. EAF, 25 Jan 2019

NDRC (2015) Vision and actions on jointly building SREB and 21st century MSR (Chap. 1). State Council, Beijing, Mar 2015

NDRC (2017) Vision for maritime cooperation under the BRI. State Council, Beijing, 20 June 2017, Chap. 1

NDRC, MoFA, MofCOM (2015) Vision and actions on jointly building SREB and 21st-century MSR. State Council, Beijing, Mar 2015

NDRC, State Oceanic Administration (2017a) Vision and actions on energy cooperation in jointly building SREB and 21st-Century MSR. State Council, Beijing, May 2017

NDRC, State Oceanic Administration (2017) Vision for maritime cooperation under the BRI. State Council, Beijing, 20 June 2017

Nehru J (2012) In: Chunhao L (ed) US-India-China Relations in the Indian Ocean: a Chinese perspective. Strat Anal 36(4):627

Nikkei (2018) China's Silk Road initiative makes prey of Oman, Maldives. Asian Rev, 26 July 2018

NSC (2019) Twitter feed—'endorsing BRI lends legitimacy to China's predatory approach to investment and will bring no benefits to the Italian people'. White House, Washington, 9 Mar 2019

O'Rourke R (2010) China naval modernization: implications for US navy capabilities. CRS, Washington, 1 Dec 2010, p i

O'Rourke R (2019) China naval modernization: implications for US navy capabilities. CRS, Washington, 2 Oct 2019, p ii

Office of the Spokesman (2011) Statement on Defence Minister Liang Guangli's visit to the Seychelles. MND, Beijing, 13 Dec 2011

Osborn K (2018) The US military is worried about China's moves in Africa. NI, 15 Aug 2018

OSD (2018a) MSDPRC. DoD, Washington, 16 May 2018, pp i, 1, 50–51, 111–112

OSD (2018b) Assessment on US defense implications of China's expanding global access. DoD, Washington, Dec 2018, pp 1-3

Panda (2015) India Plucks a Pearl from China's 'String' in Bangladesh? Diplomat, 7 June 2015

Panda J (2017) The AAGC: an India-Japan arch in the making? Focus Asia, No. 21, Aug 2017

Panda A (2018a) India gains access to Oman's Duqm Port, putting the Indian Ocean geopolitical contest in the spotlight. Diplomat, 14 Feb 2018

Panda J (2018b) India-Japan embrace should stretch out to Eurasia. EAF, 13 Nov 2018

Panda A (2019) US, India, Japan, Philippine Navies Demonstrate Joint Presence in SCS. Diplomat, 11 May 2019

Panday P (2012) Bangladesh and Myanmar resolve longstanding maritime dispute. EAF, 26 Apr 2012

Panikkar K (2012) In: Chunhao L (ed) US-India-China relations in the Indian Ocean: a Chinese perspective. Strat Anal 36(4):627

Pant H (2018) India's Indian ocean challenge. Yale Global, 3 May 2018

Parashar S (2014) Chinese submarine docking in Lanka 'inimical' to India's interests: Govt. ToI, 4 Nov 2014

Parmar T (2016) China's President Xi Jinping makes 'historic visit' to Bangladesh. Time, 14 Oct 2016

Patranobis S (2018) Wary India looks on as Chinese president talks of strategic cooperation with Sri Lanka. Hindustan Times, 5 Feb 2018

Paul R (2014) UN tribunal rules for Bangladesh in sea border dispute with India. Reuters, Dhaka, 8 July 2014

Paul R (2016) China signs deals worth billions with Bangladesh as Xi visits. Reuters, Dhaka, 14 Oct 2016

Pharis W (2009) China's Pursuit of a Blue-Water Navy. Marine Corps University, Quantico, p ii

Pilling D (2017) Chinese investment in Africa: Beijing's testing ground. FT, 13 June 2017

Pilling D (2018) India's plans on archipelago in Mauritius cause unease. FT, 25 Oct 2018

Pillsbury M (2015) The hundred-year marathon: China's secret strategy to replace America as the global superpower. Henry Holt, New York, pp 7–14

Pompeo M (2018) America's Indo-Pacific economic vision: remarks at the Indo-Pacific business forum. DoS, Washington, 30 July 2018

Port of Zeebrugge (2018) COSCO shipping ports signs Concession Agreement with Port of Zeebrugge and reaches MOU with CMA for strategic partnership. Zeebrugge, 22 Jan 2018

References

Prakash A (2011) China's maritime challenge in the Indian Ocean. Maritime Affairs 7(1):1–16
Prakash A (2016) In: Raghuvanshi V (ed) Purchase of Chinese Subs by Bangladesh 'An Act of Provocation' toward India. Defense News, 23 Nov 2016
Press Office (2017) Readout of President Donald J. Trump's meeting with Prime Minister Narendra Modi of India. White House, Manila, 13 Nov 2017
PTI (2012) GMR suspects foreign hand in Male airport crisis. Hindu, 5 Dec 2012
PTI (2014) Pro-China Maldives president Yameen loses election, India welcomes result. Hindustan Times, 24 Sept 2018
PTI (2018a) Indian Ocean cannot be backyard of India: China. Economic Times, 12 July 2018
PTI (2018b) Pro-China Maldives president Yameen loses election, India welcomes result. Hindustan Times, 24 Sept 2018
PTI (2018c) China, Japan pledge new economic collaboration as Abe backs BRI. Financ Exp, 26 Oct 2018
PTI (2018d) India has nothing to worry about China-Bangladesh ties, says Sheikh Hasina. Indian Express, 21 Feb 2018
Public-Private Partnership Authority (2017) 3rd sea port: Payra Port coal terminal. PMO, Dhaka, 4 Oct 2017
Quadir S (2018) China to develop Bangladesh industrial zone as part of South Asia push. Reuters, Dhaka, 4 Apr 2018
Radhakrishnan R (2012) Maldives denies China role in GMR row. Hindu, 10 Dec 2012
Rahman M (2015) China to build deep-sea port for Bangladesh. Gulf Times, 16 Apr 2015
Raina H (2014) An Indian view of sea power. Strategist, 26 Mar 2014
Reddy V (2016) Reinforcing China's Malacca Dilemma. CIMSEC, Washington, 15 Sept 2016
Report (2010) Bangladesh to persuade China to use Chittagong port. Reuters, Dhaka, 14 Mar 2010
Report (2013a) Maldives holds fresh election for president. BBC, 9 Nov 2013
Report (2013b) Maldives elects new leader. UNDP, Male/Geneva, Nov 2013
Report (2014) Chinese firm to construct Sonadia seaport. Prothom Alo, 4 June 2014
Report (2016) Bangladesh, China sign 27 deals as President Xi visits Dhaka. BDNews24, Dhaka, 14 Oct 2016
Report (2017a) How much trade transits the SCS. China Power, CSIS, Washington, 27 Oct 2017
Report (2017b) Master plan for Payra deep-sea port on the cards. Star, 20 Sept 2017
Report (2018a) In first, Japanese submarine conducts drills in disputed SCS. JT, 17 Sept 2018
Report (2018b) China grants $66m for Jamestown Port, others. Daily Guide Africa, 10 Apr 2018
Report (2018c) China strengthens investments in African ports. Undercurrent News, 23 Apr 2018
Report (2018d) Sri Lanka political crisis deepens with parliament suspended after prime minister sacked. ABC, 28 Oct 2018
Report (2018e) Sri Lanka in political turmoil after prime minister Wickremesinghe sacked. Guardian, 27 Oct 2018
Report (2018f) Mahinda Rajapakse sworn in as new Prime Minister. Presidential Secretariat, Colombo, 26 Oct 2018
Report (2018g) Maldives election: opposition defeats China-backed Abdulla Yameen. BBC, 24 Sept 2018
Report (2018h) Operation cactus: how Indian troops went to Maldives and helped quell a coup. ToI, 7 Feb 2018
Report (2018i) India Looks at the Maldives and Sees China. Stratfor, Washington, 8 Feb 2018
Report (2018j) Opposition candidate wins Maldives presidential polls: Election Commission. Xinhua, Male, 24 Sept 2018
Report (2018k) Real reason Mahathir vague about cancelling China projects. Free Malaysia Today, 29 Aug 2018
Report (2018l) The Prime Minister in action: visit to China. Kantei, Beijing/Tokyo, 25 Oct 2018
Report (2019a) Europe continues to welcome Chinese investment in its ports. Belt & Road News, 22 June 2019

Report (2019b) Agreement signed for $800m line of credit from India. Maldives Indep, 19 Mar 2019

Report (2019c) China GDP. Trading Econ. https://tradingeconomics.com/china/gdp. Accessed 27 Aug 2019

Report (undated) Hambantota port development project, Sri Lanka. Environ Justice Atlas. https://ejatlas.org/conflict/hambantota-deep-seaport-and-the-projected-industrial-zone-sri-lanka. Accessed 18 Sept 2018

Reuters (2015) India's Modi hopes to tamp down China's influence in Bangladesh. VoA, 27 May 2015

Reuters (2016) Chinese investors to build industrial park at Oman's Duqm port. Gulf News, 23 May 2016

Reuters (2017) India's concerns addressed? Sri Lanka says no military use of Chinese-built port. Hindustan Times, 25 July 2017

Reuters (2018a) Maldives opposition candidate says he has won presidential election. CD, 24 Sept 2018

Reuters (2018b) Maldives seeks scaling back of Indian presence as it woos China. Economic Times, 10 Aug 2018

Riedel B (2018) Maldives democracy prevails for now. Brookings, Washington, 27 Sept 2018

Robinson J (2012) Maldives cancels GMR's $511m airport project. Reuters, Male, 28 Nov 2012

Rolland N (2019) A concise guide to the BRI. NBR, Seattle, 11 Apr 2019

Rolland N, Boni F, Bouwens M, Samaranayake N, Khurana G, Tarapore A (2019) Where the Belt meets the Road: security in a contested South Asia. Asia Policy 14(2):1–41

Roy S (2018) India gets access to strategic Oman port of Duqm for military use, Chabahar-Gwadar in sight. Indian Express, 13 Feb 2018

Rumsfeld D (2004a) The global posture review of US military forces stationed overseas: SASC testimony. DoD, Washington, 23 Sept 2004

Rumsfeld D (2004b) Positioning US forces for the 21st century. DoD, Washington, 5 Aug 2004

Sahgal A (2010) In: Krishnan A (ed) China offers to develop Chittagong port: Dhaka allays New Delhi's concerns over ties with Beijing. Hindu, 20 Sept 2010

Saieed Z (2018) Malaysia to gain from cancellation of controversial China projects. Star, 17 Sept 2018

Sen S (2017) China deploys a submarine in Indian Ocean as tensions over borders flare. Mail, 3 July 2017

Shams S (2018a) Sri Lanka's political crisis: in the shadows of China and India. DW, 18 Nov 2018

Shams S (2018b) Ex-Maldives President Nasheed: 'opposition victory resets ties with India'. DW, 24 Sept 2018

Shankar A (2019) Govt still in talks over Trans-Sabah gas pipeline, says Guan Eng. Edge Mark, 17 July 2019

Shepard W (2017) Why China is building a new city out in the desert of Oman. Forbes, 8 Sept 2017

Shibasaki R, Watanabe T (2011) Risk assessment of blockade of the Malacca Strait using international cargo simulation model. Int Assoc Marit Econ, Chios, pp 1–21

Shinn D (2018) FOCAC meets the Belt and Road. EAF, 18 Oct 2018

Shriver R, Gertz B (2018) PLA expanding power through BRI. WFB, 11 Sept 2018

Simon S (2017) US-Southeast Asia relations: mixed messages. Comp Connect 19(1):41–50

Singh A (2014) China's MSR: implications for India. IDSA Comment, Delhi, 16 July 2014

Singh A (2015) China's 'maritime bases' in the IOR: a chronicle of dominance foretold. Strat Anal 39(3):293–297

Singh S (2018a) How China-consciousness defines Malabar-2018. Asia Times, 5 June 2018

Singh A (2018b) India needs a better PLAN in the Indian Ocean. EAF, 12 May 2018

Singh K (2018c) At Modi-Abe summit, China will loom large. Asian Age, 17 Oct 2018

Singh A (2019) India bides its time in the Indian Ocean. EAF, 8 Feb 2019

SLPA (2018) Hambantota MRMR port. Colombo/Hambantota, 2018. http://www.slpa.lk/port-colombo/about-mrmr. Accessed 18 Sept 2018

Smyth J, White E (2018) US to make 'counter offer' to stop Huawei in Pacific. FT, 28 Sept 2018

Solomon S (2017) In trade with Africa. US playing catch-up. VoA, 18 Jan 2017

Special Correspondent (2018) Bangladesh resists pressure to allow Indian exports. JoC, 15 Oct 2018

Spinetta L (2006) 'The Malacca Dilemma': countering China's 'string of pearls' with land-based airpower. Air University, Maxwell AFB, June 2006

Spokesperson (2018) Joint statement on the inaugural US-India 2 + 2 ministerial dialogue. DoS, Delhi, 6 Sept 2018

Srinivasan R (2014) China's maritime threat: how India let its best bet Vizhinjam be sabotaged. First Post, 10 Nov 2014

Srinivasan M (2018a) Sri Lankan President Sirisena alleges that RAW is plotting his assassination. Hindu, 16 Oct 2018

Srinivasan M (2018b) Nasheed seeks Indian military intervention in Maldives. Hindu, 7 Feb 2018

Stacey K (2017) China signs 99-year lease on Sri Lanka's Hambantota port: critics denounce move as an erosion of country's sovereignty. FT, 11 Dec 2017

Stacey K (2018a) China investment in Bangladesh worries India. FT, 14 Aug 2018

Stacey K (2018b) Chinese investment in Bangladesh rings India alarm bells. FT, 7 Aug 2018

Stewart P (2019) With an eye on Iran, US clinches strategic port deal with Oman. Reuters, Washington, 24 Mar 2019

Storey I (2006) China's 'Malacca Dilemma'. China Brief 6(8), 12 Apr 2006

Subramanian S (2018) 'China factor' deters India from Maldives intervention. National, Abu Dhabi, 18 Feb 2018

Sullivan J (2019) US Africa strategy speech at trade and investment luncheon. DoS, Luanda, 20 Mar 2019

Szechenyi N (ed) (2018) China's MSR: strategic and economic implications for the Indo-Pacific region. CSIS, Washington

Taneja K (2016) Why India is worried about China's ambitious OBOR initiative. Scroll, 28 Mar 2016

Tarapore A (2018) The geopolitics of the Quad. NBR, Seattle, 16 Nov 2018

TBP (2014) Maldives, China eye bridge project for Male airport. BRICS Post, 2 Dec 2014

Teoh S (2017) ECRL: Malaysia touts rail trade route as rival to Singapore. Straits Times, 10 Aug 2017

Tiezzi S (2018) FOCAC 2018: rebranding China in Africa. Diplomat, 5 Sept 2018

Tisdall S (2012) Maldives: Coni report causes predictable outrage. Guardian, 9 Sep 2012

Todd L, Slattery M (2018) Impact of Investment from China in Malaysia on the local economy. Institute for Democracy and Economic Affairs, Kuala Lumpur, Oct 2018, p 5

Trump D, Modi N (2017) US and India: prosperity through partnership. White House, Washington, 26 June 2017

Tuegno (2018) FOCAC 2015 Summit: China fulfills all commitments on schedule. Cameroon Tribune, 7 June 2018

Tusk D (2018) Global partners for global challenges: Chair's statement. EC ASEM-12, Brussels, 19 Oct 2018, pp 1, 6. Emphasis in original

UNB (2010) China to help build deep-sea port. Star, 15 June 2010

UNCTAD STAT (2018) China Merchandise Trade Annual: 1948–2017. UNCTAD, Geneva

Under Secretary of Defence (Policy) (1999) 1999 summer study final report: Asia 2025. ONA/DoD, Newport, 4 Aug 1999

US Embassy (2019) Statement on the signing of the strategic framework agreement. DoS, Muscat, 24 Mar 2019

UNDP (2018) About Maldives. UN, New York/Geneva

Usmani A (2016) GMR to get $270m compensation from Maldivian government. Bloomberg, 28 Oct 2016

Usui K (2005) Stronger economic ties between Japan and the booming Indian economy. JBIC Today, JBIC, Tokyo, June 2005, pp 7–11

van Tol J, Gunzinger M, Krepinevich A, Thomas J (2010) AirSea Battle: a point-of-departure operational concept. CSBA, Washington, 18 May 2010

Wang C (2018a) China-Maldives 'project of the century' bridge opens, reflects China's BRI vision. GT, 31 Aug 2018

Wang C (2018b) Maldives celebrates opening of China-built bridge as the two countries vow to strengthen ties. GT, 31 Aug 2018

Wardell J, Packham C (2018) Australia, US, India and Japan in talks to establish belt and road alternative. Reuters, Sydney, 19 Feb 2018

Washburne R (2018a) In: Churchill O (ed) China hasn't changed Belt and Road's 'predatory overseas investment model', US official says. SCMP, 13 Sept 2018

Washburne R (2018b) In: Delaney R (ed) India in talks to join US global development partnership countering China's Belt and Road plan. SCMP, 25 Sept 2018

Washburne R (2018c) OPIC President and CEO's statement as president signs BUILD Act into Law. OPIC, Washington, 5 Oct 2018

Web Desk (2018) Maldives crisis: India's stand, the China angle. Indian Express, 9 Feb 2018

Wey A (2018) Foreign policy concerns swayed Malaysia's voters. EAF, 28 July 2018

WBG Data (2019) https://data.worldbank.org/country/china. Accessed 27 Aug 2019

Wickremesinghe R (2017) In: Dutta P (ed) Sri Lanka leases Hambantota port to China for non-military use. Should India worry? India Today, 29 July 2017

Wing C (2015) Fujian to explore 21st century MSR opportunities. HKTDC Research, Hong Kong, 2 Dec 2015

Withnall A (2018) Maldives election: pro-China President Yameen concedes shock defeat as rival India celebrates 'triumph of democracy'. Independent, 24 Sept 2018

Wolfrum R, Cot J, Mensah T, Rao P, Shearer I (2014) Bay of Bengal maritime boundary arbitration between Bangladesh and India. PCA, the Hague, 7 July 2014

Wong S (2018) ASEAN Braces For a Risen China. ASEAN Focus 25(6), 25 Dec 2018, pp 2–3

Xi J (2014) Remarks during the Joint Press Statement with President Yameen. MoFA, Male, 15 Sept 2014

Xi J (2018a) Keynote speech at 2018 FOCAC opening ceremony. Xinhua, Beijing, 4 Sept 2018

Xi J (2018b) Message to 2nd ministerial meeting for the China and the Community of Latin American and Caribbean States Forum. Xinhua, Beijing, 23 Jan 2018

Xinhua (2005a) Zheng He anniversary harvests peace, prosperity. CD, 7 Aug 2005

Xinhua (2005b) Zheng He: a peaceful mariner and diplomat. State Council, Beijing, 12 July 2005

Xinhua (2014) China to help Maldives build 1,500 homes. CD, 16 Sept 2014

Xinhua (2017) Fujian: a new star on ancient MSR. CD, 8 Mar 2017

Xinhua (2018a) Beijing declaration, action plan adopted as FOCAC summit. MoFA, Beijing, 5 Sept 2018

Xinhua (2018b) China's maritime economy expands by 7.5 pct in recent five years. China Focus, Beijing, 21 Jan 2018

Xinhua (2018c) President Xi meets Malaysian PM, calling for better ties in new era. State Council, Beijing, 21 Aug 2018

Xinhua (2018d) China pledges to lift ties with Malaysia to new high. State Council, Beijing, 21 Aug 2018

Xinhua (2018e) Xi meets Japanese Prime Minister, urging effort to cherish positive momentum in ties. PD, 28 Oct 2018

Yameen A (2014) Press statement on the state visit by H.E. Xi Jinping, President of the PRC. MoFA, Male, 15 Sept 2014

Yang W (2014) Admiral Zheng He's voyages to the 'West Oceans'. Educ About Asia 19(2):26–30

Yang X (2018) When India's strategic backyard meets China's strategic periphery: the view from Beijing. WotR, 20 Apr 2018

Ye Q, Han S (2019) China contributes to Italian port construction for win-win results. PD, 22 Mar 2019

Yong Y (2018) Azmin: FoN in SCS must remain unencumbered. Star, 11 Oct 2018

References

Yoshida R (2014) Bangladesh, Japan seal aid deal, agree to nuclear power talks. JT, 26 May 2014

You J (2007) Dealing with the Malacca Dilemma: China's effort to protect its energy supply. Strat Anal 31(3):467–489

Yu Z (2016) The importance of the Malacca Dilemma in the BRI. J Policy Sci 10:85–109

Zhang Y (2018) President: Belt, Road needs Latin Flavor. CD, 23 Jan 2018

Zhen S (2016) Chinese company Landbridge wins 99-year lease on northern Australia's Darwin port. SCMP, 11 May 2016

Zhou B (2014) The string of pearls and the MSR. China-US Focus, Hong Kong, 11 Feb 2014

Zhu L (2005) New corridor opens to boost Sino-Bangladesh trade. Xinhua, Beijing, 17 Nov 2005

Zhu R, Tang L (2019) Maldives, China set example of friendship between small and big countries in int'l community: Maldivian FM. Xinhua, male, 21 July 2019

Zielinski S (2013) The giraffes that sailed to medieval China. Sci News, 1 Oct 2013

Chapter 7
Conclusion

7.1 The Belt and Road Initiative's Volatile Strategic Backdrop

In late-April 2019, Xi Jinping welcomed 37 heads of state and -government,[1] leaders of major multilateral organisations and around 5000 delegates from more than 150 countries, to the 2nd Belt and Road Forum for International Cooperation (BRF) in Beijing. While much larger than its 2017 forerunner, the 2nd BRF was a more sombre gathering. China was in the throes of an intensifying tariff war with the United States, whose multi-dimensional and openly adversarial focus, increasingly described as a new Cold War, was causing serious economic, diplomatic and strategic stress, both bilaterally and in the wider geopolitical and international security realms.[2] Participation by the leaders of several US-allied countries may have been the major surprise.

Xi's assurances that China harboured no hegemonic aspirations to *primacy* made little impact on his critics. He vowed, 'the BRI may be China's idea, but its opportunities and outcomes are going to benefit the world. China has no geopolitical calculations, seeks no exclusionary blocs and imposes no business deals on others.'[3] He had marked the 40th anniversary of Deng Xiaoping's transformative reforms by noting, 'there is no textbook of golden rules to follow for reform and development in China, a country with over 5000 years of civilization and more than 1.3 bn people. No one is in a position to dictate to the Chinese people what should or should not be done.'[4] Four months later, on the 70th anniversary of NATO's founding, US Vice

---

[1] Austria, Azerbaijan, Belarus, Brunei, Cambodia, Chile, Cyprus, Czech Republic, Djibouti, Egypt, Ethiopia, Greece, Hungary, Indonesia, Italy, Kazakhstan, Kenya, Kyrgyzstan, Laos, Malaysia, Mongolia, Mozambique, Myanmar, Nepal, Pakistan, Papua New Guinea, Philippines, Portugal, Russia, Serbia, Singapore, Switzerland, Tajikistan, Thailand, Uzbekistan, UAE, Vietnam.
[2] Watts (2019), Pieraccini (2019), Khanna (2019) and Dodwell (2019).
[3] Xi (2018a).
[4] Xi (2019a).

President Mike Pence aimed the Atlantic alliance's power at China, targeting BRI specifically.[5]

Washington, most vocal among Beijing's critics, led a growing band. An expanding body of opinion among US allies, especially Quad-members, but also including several European states, pointed an accusatory finger at China as the source of serious, even existential, threats to good global order, stability and prosperity. Politicians of varying ideological hues, national security officials and scholars and analysts from these lands built up a large corpus of policy-documents and academic research purportedly presenting evidence of Xi-led China's subtly malign attempts at covertly usurping global power to supplant the USA as the *systemic primate*. BRI was highlighted as a key element of that dangerous stratagem. The World Bank disagreed.

7.1.1 Economic Outcomes

WBG economists reported that BRI would create 'new land and maritime connections' across Eurasia, and reduce trade-time by between 2.4% and 4.4%. When infrastructure improvements were combined with reduction in cross-border delays, trade-time would be reduced by between 7.4% and 10.9% respectively. WBG analysts inferred, if BRI partners reformed trade-facilitation while improving connectivity, BRI's 'positive impact on trading time will be magnified.'[6] They concluded, 'deeper trade agreements and improved market access would magnify the trade impact of BRI infrastructure projects.' Easier access and movement, supportive regulatory regimes, and state-level cooperation would enable BRI infrastructure projects to increase 'total exports by 11.2% and 12.9%' respectively, trade gains being 'larger if trade cooperation complemented infrastructure cooperation', as urged by Beijing.[7]

Even the US-government funded RAND Corporation, a think-tank traditionally associated with the intellectual bases underpinning much DoD strategy,[8] concluded BRI made a large contribution to economic development among partner-states. RAND analysts found multimodal transport infrastructure and connectivity were key to boosting international trade and growth. Specifically, efficient transport infrastructure reduced transport costs and—times, improving delivery reliability. Empirical evidence showed cost of transport, in both time and money, significantly impacted on trade flows. Good transport infrastructure facilitated trade expansion. Efficient transport systems encouraged industrialisation while enabling efficient regional and global production networks. This generated more employment, positively affecting economic development.[9]

---

[5]Pence (2019).
[6]Baniya et al. (2018, p. 28).
[7]Baniya et al. (2018, p. 29).
[8]See, for instance, Ochmanek et al. (2015).
[9]Lu et al. (2018, pp. 2–3).

7.1 The Belt and Road Initiative's Volatile Strategic Backdrop

Better transport infrastructure improved regional connectivity, enabling economic integration. This could lower transport and trade costs, accelerate industrial agglomeration, increase labour productivity and foster development and regional- and national welfare. Improving transport infrastructure across the BRI alignments 'will result in a win-win situation for trade, both for the BRI region and areas further afield.' Trade volumes could increase beyond the BRI region, e.g., the EU. RAND recommended, 'Countries should work together to ensure the initiative delivers sustained economic, social and environmental benefits.'[10]

BRI-managers at the Leading Group established to supervise implementation, and NDRC executives monitoring execution, acknowledged implementation faced numerous challenges, but not necessarily of the variety promoted by China's US-led critics. The former reminded BRF participants that BRI 'originated in China, but it belongs to the world. It is rooted in history, but oriented toward the future. It focuses on Asia, Europe and Africa, but is open to all partners. It spans different countries and regions, different stages of development, different historical traditions, different cultures and religions, and different customs and lifestyles.'

Significantly, BRI was 'an initiative for peaceful development and economic cooperation, rather than a geopolitical or military alliance. It is a process of open, inclusive and common development, not an exclusionary bloc or a "China Club". It neither differentiates between countries by ideology nor plays the zero-sum game. Countries are welcome to join in the initiative if they so will.' Quoting Xi's remarks at BRI's 5th Anniversary symposium, BRI managers noted that the initiative would advance via 'transition from making high-level plans to intensive and meticulous implementation, so as to realize high-quality development, bring benefits to local people, and build a global community of shared future.'[11] Xi and his 37 peers, pledging to advance BRI-implementation via 'extensive consultation, joint efforts, shared and mutual benefits', emphasizing the import of 'the rule of law and equal opportunities for all', asserting that 'all states are equal partners for cooperation', and affirming they respected 'sovereignty and territorial integrity of each other', endorsed this perspective.[12]

Data presented by BRI's NDRC managers informed this consensus: in 2013–2018, goods trade between China and its BRI partners reached $6975.62 bn; China-BRI trade grew faster than China's foreign trade generally; private enterprises from across BRI-states accounted for 43% of the trade volume; China chiefly exported electromechanical items while it mainly imported electrical equipment and fossil-fuels; China's trade with Kazakhstan, Mongolia, Montenegro and Qatar grew by over 35% in 2017, showing the potential; by the end of February 2019, SREB's Europe-China freight rail-service completed 14,000 trips linking 48 Chinese cities to 42 cities in 14 European countries. MSR initiatives contributed to links established among more than 600 ports in 200 countries; civil aviation agreements with 62 BRI-members led to about 5100 direct flights weekly to 45 countries; China signed 171 agreements with 123 countries and 29 international organisations enabling and operationalising these

---

[10]Lu et al. (2018, pp. 3–8).

[11]Leading Group for Promoting the BRI (2019).

[12]FMPRC (2019).

terrestrial and maritime networks, 50 agreements being signed in 2017. Financing, tourism, cultural exchanges and scholarly interaction, too, significantly expanded.[13]

BRI's dynamic project-portfolio, progressively expansive catchment area, and resultant complexity, did pose substantial challenges. The ADB's experience was instructive. Confronting widespread critique, in the spring of 2016, Beijing asked the ADB to conduct a 'technical assistance' assessment of the BRI blueprint. ADB and NDRC officials reached accord on the evaluation's objectives, framework and methodology on 22 June 2016, and signed a formal agreement on 22 July, when the TA became 'effective'. The contract set out the completion date for the assessment on 31 December 2017. However, as the complexity of the evaluation process dawned on the ADB team, the deadline was first delayed to 2018, then to 30 June 2019 and, then again, to 30 June 2020.[14] In short, comprehensively evaluating BRI's outcomes in contributions and liabilities would challenge the talents of even the ablest teams of professionals. For others, the difficulties would likely be impossible to meaningfully master.

The BRF's displays of mutuality driving BRI-implementation made little impression on the downward trajectory defining Sino-US relations or US perceptions of Chinese policy. As the US trade war against China advanced, any residual misconceptions Beijing may have nurtured about the nature of Beltway angst were corrected by US leaders, e.g., Vice President Mike Pence, Secretary of State Mike Pompeo, acting-Secretary of Defence Patrick Shanahan, and his successor, Mark Esper, in formal statements. Pence's October 2018 address at the Hudson Institute in Washington was effectively a declaration of Cold War with China. Pompeo accused Beijing of threatening US 'national interests' with corrupt practices allegedly underpinning BRI: 'When China shows up with bribes to senior leaders in countries in exchange for infrastructure projects that will harm the people of that nation, then this idea of a treasury-run empire build is something that would be bad for each of those countries and certainly presents risks to American interests, and **we intend to oppose them at every turn**.'[15]

7.1.2 The Geopolitics of Geoeconomics

Sinophobia gripped not just the Washington Beltway, but also allied capitals in Europe and Asia. This was seen in several governments barring Chinese firms, notably Huawei and ZTE, from building elements of the 5th-generation (5G) ICT networks on national security grounds. The USA's 'five-eyes' allies openly or tacitly joined Washington's vigorous campaign to insulate 'free-world' networks from Chinese influence. US allies or 'partners' engaging with Chinese providers were

---

[13] Bianji (2019), NDRC (2019) and Zhang (2018).
[14] Bo (2016).
[15] Pompeo (2018). Emphasis added.

7.1 The Belt and Road Initiative's Volatile Strategic Backdrop

threatened with restrictions on classified information exchanges.[16] 5G systems would form the core of data-processing-and-sharing pathways powering the future global economy and security ecology. Huawei and other Chinese firms, advancing their 'Chinese' framework, promised to erect China's 'digital Silk Road' to planetary pre-eminence, putatively threatening US scientific-technological dominance.

Constructs like the 'Arctic Silk Road' and 'Polar Silk Road', extending the BRI framework into realms hitherto unused to substantial Chinese presence, consolidated the notional 'China challenge'.[17] Perceptually formidable threats to the US-led international order posed by Xi's supposedly increasingly centralised, assertive, authoritarian and authoritative administration elicited robust counteraction. This dialectic dynamic pushed the progressively bipolar *systemic core* into a definably adversarial framework—a 'new Cold War'.[18] This was the landscape on which BRI's Western critics fashioned the counter-BRI discourse. And yet, empirical examination of the Chinese reality demonstrated that contrary to the BRI literature, tighter centralisation under Xi notwithstanding, 'China's complex, multilevel governance system still makes it extremely difficult for Beijing to pursue a coherent, consistent grand strategy.'[19]

With Donald Trump expanding punitive tariffs, threatening to tax all Chinese-made imports, and Xi Jinping refusing to compromise, the two leaders met on the sidelines of a G20 summit in Argentina. Not surprisingly, this 'last opportunity' to negotiate a resolution proved to be anything but. Even though US and Chinese officials walked back from a complete commercial breach, talks collapsed amidst mutual recriminations. Trump and Xi made another attempt at reviving trade negotiations during a subsequent G20 summit in Japan in June 2019, and after 13 rounds of talks, Trump announced a 'Phase-I' truce in October,[20] but asymmetrically offensive-defensive competitive dynamics deepening the divide could not be bridged.

On assuming the role of Acting Secretary of Defence, Shanahan, articulating US strategic objectives, told DoD leaders, 'While we're focused on ongoing operations, remember China, China, China.'[21] Conflating BRI's geoeconomic bases with its purported geopolitical threats to US *primacy*, he rationalised strategic responses: 'Starting in the Indo-Pacific, our priority theatre, we continue to pursue many belts and many roads by keeping our decades-old alliances strong and fostering growing partnerships.'[22] Vice President Pence, spearheading the United States' China mission, warned NATO-members at the alliance's 70th anniversary, 'perhaps the greatest challenge NATO will face in the coming decades is how we must all adjust the People's Republic of China. And adjust we must.' BRI topped in Pence's wake-up call

---

[16] Dow Jones (2019), Reuters (2019) and Sanger et al. (2019).
[17] Sukhankin (2018), Eiterjord (2018), Scrafton (2018) and Wen (2018).
[18] Edel and Brands (2019), Pei (2019a) and Wu (2019).
[19] Jones and Zeng (2019).
[20] Politi (2019) and Swanson (2019).
[21] Shanahan et al. (2019).
[22] Shanahan (2019a).

alerting Europe to 'the challenge of Chinese 5G technology, meet the challenge of the easy money offered by China's BRI, (it) is a challenge European allies must contend with every day.' BRI highlighted China's 'global threat': 'The implications of China's rise will profoundly affect the choices NATO members will face, individually and collectively … our European allies must do more to maintain the strength and deterrence of our transatlantic alliance with their resources.'[23]

Defending DoD's draft NDAA 2020 budgetary allocations, CNO, Admiral John Richardson, told Senators that China (and Russia) determinedly drove 'to replace the current free and open world order with an insular system. They are attempting to impose unilateral rules, redraw territorial boundaries, and redefine exclusive economic zones so they can regulate who comes and who goes, who sails through and who sails around.' These actions were 'undermining international security. This behaviour breeds distrust and harms **our most vital national interests**.'[24] BRI loomed large in Washington's strategic insecurity calculus against that threatening backdrop: 'China's BRI in particular is blending diplomatic, economic, military, and social elements of its national power in an attempt to create its own globally decisive naval force.' BRI built commercial ports, upgraded domestic facilities and invested in infrastructure; 'slowly, as the belt tightens, these commercial ports transition to dual uses, doubling as military bases that dot strategic waterways. Then, the belt is clinched as China leverages debt to gain control and access…like an anaconda enwrapping its next meal.'[25]

The message: DoD must robustly respond. In a symbolic tightening of the planetary noose around Beijing, Shanahan warned China's neighbours that while Washington led the effort to box Beijing in, they must reinforce the counter-China coalition. Addressing counterparts and military commanders from across the Pacific rim and Europe, Shanahan pointed to the United States' Better Utilization of Investment Leading to Development (BUILD) Act, and the US International Development Finance Corporation it established, with development-finance funds of $60bn, 'to better service infrastructure needs across the region', in a clear counterpoint to BRI and the AIIB. Signalling US determination to retain dominance, Shanahan urged, 'no nation can go it alone. No one nation can or should dominate the Indo-Pacific. It is in all our interests to work together to build a shared future, one that is better than anything any of our nations could achieve on our own.'

7.1.3 Dialogue de Sourds

That appeal for collaboration preceded an attack on 'the greatest long-term threats to the vital interests' of regional states 'from actors who seek to undermine rather than uphold the rules-based international order.' Shanahan's focus on 'actors (who)

---

[23] Pence (2019).

[24] Richardson (2019, p. 3). Emphasis added.

[25] Richardson (2019, p. 4).

undermine the system by using indirect, incremental action and rhetorical devices to exploit others economically and diplomatically and coerce them militarily', left little to the imagination. He summarised China's political, economic, territorial and military sins:

- Deploying advanced weapons systems to militarise disputed maritime areas, destabilising the peaceful status quo by threatening the use of force to compel rivals into conceding claims
- Using influence operations to interfere in the domestic politics of other nations, undermining the integrity of elections and threatening internal stability
- Engaging in predatory economics and debt sovereignty deals, lubricated by corruption, which take advantage of pressing economic needs to structure unequal bargains that disproportionately benefit one party, and
- Promoting state-sponsored theft of other nations' military and civilian technology.[26]

The charges covered the spectrum of China's alleged malfeasance repeatedly cited in myriad US official and congressional reports, testimonies and statements. BRI formed a part of the threatening landscape on which the USA plotted coalition-warfare against the allegedly revisionist China. Shanahan warned, 'behaviour that erodes other nations' sovereignty and sows distrust of China's intentions must end.' Washington did 'not seek conflict', and China could return to the US-led order for a cooperative, and subordinated, future, 'but we know that having the capability to win wars is the best way to deter them.' DoD sought to ensure this with the highest ever R&D investment 'in emerging technologies like artificial intelligence, hypersonics and directed energy, much of which is aimed at the unique challenges in this theatre.'[27] China's 'challenges' were clearly exercising the US policy-elites.

To meet these, Indo-PACOM deployed over 370,000 personnel, more than four times the strength of any other combatant command, 2000-plus aircraft, and more than 200 ships and submarines, to the Indo-Pacific. Forces were boosted, alliances/partnerships deepened, and interoperability reinforced. Shanahan urged partners to take their own responsibilities seriously. His remarks introduced the latest iteration of DoD's Indo-Pacific Strategy Report launched simultaneously. In it, Shanahan accused China's CPC leadership of seeking 'to reorder the region to its advantage by leveraging military modernization, influence operations, and predatory economics to coerce other nations' while, in contrast, the United States' military activities supported 'choices that promote long-term peace and prosperity for all.' He warned, 'We will not accept policies or actions that threaten or undermine the rules-based international order… We are committed to defending and enhancing these shared values.'[28]

Shanahan's Chinese counterpart, General Wei Fenghe, at the same forum, issued a robust rebuttal: at the crossroads where humanity found itself, 'building a community

---

[26] Shanahan (2019b).
[27] Shanahan (2019b).
[28] Shanahan (2019c).

with a shared future for mankind' was 'the right path forward'. His rhetorical salvo painted the *primate* in narrow-minded, selfish colours—'which should we choose, peace and development or conflict and confrontation? openness and inclusiveness, or isolation and exclusiveness? win-win cooperation or zero-sum game? mutual learning among civilizations or arrogance and prejudice?'[29] Slamming 'some' for 'recklessly hyping up, exaggerating and dramatizing the "China threat theory",' and pledging China would never attack another state, acquire foreign land, or seek hegemony, Wei vowed, specifically regarding Taiwan and SCS maritime claims, never to give up 'an inch' of China's territory, definitely counter-attack if attacked, and defend the national interest 'until the end'. While stressing China's pacific and defensive approach to the world, Wei cautioned, 'No country should ever expect China to allow its sovereignty, security and development interests to be infringed upon.'[30] The status quo-vs.-revisionist line dividing the self-proclaimed *systemic primate* and its presumed near-peer rival was drawn.

Reinforcing Wei's rejection of US allegations that Beijing's reneging on agreed compromises triggered the collapse of earlier trade-talks, China's cabinet accused Washington of repeatedly doing just that, while progressively adding new demands encroaching on China's sovereign jurisdiction. 'China's only intention' was 'to reach a mutually acceptable deal' that would be 'in accordance with the principle of mutual respect, equality and mutual benefit.' Beijing explained: 'mutual respect means that each side should respect the other's social institutions, economic system, development path and rights, core interests, and major concerns.' Additionally, 'one side should not cross the other's "red lines". The right to development cannot be sacrificed, still less can sovereignty be undermined.'[31] Each side thus accused the other of committing fundamental breaches of long-established norms of inter-state conduct.

By demanding fundamental changes to the PRC's political-economic structure so as to conform to the US model of free-market capitalism, Washington, in Chinese eyes, violated these basic principles. Beijing's insistence on equality and mutual respect indicated China's rejection of its permanent subordination within an order designed and enforced by others; the reverse of that defensive insecurity was the US perception of Chinese action as an intolerable threat to the USA's perpetual *primacy*. That openly adversarial dialectic exchange, while clarifying any residual ambiguity over the virtually complete re-bipolarisation of the *systemic core*, triggered much regional concern. Anxiety was writ large in Malaysia's Defence Minister Mohamad Sabu's somewhat plaintive response, 'We love America. But we also love China.'[32]

---

[29] Wei (2019, pp. 1–3).
[30] Wei (2019, pp. 3–5).
[31] State Council (2019).
[32] Sabu et al. (2019).

7.2 Empiricism's Rational Sobriety

In his keynote address inaugurating the 18th Shangri-La Dialogue, host-state Singapore's Prime Minister, Lee Hsien Loong, underscored the strategic turbulence precipitated by deepening US-Chinese rivalry. Singapore's status as a 'steadfast US partner', the United States' 'only major Security Cooperation Partner' in Southeast Asia providing 'valuable access to US Navy ships and US military aircraft', and one whose fighter squadrons regularly trained with US counterparts, was acknowledged by Secretary of Defence Shanahan.[33] That unique security linkage, manifest in Singaporean naval participation in the Quad's first major Indian Ocean drill, Malabar-07-2, in September 2007, and close association since, placed Lee in a strong position to balance the varied perspectives informing national, regional and global strategic discourses from a US-friendly standpoint. His remarks sympathetically illuminated the dangerously combustible flux.

Reflecting upon regional anxiety, Lee noted, 'China's growth has shifted the strategic balance and the economic centre of gravity of the world—and the shift continues. Both China and the rest of the world have to adapt to this new reality.' Given these changes, China could 'no longer expect to be treated in the same way as in the past, when it was much smaller and weaker.' In the security realm, too, 'China is a major power … To protect its territories and trade routes, it is natural that China would want to develop modern and capable armed forces and aspire to become not just a continental power, but also a maritime power.' Still, 'to grow its international influence beyond hard power, military strength, China needs to wield this strength with restraint and legitimacy.' In that context, 'the US/China bilateral relationship is the most important in the world today. How the two work out their tensions and frictions will define the international environment for decades to come….China must now convince other countries through its actions that it does not take a transactional and mercantilist approach.'[34]

Realistic about the inevitability of occasional frictions between China and other countries, Lee urged Beijing to address these with diplomacy and within the bounds of international law, e.g., UNCLOS, 'rather than force or the threat of force', so that it gained respect and trust as a reliable and peaceful power. Reciprocity was key: 'The rest of the world, too, has to adjust to a larger role for China. Countries have to accept that China will continue to grow and strengthen and that it is neither possible nor wise for them to prevent this from happening.' Indeed, 'China should be encouraged to play commensurate and constructive roles in super-national institutions, like the IMF, the World Bank and the WTO.' The risk was, 'If China cannot do so, it will create its own alternatives.' That was presumably the framework for assessing BRI.

'Amid the geopolitical shifts, new concepts and platforms for regional cooperation have emerged, notably China's BRI. Singapore supports a Belt and Road Initiative. We see it as a constructive mechanism for China to be positively engaged with the region and beyond. That is why we are active participants.' Lee noted, Singapore

---

[33] Shanahan (2019b).
[34] Lee (2019).

worked with the World Bank to promote financial and infrastructure connectivity, and provided supporting professional and legal services to BRI-partners. Singapore also partnered with China in fashioning the new International Land-Sea Trade Corridor, connecting western China to Southeast Asia under the China-Singapore Chongqing Connectivity Initiative. Lee insisted, 'the substance of the BRI and the way in which the BRI is implemented are very important. The specific projects must be economically sound and commercially viable and must bring long-term benefits to its partners. This has not always been the case.' His conclusion: 'BRI must be open and inclusive and must not turn the region into a closed bloc centred on a single major economy.'[35] This jibed closely with Xi Jinping's own assurances to his 37 BRI counterparts at the 2nd BRF.

Lee's Malaysian counterpart, Mahathir Mohamad, addressing another international gathering three weeks later, refined the proposition: 'nations which once may be opposed to each other should find common grounds and become partners in ventures and endeavours that would benefit both. Hostility and belligerence benefits no one except those arms traders and, of course, war profiteers.' Focusing on the impact of US-China tensions on Southeast Asia, Mahathir added, 'rejection of wars serves our region well as it had enjoyed relative peace and had prospered on that very peace.' His critique of great powers was direct: 'At a time when we pride ourselves as being civilised, we find leading nations still bent on killing people in the pursuit of their national interest and agenda. There is nothing civilised nor advanced when war is an option to solve problems. We should all work towards criminalising wars.' Against that backdrop, Mahathir believed, 'There should be no doubt that the BRI with the full and fair participation of all stakeholders, can be a win-win proposition.'[36]

7.2.1 The Continuity and Disruptions Marking Xi's China

Lee Hsien Loong, Mahathir Mohamad, the World Bank and the ADB could not be accused of serving Beijing's covert design to subvert the US-led post-War order and the post-Soviet security *system*. So, what triggered Washington's abrupt switch from restrained engagement balanced with Obama's Pacific *Pivot* to Trump's neo-containment campaign? Western analysts identified myriad 'threats' emanating from China, but a consensus emerged on the personality of, and policy goals advanced by, the Xi Jinping-led CPC. Some commentators insisted 'emperor Xi', centralising authority, abandoning collective leadership procedures, solidifying Party control over the Chinese state, removing presidential term-limits and replacing Deng Xiaoping's 'hide and bide' dictum with an assertive approach to allegedly time-bound plans to supplant US *primacy*, thus existentially threatening US 'national security' interests, was the problem.[37]

---

[35] Lee (2019).

[36] Mahathir (2019).

[37] Thayer and Han (2019), Lian (2010), Economy (2019) and Fukuyama (2018).

The charges reflected China's reality.[38] However, an apparent neglect of contextual complexity weakened the logical merit of the more extreme commentaries. As Chap. 2 demonstrates, Xi inherited the legacy handed him by Hu Jintao who, rather than Xi, had initiated China's robust responses to perceived threats to China's maritime economic life-line, phrased as Beijing's 'Malacca dilemma', to be addressed with expanded duties Hu tasked the PLA, especially a fast-modernising PLAN, to perform. Hu also boosted ties to China's Eurasian neighbours so as to both reinforce security ties and expand commercial and transport linkages. The foundations of much of what Xi did in his first term was laid by his predecessors, but action to counter challenges to CPC-centrality manifest in the conduct of regional leaders, e.g., Bo Xilai, and some senior PLA officers, had only begun to be taken as Hu's term neared its end.

Xi's anti-corruption campaign formalised and dramatically expanded that endeavour, but insecurity triggered by its scale and consequences might not have been fully appreciated either within China or without. Even Deng, conscious of corrupt practices among CPC cadres flowing from economic liberalisation, railed against these, but only took limited action. Deng also harangued CPC leaders on the need for regional officials to obey Central Committee- and State Council instructions, which some, betraying a 'mountain-stronghold mentality', did not.[39] Successive General Secretaries and Premiers, including Hu Jintao and Wen Jiabao, later expressed frustration over the lack of speed or vigour with which central decisions were implemented, and occasionally ignored. Rapidly gathering challenges to domestic coherence[40] appeared significant enough to demand predictability enforced by unifying PBSC leadership, CMC chairmanship—both constitutionally indefinite—and the presidency. Anxious insecurity may have played as big a role as confident over-reach in shaping those decisions.

Xi's authoritarian bent and the concentration of power in his own hands and those of trusted acolytes were seen to have contributed to the perception of threats originating in China's growing strength and assertiveness. However, incipient and perhaps ineradicable factionalism within the CPC mainstream, and Xi's record of leavening strictly-phrased policy proclamations targeting domestic and external audiences with flexible interpretations and implementation indicated rhetoric's role in mobilisational initiatives and tactics in the pursuit of 'the ultimate goal of growing Xi's own power and that of China.' The 'reality' unveiled by analysts was that Xi 'faces challenges both at home and abroad.'[41] Passionate examinations pointed to axiomatic assumptions of greater capability acquired, and consequently, a bigger threat potentially posed, by the CPC-led China than would be revealed by rigorous empirical-rational scrutiny.

---

[38] Pei (2019a) and Wu (2019).

[39] Deng (1989a).

[40] Lian (2010), Economy (2019), Yin (2019), Tang (2019) and Ming (2018).

[41] Li and Liang (2019).

7.2.2 Beijing's Insecurity Angst

Chapters 3 and 6 have examined the post-Cold War strategic imbalances facing China following the Soviet bloc's evisceration and emergence of the US-led *unipolar systemic core*. The loss of a socialist bulwark and the advent of a wilful 'sole-superpower' willing to eliminate all opposition confronted Beijing with grim choices. A dialectic dynamic pitted China, for nearly two decades a silent US ally against the Soviet Union, and the United States, in asymmetric contestation. The *primate* acted to reshape the world in its own image; Deng Xiaoping's China, suddenly isolated from the global mainstream, stood apart in defensive insecurity. Deng, leading architect of modern China's ideological, party-political, socio-economic, scientific-technological and military-strategic parameters, set out the framework successive leaders then built upon. Coloured by China's relative frailty exposed to a threatening milieu dominated by an unfriendly *primate*, Deng's design for the next five decades aimed at fashioning a protective carapace which would enable China to secure its independence, socialism with CPC paramountcy, growing prosperity, and external power and influence.

Deng's US-rooted foreboding came to the fore after Tiananmen Square demonstrators prevented the official reception planned for President Mikhail Gorbachev, in Beijing on a crucial visit, forcing the ouster of CPC General Secretary Zhao Ziyang, and leading to much violence. After ordering the PLA to crush what he saw as 'counter-revolutionary turbulence' in which hundreds, perhaps thousands, of soldiers and civilians were killed or wounded, Deng noted the risks untrammelled democratic challenges posed to the PRC. Paying tribute to security personnel killed and injured in the clashes, Deng explained the potent brew of internal and external threats he saw, to the PBSC, CMC and PLA commanders enforcing martial law:

'We were dealing not only with people who merely could not distinguish between right and wrong, but also with a number of rebels and many persons who were the dregs of society. They tried to subvert our state and our Party … The handful of bad people had two basic slogans: overthrow the Communist Party and demolish the socialist system. Their goal was to establish a bourgeois republic, an out-and-out vassal of the West…Their ultimate aim was to overthrow the Communist Party and demolish the socialist system. This is the crux of the matter.'[42]

Against that perceptibly existential threat, the US role, allegedly covert, still stood out: 'The United States has blamed us for suppressing the students. But didn't the U.S. itself call out police and troops to deal with student strikes and disturbances, and didn't that lead to arrests and bloodshed? It suppressed the students and the people, while we put down a counter-revolutionary rebellion. What right has it to criticize us?' Conscious of the fallout from the violence, Deng told his successors, 'In future, however, we must make sure that no adverse trend is allowed to reach that point.'[43] Pre-emption was thus codified into the PRC's insistence on 'stability'.

---

[42] Deng (1989b).
[43] Deng (1989b).

Admitting to 'serious mistakes', although offering no explanation of what these were, Deng verbally resigned from all official posts, installing Jiang Zemin as the PBSC 'Core'. Despite his supposed withdrawal into an 'advisory' role, the relative lack of authority afflicting Jiang and fellow-PBSC members meant Deng remained China's 'paramount leader'. In that capacity, he warned shaken CPC elites, 'Western imperialists are trying to make all socialist countries abandon the socialist road, to bring them in the end under the rule of international monopoly capital and set them on the road to capitalism. We have to take a clear-cut stand against this adverse current. Because if we did not uphold socialism, we would eventually become, at best, a dependency of other countries, and it would be even more difficult for us to develop.'[44]

Deng encouraged his acolytes, notwithstanding growing pressure, to hold fast on to their socialist beliefs and faith in the PRC's future: 'The international market has already been fully occupied, and it will be very hard for us to get in. Only socialism can save China, and only socialism can develop China.' He removed all ambiguity in his 'reform and opening up' initiative: 'The highest objective of our political restructuring is to keep the environment stable. I have told Americans that China's overriding interest is in stability. Anything that helps maintain stability is good. We have never backed down from our position of upholding the Four Cardinal Principles.[45] The Americans' criticisms and rumors are nothing.'

Deng urged resistance to US pressure: 'China is an independent country. Why do we say we are independent? Because we are trying to build socialism, a socialism suited to our own conditions. Otherwise, we should have to act in accordance with the will of the Americans, or of people in other developed countries or in the Soviet Union. How much independence would we have then? At the moment, the media worldwide are putting pressure on us; we should take it calmly and not allow ourselves to be provoked.'[46] Even when formally resigning, Deng warned his successors, 'There is no doubt the imperialists want socialist countries to change their nature. The problem now is ... whether the banner of China will fall. The most important thing is that there should be no unrest in China and that we should continue to carry on genuine reform and opening to the outside.' He advised, 'First, we should observe the situation coolly. Second, we should hold our ground. Third, we should act calmly.'[47]

Deng simultaneously cautioned and assured a US emissary: 'The West really wants unrest in China. It wants turmoil not only in China but also in the Soviet Union and Eastern Europe. The United States and some other Western countries are trying to bring about a peaceful evolution towards capitalism in socialist countries.

---

[44] Deng (1989c).

[45] These being adherence to the socialist road, adherence to the people's democratic dictatorship, adherence to the leadership of the CPC, and adherence to Marxism-Leninism and Mao Zedong Thought. State Council (1979) Four Cardinal Principles. Foreign Language Press, Beijing, March 1979.

[46] Deng (1989c).

[47] Deng (1989a, d).

The US has coined an expression: waging a world war without gunsmoke. We should be on guard against this. Capitalists want to defeat socialists in the long run… but we have to look after our own… The Chinese people will not be intimidated. Anyone who tries to interfere in our affairs and bully us will fail.'[48] His insistence that no one could 'shake socialist China' was to instil confidence among untested neophytes taking over at a time of *systemic transition*, while deterring adversaries: 'Foreign aggression and threats arouse the Chinese people's sense of unity, their patriotism, and their love for socialism and the Communist Party and only make us clearer in our thinking…It is not wise for foreigners to resort to aggression and threats; that only works to our advantage.'[49]

That schizophrenic mix of warnings to an adversarial USA, and reassurances to anxious successors at home, was best evidenced in Deng's expositions on post-Tiananmen China-US relations. When former President Richard Nixon brought messages from George H.W. Bush, Deng told him, 'the recent disturbances and the counter-revolutionary rebellion that took place in Beijing were fanned by anti-communism and anti-socialism. It's a pity that the US was so deeply involved in this matter and that it keeps denouncing China; actually China is the victim. China has done nothing to harm the US.' Assuring Nixon that Beijing was focused on domestic development, Deng insisted China would do what was necessary to ensure it: 'We cannot tolerate turmoil, and whenever it arises in future we shall impose martial law. This will do no harm to anyone or to any country. It is our internal affair. The purpose … is to maintain stability; only with stability can we carry on economic development … In China, which has a huge population and a poor economic foundation, nothing can be accomplished without good public order, political stability and unity.' Deng stressed, 'Stability is of overriding importance.'[50]

Rejecting suggestions that Beijing initiate conciliatory measures, Deng said, 'The US should take the initiative in putting the past behind us, because only your country can do that … it is not possible for China to do so, because the US is strong and China is weak. China is the victim. Don't ever expect China to beg the US to lift the sanctions. If they lasted a hundred years, the Chinese would not do that.' Deng's message to Bush: 'If China had no self-respect, it could not maintain its independence for long and lose its national dignity. Too much is at stake … The PRC will never allow any country to interfere in its internal affairs.'[51] When Bush sent his NSA, Brent Scowcroft, to Beijing for a second time, Deng's tone was softer, but the message was unchanged: 'China cannot be a threat to the US, and the US should not consider China as a threatening rival. We have never done anything to harm the US.' He urged moderation: 'China and the US should not fight each other—I'm not talking just about a real war but also about a war of words. We should not encourage that.'

As for political reforms, 'China cannot copy the system of the US. It is up to the Americans to say whether their system is good or bad, and we do not interfere.' Deng

---

[48] Deng (1989e).
[49] Deng (1989f).
[50] Deng (1989g).
[51] Deng (1989g).

insisted, 'In relations between two countries, each side should respect the other and consider the other's interests as much as possible. That is the way to settle disputes. Nothing will be accomplished if each country considers only its own interests. It will require efforts by both China and the US to restore good relations. This must not be put off too long, or it would be damaging for both sides.'[52] Deng's views informed a 1993 CPCCC/CMC joint-analysis of China's strategic challenges which, six years before DoD's ONA described China as an 'emerging constant competitor', concluded that the United States, supported by Japan and South Korea, posed the most severe threats to China's security and prosperity.

Beijing's threat perceptions under Jiang, Hu and Xi were notably consistent. BRI-phobia thus reflected and reinforced well-established patterns of elemental mutual insecurity culminating in Washington declaring China a 'strategic rival'. Beijing's response, while more subtle, was nonetheless clear. Celebrating the PRC's 70th anniversary with a large military parade displaying both DF-41 MIRVed ICBMs capable of reaching any target in the USA, and DF-17 hypersonic glide vehicles analysts agreed could be hard to stop, Xi pledged to pursue 'peaceful development' but ordered the PLA to 'resolutely defend' China's sovereignty and security.[53] Visibly securing assured second-strike deterrence appeared to be the primary objective.

7.2.3 Belt-and-Road's Dialectic Dynamics

Under Deng's successors, Beijing's US-rooted defensiveness persisted, indeed deepened, even as they sought to work out a *modus vivendi* with the 'sole *superpower*'. Despite difficulties, Jiang collaborated with Bill Clinton in the aftermath of the May 1998 Indian- and Pakistani nuclear tests which not only threatened the peace across Asia, but also destabilised the global non-proliferation regime and rendered the frightening prospects of a deliberate nuclear exchange a realistic possibility. But the following year, as noted, Washington's bombing of the Chinese Chancery in Belgrade, the ONA's description of China as an 'emerging constant competitor', and the openly anti-PRC provisions of the 2000 NDAA, made clear for Beijing Washington's China perspective. Clinton's valedictory actions—forcing Israel to cancel the contract to supply two *Phalcon* AWACS platforms to the PLA, and visits reviving relations with India and Vietnam, two countries China had engaged in combat with, resulting in profound PRC-phobia in both—set the scene for hardening relations under George W. Bush.

The April 2001 aerial collision between a US ISR-aircraft and a Chinese fighter-jet, the latter crashing into the sea, killing the pilot, and the detention of the US crew in Hainan for 11 days, triggered a crisis which, although resolved diplomatically, further embittered mutual perceptions. The 2004 US Global Posture Review, authorising reinforcement of US troops and *materiel* forward deployed close to China, reinforced

---

[52] Deng (1989h).
[53] Kosaka (2019), Tillett (2019), Liu (2019a) and Martina (2019).

the trend towards building pressure, eliciting Chinese responsive efforts to build a 'counter-intervention' bubble around the PRC's periphery. What was designed in Washington as prudent, status quo-orientated, policy, looked like strategic encirclement to Beijing. Instead of reinforcing stability, the dialectics aroused profound Chinese insecurity, catalysing defensive/deterrent action which, in turn, reinforcing US suspicions, precipitated a vicious circle of deepened mistrust and spiralling counter-action.

China's anxiety over threats to its political-ideational institutions generally and its commercial- and energy lifelines in particular informed successive defence white papers issued by the CMC under Hu. Beijing averred, 'Economic globalization and world-multipolarization are gaining momentum.' Major powers continued 'to compete with and hold each other in check', but 'new emerging developing powers are arising,' leading to 'a profound readjustment brewing in the international system.' In addition to transitional fluidity, 'the US has increased its strategic attention to and input in the Asia-Pacific region, further consolidating its military alliances, adjusting its military deployment and enhancing its military capabilities.' In sum, 'China is faced with the superiority of the developed countries in economy, science and technology, as well as military affairs. It also faces strategic maneuvers and containment from the outside while having to face disruption and sabotage by separatist and hostile forces from the inside.' China would 'never seek hegemony or engage in military expansion', but US arms sales to Taiwan caused 'serious harm to Sino-US relations' and to 'peace and stability.'[54]

Looking to the second decade of the twenty-first century as an 'important period of strategic opportunities for national development', China planned to 'apply the Scientific Outlook on Development in depth, persevere on the path of peaceful development, pursue an independent foreign policy of peace and a national defense policy that is defensive in nature.' It sought to build 'a society that is moderately affluent on a general basis' and 'realize the united goal of building a prosperous country and strong military.'[55] Aspirations confronted 'profound and complex changes' roiling the *system*: 'International strategic competition and contradictions are intensifying, global challenges are becoming more prominent, and security threats are becoming increasingly integrated, complex and volatile.' Notably, 'international military competition remains fierce. Major powers are stepping up the realignment of their security and military strategies, accelerating military reform, and vigorously developing new and more sophisticated military technologies' dominating 'outer space, cyber space and polar regions', prompt global strikes and BMD, 'to occupy the new strategic commanding heights.'[56] US deepening of 'regional military alliances' and increased 'involvement in regional security affairs', including enabling 'Taiwan independence' tendencies, posed grave threats.[57]

---

[54] MND (2009, Chap. 1).
[55] MND (2011). Preface.
[56] MND (2011, Chap. 1).
[57] MND (2011, Chap. 1).

7.2 Empiricism's Rational Sobriety

When Hu Jintao transferred CMC Chairmanship to Xi Jinping, strategic assessments reflected persistent anxieties flowing from 'more diverse and complex security challenges'. China faced 'heavy demands in safeguarding national security', particularly from Taiwanese 'separatist force' receiving US military aid, and similar tribulations in Xinjiang and Tibet. In that complex milieu, Beijing set out 'fundamental policies and principles' guiding military modernisation, reforms and deployments: safeguarding national sovereignty, security and territorial integrity, and supporting China's peaceful development; aiming to win local wars under conditions of informationization and expanding and intensifying military preparedness; conceptualising comprehensive security and effectively conducting military operations other than war (MOOTW); deepening security cooperation and fulfilling international obligations; and 'acting in accordance with laws, policies and disciplines.'[58]

As Xi embarked on a quest to attain the 'two centenarian goals', 'profound changes' transformed the *system*: 'historic changes in the balance of power, global governance structure, Asia-Pacific geostrategic landscape, and international competition in the economic, scientific and technological, and military fields.' Although 'forces for world peace' were ascendant, China faced 'new threats from hegemonism, power politics and neo-interventionism.' Rivalry over 'the redistribution of power, rights and interests' intensified. Hence, the 'immediate and potential threats of local wars.' US 'rebalancing', Japan's overhauling of its 'military and security policies', neighbours' 'provocative actions' and military reinforcements on 'China's reefs and islands', 'external countries' meddling' in SCS affairs, and Taiwan, posed grim tests.

China's growing 'national interests' rendered its national security even 'more vulnerable to international and regional turmoil'. The security of China's 'overseas interests concerning energy and resources, strategic SLoCs, as well as institutions, personnel and assets abroad', became 'an imminent issue'.[59] Beijing tasked the PLA to, among other things, 'safeguard the security of China's overseas interests', alongside more traditional military assignments.[60] This was the landscape on which Beijing fashioned its BRI, especially the MSR, blueprints.

7.3 Concluding Observations

This work has traced the evolution of China's regional developmental asymmetries challenging national integration and stability, consequent elite focus on planning and resourcing inland regional development schemes as significant elements of national developmental programmes, extending trans-frontier economic-commercial linkages between poorer border regions, especially Xinjiang and Yunnan, and neighbouring states, thus laying the foundations of the SREB network. It has examined contributions made to both the conceptual definition and practical implementation of the BRI

---

[58] MND (2013).
[59] MND (2015).
[60] Yao (2016).

blueprint by external actors, notably multilateral bodies e.g., WBG, IMF and UN organs; US visionaries, academics and multinationals; authorities in Tokyo, Washington and Brussels; and a large number foreign corporations driving China-EU trade. The combined impact of the palimpsest of these domestic and foreign actors' varied and coincidentally synergistic efforts nurtured BRI's SREB enterprise.

The work has also traced the fundamental continuity in the core beliefs, priorities and policies animating Deng Xiaoping and his successors, especially in the aftermath of China's Tiananmen Square catharsis and the near-simultaneous fission of the USSR and the Soviet *bloc*, replacing the Cold War's bipolar certitudes with the USA's unipolar *primacy*, seen to existentially threaten the CPC-led PRC. The study charted the adversarial dialectic dynamic catalysed by the US determination to strengthen, indeed indefinitely extend, its post-Soviet *primacy* by expanding the capitalist-democratic sphere, even by applying the force of arms, and an insecure China responding by marshalling all its resources to boost its national power and secure a defensive carapace aspirationally impermeable to US- and allied penetration.

Given the contradictions flowing from a globally open economic system sustained by a relatively closed political framework, Beijing's perception of threats from a semi-hostile *primate,* and Washington's identification of China as the one actor capable of obstructing the United States' force-projection capacity across the Asia-Pacific and, with Russia, across Eurasia, fed off and deepened mutual insecurity. Economic interdependence and pro-engagement commercial interests sustained ties until Donald Trump's multi-pronged assault on China's much-feared revisionist potential. BRI, with MiC2025, irrespective China's original drivers, thus assumed primarily geopolitical imagery, visage and import in US- and allied narratives.

7.3.1 *Displacement Anxiety and Terminal Fears*

In April 2019, after former President Jimmy Carter wrote to Trump expressing concern that the United States and China were sliding into a 'modern Cold War', Trump called Carter on a private line 'to say very frankly to me that the Chinese were getting way ahead of the United States.' Carter's own view—a superpower was defined by 'not just who has the most powerful military, but who is a champion of the finer things in life', apparently had little impact.[61] Trump escalated the tariff war almost directly afterwards by hiking duties on $200bn of imports from China from 10 to 25%, and threatening to impose fresh duties on the remainder of China's exports. US imposition of sanctions on five Chinese entities linked to Beijing's supercomputing advances in June 2019 reflected anxiety mirroring a general, bipartisan, insecurity afflicting US policymakers, legislators, think-tank and media analysts, and even businessmen, the latter hitherto vocal supporters of US-China engagement.

---

[61] Liu (2019b).

Fear that China was 'getting way ahead' in areas crucial to sustaining US *primacy* drove Washington's overt shift to challenging, even containing, China across the spectrum of policy-options. Washington's national security community was convinced of the existential nature of the purported Chinese threat to the USA's *primacy*. Assistant Secretary of State, Christopher Ford, laid out the official version of the challenge 'China's wealth and power' posed in the form of Beijing's pursuit of 'Indo-Pacific regional hegemony in the near-term and displacement of the United States to achieve global pre-eminence in the future.'[62] Congress was advised, China's military modernisation, 'including its naval modernization effort, has become the top focus of US defense planning and budgeting.'[63] A consensus united the politically polarised US cognoscenti—China's actions in the SCS and the ECS, especially the start of its massive reclamation projects in 2013–2014 and militarisation of these artificial bases in 2016–2018, and Russia's annexation of Crimea in March 2014, marked the end of the United States' post-Soviet *unipolar* pre-eminence, and the beginning of great power competition pitting the USA against the occasionally co-ordinated challenges posed by China and Russia.

Key features of the resumption of great power rivalry included authoritarian ideological threats to liberal democracy; Chinese and Russian narratives of historical grievances, injustice, humiliation and victimisation by Western powers, and support for revanchist or irredentist goals; Chinese and Russian use of 'hybrid operations' and 'gray-zone warfare' for which conventional military responses appeared inappropriate or inadequate; and Chinese and Russian deployment of coercion as an instrument of dispute-resolution, eroding the efficacy of the US-led 'rules-based order'.[64] US legislators were urged to significantly expand defence budgets as the most urgent measure to meet these challenges to the US 'national' security.

Xi Jinping, centralising CPC- and state authority unto himself, removing term-limits to the presidency, initiating and sustaining 'militarisation' of reclaimed SCS islets, hastening PLA modernisation, reforms and growth, and permitting a personality cult, apparently personified revisionist challenges. Xi's instructions to China's diplomats to raise Beijing's role in global governance deepened US concerns: China should 'actively participate in leading the reform of the global governance system, build a more comprehensive global partnership network, and strive to create a new diplomatic situation with big powers with Chinese characteristics.' It was 'necessary to grasp the general trend of accelerating the multi-polarization of the world and attach importance to the in-depth adjustment of relations between major powers.' Hence the need 'to plan for the relationship between the major powers and promote the construction of a major stable and balanced development framework for major powers.' In that context, 'It is necessary to persist in co-construction and sharing, and promote the construction of the "Belt and Road" to achieve a deeper and deeper pace, and push the opening up to a new level.'[65]

---

[62] Ford (2019).
[63] O'Rourke (2019).
[64] O'Rourke (2018, pp. 2–7).
[65] Xi (2018b).

Although Xi had insisted on BRI's economic focus, and reiterated successive predecessors' pledges abjuring hegemonic aspirations, his emphasis on active encouragement of *systemic multipolarisation,* by definition antithetical to the USA's determination to effect perpetual *primacy,* and linking BRI implementation to that structural aspiration, seemingly confirmed suspicions that BRI indeed concealed geopolitical, indeed grand-strategic, objectives. US analysts acknowledged that Beijing viewed Barak Obama's 'Asian Pivot/Rebalance' as strategic encirclement from which China must break out. An obvious response was Xi's attempt to rebuild the PLA into a modern military that could 'fight and win battles'.[66] If Xi's reforms succeeded, 'US forces operating throughout the Indo-Pacific region will face a PLA' able to rapidly respond to regional crises and 'conduct counter-intervention operations more effectively.' If the PLA harnessed civilian scientific innovation to military purposes, China would become 'an even more formidable strategic adversary'. To sustain 'a favourable regional balance of power' would require the USA to 'regain its technological edge.'[67] Conflation of structural concerns and BRI's presumed catalytic potential explained the depth of US anxiety.

Mirroring US concerns over what China perceived as its own defensive response to significant threats from the former but Washington viewed as unacceptable revisionism, Beijing's 2019 Defence White Paper illustrated the persistently cyclical feature of US-Chinese strategic insecurity dialectic dynamics: 'The international strategic landscape is going through profound changes. As the realignment of international powers accelerates and the strength of emerging markets and developing countries keeps growing, the configuration of strategic power is becoming more balanced.' That broadly positive trend was negated, however, by the fact that 'international security system and order are undermined by growing hegemonism, power politics, unilateralism and constant regional conflicts and wars.' Notably, 'International strategic competition is on the rise. The US has adjusted its national security and defense strategies, and adopted unilateral policies.' Specifically, 'It has provoked and intensified competition among major countries, significantly increased its defense expenditure, pushed for additional capacity in nuclear, outer space, cyber and missile defense, and undermined global strategic stability.' In short, China accused the USA of doing much of what the USA accused China of doing.[68] This mutually reinforcing dialogue of the deaf characterised the *system.*

What Beijing saw as its deterrent imperative and a natural function of 'national rejuvenation', appeared dangerously revisionist in US eyes. DIA analysts agreed, China's 'contemporary strategic objectives' were largely domestic and defensive, i.e., to perpetuate CPC rule, maintain domestic stability, sustain economic growth and development, defend national sovereignty, and secure China's status as a great power.[69] DIA Director, Lt. General Robert Ashley, still insisted, 'China is building a robust, lethal force with capabilities spanning the air, maritime, space and information

---

[66] Report (2014) and Saunders et al. (2019, p. 711).

[67] Saunders et al. (2019, p. 723).

[68] MND (2019).

[69] Ashley (2019, p. 12).

7.3 Concluding Observations

domains which will enable China to impose its will in the region. As it continues to grow in strength and confidence, our nation's leaders will face a China insistent on having a greater voice in global interactions, which at times may be antithetical to US interests.'[70]

Admiral Philip Davidson, Commander, Indo-PACOM, the combatant-command responsible for containing China's military and diplomatic prowess, insisted that Beijing represented 'the greatest long-term strategic threat to a Free and Open Indo-Pacific and to the United States.' China used 'fear and economic pressure' to 'expand its form of Communist-Socialist ideology in order to bend, break, and replace the existing rules-based international order.' Beijing allegedly sought 'to create a new international order led by China and with "Chinese characteristics"—an outcome that displaces the stability and peace of the Indo-Pacific that has endured for over 70 years.'[71] US *primacy* and global well-being were thus conflated.

Davidson and his peers ignored the fact that in the preceding 70 years, millions of Indo-Pacific residents had been killed or wounded or rendered destitute in conflicts, e.g., the Korean War, the 1st Indo-China War, the 2nd Indo-China War, several Indo-Pakistani wars, a Sino-Indian war, and another involving China, Vietnam and Cambodia. Orwellian false assertions reflecting mass-amnesia, reinforced by authoritative repetitions, established China as axiomatically posing severe revisionist threats to an allegedly benign order. Trade, technology, financial leverage, regulations, diplomacy, military deployments, even education and research, were vectors for an 'all of government' push-back against China's growing capabilities.

Executive branch policy documents, official remarks and statements by Trump and Pence, cabinet-level principals and their subordinates, senior legislators and their USCC advisors, and justices administering a growing tide of espionage cases against alleged Chinese 'agents' and organisations, manifested fears of rapidly eroding global stature and impending loss of control. Shanahan's Congressional testimonies bore witness to deepening unease. China had invested in programmes 'specifically intended to offset US advantages, including robust A2/AD networks, more lethal forces, and new strategic capabilities,' which 'could seek to achieve a "fait accompli" that would make reversing China's gains more difficult, militarily and politically.'[72] DoD openly targeted China in its 2019 Indo-Pacific Strategy: 'the PRC, under the leadership of the CPC, seeks to reorder the region to its own advantage by leveraging military modernization, influence operations, and predatory economics to coerce other nations.' Shanahan vowed not to 'accept policies or actions that threaten or undermine the rules-based international order.'[73] The USA must act; hence the nearly-$750bn defence budget.[74]

Shanahan's successor, Mark Esper, in his first instructions to all DoD personnel, reaffirmed the guiding role of the 2018 National Defence Strategy identifying China

---

[70] Ashley (2019, p. v).
[71] Davidson (2019).
[72] Shanahan (2019d).
[73] Shanahan (2019e).
[74] Shanahan (2019d).

(and Russia) as rival powers which US forces must prepare to fight. He instructed subordinates, 'the surest way to deter adversary aggression is to fully prepare for war.'[75] This mirrored his vows to Congress: 'China and Russia (are) attempting to match or surpass the United States in multiple domains…If confirmed, my focus will be executing the NDS, and continuing to expand the competitive space for the US Armed Forces in all domains to counter the challenges we face.'[76]

Deep discomfort arguably flowed from *systemic transitional fluidity,* catalysing general anxiety afflicting both shores of the Pacific, almost overwhelming rational responses to a fast-changing external milieu. Deng's China had feared 'imperialist' designs to subvert the PRC's socialist foundations and replace its CPC leadership by using all available means; post-*unipolar* United States was now gripped by fears of its dominance being eroded by China's growing capabilities and will to the point of it being dethroned from the crown of the *systemic* hierarchy. Beijing's terminal fears both mirrored and ironically engendered the US' displacement anxiety. The two actors' divergent aspirations and fears, focused largely on each other, rendered their respective grand-strategic insecurity virtually zero-sum, not easily amenable to alleviation.

As the dialectic spiral approached culmination under the Trump-Xi diarchy, BRI appeared to reinforce that virtually autonomous vicious circle. Washington not only challenged China, but also threatened its allies and partners with sanctions if they cooperated with Chinese (as well as Russian and Iranian) entities. The *primate*'s long-established and substantial dominance in most realms of economic, scientific-technological, diplomatic and military capabilities, now deployed in the form of tangible and intangible tools against the networked global supply-chains reinforcing economic, commercial and financial integration, generated destabilising disruptions. The debate over BRI's purpose thus both reflected and reinforced a much wider and deeper grand-strategic insecurity discourse examining the relative and absolute status of particular major powers within an evolving, dynamic, even fluid, *systemic* architecture.

7.3.2 The Belt and Road Initiative in the Systemic *Context*

Trump Administration officials not only examined ways of preventing Chinese investment and equipment from threatening what they saw as US national security interests at home, but also pressed upon their allies and partners on the need to steer clear of such instruments of Beijing's purportedly dangerous outreach outside the USA. BRI featured prominently in that endeavour. Practitioners explained to friendly states why BRI projects, especially their financing, lacked acceptable levels of transparency, and how ICT networks, hardware and software, particularly those linked

---

[75]Esper (2019a).
[76]Esper (2019b).

7.3 Concluding Observations

to 5G telecommunications, posed challenges to the US-led 'rules-based' order.[77] Persuasion succeeded most visibly with the United States' 'Five-Eyes' Anglophone intelligence-partners, but proved less effective elsewhere.

Senior US officials and their Chinese counterparts appeared to engage in a proxy exchange of accusations and counter-accusations which suggested they were talking past each other in a dialogue of the deaf. The status quo-orientated *systemic primate* found it almost impossible to countenance the possibility that a reinvigorated China could not only match its own globe-girding capabilities but, indeed, overtake them. In early 2019, having marked the 40th anniversary of the establishment of US-China diplomatic relations with warnings that the two countries risked sliding into a 'modern Cold War', former President Jimmy Carter explained to his church congregation the differences between the two powers: 'I normalized diplomatic relations with China in 1979. Since 1979, do you know how many times China has been at war with anybody? None.' In contrast, 'We have stayed at war.' Carter noted that the USA had been 'the most warlike nation in the history of the world' due to 'a desire to impose American values on other countries.' Rhetorically responding to Trump's concerns about China, Carter said, 'We have wasted, I think, $3 trillion (in recent warfare). China has not wasted a single penny on war, and that is why they are ahead of us. In almost every way.'[78] Carter's perceptive analysis did not, however, persuade policymakers or inform the Beltway perspective.

Secretary of State Mike Pompeo, a powerful voice in the Administration's 'China threat' formulation, often criticised Chinese policies in remarks made at home and overseas. Nonetheless, he felt it necessary to stress that the USA was not being aggressively reactionary—it was determined merely to uphold fairness and responsible state-behaviour, attributes of the post-Cold War US-led order that China was allegedly threatening: 'The US is not pursuing a cold war or containment policy with China. Rather, we want to ensure that China acts responsibly and fairly in support of security and prosperity in each of our two countries.'[79]

Vice President Mike Pence, a vocal champion of the 'China threat' school, persuaded NATO leaders that China, using BRI as a subtle tool, posed a grand-strategic threat not just to the USA, but to the Atlantic alliance as a whole, a threat the latter needed to address: 'The greatest challenge NATO will face in the coming decades is how we must all adjust to the rise of the PRC. And adjust we must, for determining how to meet the challenge of Chinese 5G technology, meet the challenge of the easy money offered by China's BRI.'[80] The argument proved so effective that NATO Secretary General Jens Stoltenberg, on his first China-focused visit to the Asia-Pacific, stressed his efforts were 'not about moving NATO into the Pacific, but this is about responding to the fact that China is coming closer to us, investing heavily in critical

---

[77] Strayer (2019).
[78] Carter and Hurt (2019).
[79] Spokesperson (2018) and Gehrke (2018).
[80] Pence (2019).

infrastructure in Europe, increased presence in the Arctic and also increased presence in Africa, and in cyberspace.'[81] His was a tacit but visible action manifesting NATO's presence.

Beijing questioned the legitimacy of US measures, especially against the Chinese ICT firm Huawei, and warned of strong responses.[82] More generally, senior Chinese officials, civilian as well as military, emphasised the primarily domestic focus of Beijing's policies and pledged to abjure disruptions to the *system*. Minister of National Defence Wei Fenghe assured his US counterpart, 'We will never seek hegemony, or aggression, or expansion, or an arms race. It is only for the protection of China itself, to protect China's people from war, to give them a life of peace.'[83] Xi Jinping's senior foreign policy adviser Yang Jiechi similarly reassured US leaders, 'Everything that we do is to deliver better lives for the Chinese people, to realize rejuvenation of the Chinese nation. It is not intended to challenge or displace anyone.'[84]

Xi Jinping, hosting dozens of foreign dignitaries at the 2nd BRF in April 2019, personally sought to address some of the concerns his policies, and BRI specifically, had triggered: 'We wish to work with all parties to promote in-depth and substantial development of Belt and Road cooperation to better benefit the people of all countries. We hope to work with all parties to improve the cooperation concept of high-quality development of Belt and Road cooperation.'[85] Despite these assurances issued at the highest levels of the Chinese government, allegations and counter-allegations vitiated the atmosphere. US-Chinese mutual anxiety, although widely shared, did not however monopolise either Chinese or Western thinking.

Eminent practitioners and analysts shed rational light on the conundrum simultaneously dividing and uniting the two powers in their schizophrenic quest for both advantage and stability. The senior China-specialist, former Australian leader, Kevin Rudd, noted, 'if the pillars of strategic analysis are capabilities, intentions, and actions, it is clear from all three that China is no longer a status-quo power.' Still, linearity was not in the gift of the CPC and the combination of debt, environmental damage, climate change and demography could convert China's economy into a liability. BRI confronted mixed fortunes as some partners pushed back against perceived unfair practices. However, if one state hesitated, 'another queued up'. Rudd saw the 'long-term drain' on China's resources, rather than foreign resistance, as BRI's 'biggest impediment.' On that basis, 'it would be prudent to assume, absent major and enduring policy change in Beijing or Washington', that China had 'at least some chance of success.'[86]

While noting the USA's declaratory shift to competition with China, Rudd asked, what was Washington's desired strategic end-point if Beijing refused to 'acquiesce to the demands outlined' in Pence's October 2018 speech; what were the new 'rules

---

[81] Mair and Packham (2019).
[82] Lu (2019).
[83] Xinhua (2018a).
[84] Spokesperson (2018), Xinhua (2018b) and Gehrke (2018).
[85] Xi (2019b).
[86] Rudd (2018).

7.3 Concluding Observations 315

of the game' in this era of strategic competition? Would a new common US-Chinese strategic narrative now be possible? If competition with China resulted in 'full-blooded containment', economic decoupling or 'even a second Cold War', was the underlying logic an assumption that China would self-eviscerate as the USSR did? If not, was the USA prepared 'to provide the world a strategic counteroffer to the financial commitment' BRI represented? How would Washington compete over time with Beijing in trade and investment in non-Atlantic regions? How confident could the USA be that all its allies and partners fully endorsed its robust rivalry with China? How did Washington plan to appeal to the rest of the world to join its campaign? And what would be the immediate impact on the global economy and climate change action of 'a major cleavage in US-China relations?' Rudd urged consideration of these before taking final decisions.[87]

DoD-sponsored strategic analysts took a more granular view in questioning Washington's decision to equate China and Russia as the United States' principal strategic rivals. They asserted these two actors represented 'quite distinct challenges'. Russia, in their view, was not a peer or a near-peer competitor, but rather 'a well-armed rogue state' keen to 'subvert an international order' it could never hope to dominate. China, in contrast, was 'a peer competitor' that sought to 'shape an international order that it can aspire to dominate.' Significantly, by 'employing updated versions of defense, deterrence, information operations, and alliance relations', Russia could be contained. On the other hand, simply speaking, 'China cannot be contained.'[88] China could militarily 'be contained for a while longer; economically, it has already broken free of any regional constraints.'[89] These RAND analysts, on the basis of the data they had to hand, concluded that BRI was 'on its way to becoming the central organizing principle for China and for much of the world.'[90]

7.3.3 The Outlook

Against that fluid and presumably dangerous backdrop, US policymakers viewed BRI as an insidious, cleverly camouflaged in economic cover, strategic instrument designed to subtly expand Beijing's global footprint, increase potential bases in key areas and take control over spaces and domains crucial to the sustenance of US *primacy*. Shanahan explained the threat posed by BRI: 'China is diligently building an international network of coercion through predatory economics to expand its sphere of influence.' Ignoring evidence manifest in FOCAC and BRF summits, he insisted, 'Sovereign nations around the globe are discovering the hard way that

---

[87] Rudd (2018).
[88] Dobbins et al. (2018, p. 2).
[89] Dobbins et al. (2018, p. 10).
[90] Dobbins et al. (2018, p. 12).

China's economic "friendship" via OBOR can come at a steep cost when promises of investment go unfulfilled and international standards and safeguards are ignored.'[91]

The challenge posed by 'initiatives like the Digital Silk Road, Made in China 2025, and Thousand Talents Program' was clear. 'China aims to steal its way to a China-controlled global technological infrastructure.' A particular concern: 'China's force projection inside and outside the South China Sea disrespects and undermines our rules-based international order and threatens regional stability and security.' As the defender of planetary status quo, Washington must respond: 'Our pursuit of many belts and many roads creates alternative options for nations unwilling to succumb to China's increasingly coercive methods.' So, 'we stand ready to compete where we must to ensure our military's competitive advantage for decades to come.' To counteract China's challenge, including its BRI 'stratagem', DoD's priorities were: 'outpace Chinese military modernization to deter future conflict, or win decisively should conflict occur; protect US and partner R&D of advanced technology from rampant Chinese theft, and maintain a free and open Indo-Pacific built on strong alliances and growing partnerships.'[92]

That conflation of China's defensive measures to secure its own lifelines and Beijing's alleged threats to US *primacy* was recognised by even some of the most vocal China critics. Days after Shanahan's testimony, one of them noted, 'BRI is a counter-encirclement strategy.' Beijing saw Washington as 'an oppressive global hegemon determined to prevent China's rise,' quite openly since Obama initiated his 'Asian Pivot' in 2011. The US's massive forward-deployed forces, extensive alliance network, and 'command of the global oceans, have long been seen as posing the most direct and serious challenge to China's security.' Anxious about its energy life-line traversing US-controlled SLoCs and choke-points, Beijing sought to circumvent its 'Malacca dilemma' with alternative routes. BRI's SREB/MSR energy links would bypass the SCS, 'reducing the country's vulnerability to a potential US naval blockade in case of a military conflict.'[93] That fundamentally defensive endeavour caused grave insecurity in Washington.

Officials and academic analysts from US- and US-allied governments and institutions, in remarks made at conferences, and in personal exchanges with the author, over 2017–2019, made clear a broad consensus on the normative challenge they believed China posed to both the US-led post-Soviet liberal order, and Western values usually touted as universal ones. Perhaps the most intriguing, and revealing, comment came from Kiron Skinner, Director of US State Department Policy Planning Staff, a select body that provides horizon-scanning long-term policy advice to the Secretary of State, and thus crucially shapes structural, strategic and long-term globally-relevant frameworks for the identification of US national interests, and guidelines for pursuing these to successful conclusion.

Addressing questions about Washington's assessment of the core strategic challenges the USA would confront in 2030, Skinner stated: 'China we see as the more

---

[91] Shanahan (2019d).
[92] Shanahan (2019d).
[93] Rolland (2019).

7.3 Concluding Observations

fundamental long-term threat. That's not a partisan issue … It's a long-term fight with China, or a long-term competition. It has historical, ideological, and cultural, as well as strategic factors that a lot of Americans do not understand, even in the foreign policy community.' Skinner believed trade disputes were symptomatic of the 'deeper historical and strategic' problems the USA faced with China. Controversially, she noted that the Cold War with the Soviet Union was 'a fight within the Western family.' Fundamental accord with China, in contrast, was 'not really possible. This is a fight with a really different civilization and a different ideology, and the United States has not had that before. Nor has it had an economic competitor the way that we have.' Skinner saw China as a comprehensive rival—'In China we have an economic competitor, we have an ideological competitor, one that really does seek a global reach that many of us did not expect.' Strikingly, ignoring the 1940s war with Japan, Skinner insisted, 'this is the first time that we'll have a great-power competitor that is not Caucasian.'[94]

The fact that Skinner, herself ironically an African-American, framed US-China rivalry as a function of civilizational, cultural and racial divergences, as well as *systemic,* structural and strategic contests, suggested Sino-US tensions were fundamentally beyond resolution. Her assertions, given the significant locus of her official position, starkly reflected the depth of the 'China threat discourse' animating US- and allied policy communities. It thus created a self-reinforcing echo-chamber indicating the strength and likely longevity of both normative- and operational coalitions coalescing against China. Most, though not all, of Skinner's critics targeted her linguistic formulations rather than the substantive content of her remarks, which resonated with many. But even pragmatic opinion-leaders urged sustainable, competitive, multi-directional policy shifts aimed at meaningfully responding to China-rooted challenges.[95]

The fundamental point on which there appeared little discord, was the urgency of acting to secure and extend US leadership, or *systemic primacy*, in the face of China's multi-faceted 'threats'.[96] Even then, a handful of analysts acknowledged that while competition was elemental to US-China relations, both leaderships could do more to manage its more unwanted outcomes.[97] Several DoD-funded analysts even noted that China was 'unlikely to surpass the US by virtually any measure of national power any time soon'.[98] However, even this small group, marginal to the mainstream China discourse energising the US- and US-allied policy communities, agreed that in 'countries affected' by the SREB and MSR components of BRI, 'Chinese influence could surpass that of the United States.'[99] Symbolically, on the United States' Independence Day in 2019, celebrated with an unprecedented military

---

[94] Skinner (2019).
[95] Wolf (2019), Editorial (2019), Pastreich (2019), Pei (2019b), Chen and Hu (2019) and Schell and Shirk (2019).
[96] Report (2019), O'Rourke (2018), Zakheim (2018), Mosteiro (2018) and Grossman (2018).
[97] Franck (2019) and Hass (2018).
[98] Mazarr et al. (2018).
[99] Mazarr et al. (2018).

parade in Washington, and an address by President Trump, a hundred US academics, former-diplomats, civil- and military officials, and other eminent citizens, signed an open letter to Trump and US legislators, insisting, 'China is not an enemy'.

Noting they were 'deeply concerned about the growing deterioration' in US-China relations, they asserted this did 'not serve American or global interests.' Acknowledging China's 'recent behavior' required 'a strong response', they believed 'many US actions' contributed 'directly to the downward spiral in relations.' They posited, 'the current approach to China is fundamentally counterproductive.' They did not believe China was 'an economic enemy or an existential national security threat that must be confronted in every sphere.' Since US opposition would not 'prevent the continued expansion of the Chinese economy', a larger share of the global market going to Chinese firms, and 'an increase in China's role in world affairs', the United States could not 'significantly slow China's rise without damaging itself.'[100] Despite the eminence of the authors, and the weighty logic of their argument, a rational response was not guaranteed. The risk of passionate subjectivity overwhelming reason apparently grew.

All incoming DoD leaders, civilian and uniformed, emphasised their determination to improve and expand the USA's military capability to combat China (and Russia) as directed in Trump's National Security Strategy (NSS) and National Defence Strategy (NDS). Defence Secretary Mark Esper told Congress, 'the level of risk is increasing as the threat environment evolves. This is particularly the case regarding China's and Russia's growing ability to contest US military advantages; we cannot allow that trend to continue.'[101] CJCS, General Mark Milley, noted: 'Today we face a complicated global security environment. We rely heavily on our allies and partners to address various security challenges. Specifically, we are in the midst of a great power competition with Russia and China…our adversaries have made great strides in narrowing the (capabilities) gap and we must remain ever vigilant.'[102]

CNO, Admiral Michael Gilday, emphasised consonance with the Administration's seminal assessment that 'the central challenge to US prosperity and security is the re-emergence of long-term, strategic competition from China and Russia, amid persistent challenges to international order.' As to the US response to these challenges, Gilday asserted, 'We need to get faster—across the entire service—in order to inject uncertainty into our competitors' decision cycle and become better at competing across the full spectrum of competition.'[103] The US military leadership appeared determined to pursue the DoD's February 1992 Defence Policy Guidance to deter, or defeat any perceived threats to extending the USA's post-Soviet *unipolar primacy* into the indefinite future. Since actors within the system sustaining growth over long periods accreted capabilities that could potentially erode US *primacy,* e.g., China, in this instance, might disagree, the risk of tensions escalating appeared unavoidable.

---

[100] Fravel et al. (2019).
[101] Esper (2019c).
[102] Milley (2019).
[103] Gilday (2019).

7.3 Concluding Observations

Beijing's views, usually much more restrained in tone than that emanating from the Beltway, was only made public in statements issued by official spokesmen, state-controlled media editorials and occasionally, cabinet-level remarks. The latter were rarely reported outside China by foreign media, and made little impact on the West's China discourse. A fortnight after Kiron Skinner's remarks, Xi Jinping inaugurated a Conference on Dialogue of Asian Civilizations in Beijing. Addressing senior delegates from 47 mainly Asian countries, Xi urged his guests to 'create conditions for other civilizations to develop while keeping your own vibrant. We should leave various civilizations in the world in full blossom.' His conclusion, 'there would be no clash of civilizations as long as people are able to appreciate the beauty of them all.'[104] Although Xi did not name any country or individuals, his comment, 'It is foolish, disastrous to reshape or replace other civilizations', spoke to Beijing's assessment.

Tracing ancient commercial and intellectual links between China and distant lands and cultures, Xi noted, 'The modern day BRI is built on these channels for exchange and mutual learning, and Asian civilizations have grown stronger thanks to such interactions.'[105] However reasonable and reassuring Xi's remarks, this centrality of BRI in Beijing's global pursuits must have deepened already profound anxieties afflicting those viewing China with growing trepidation. This ongoing and perhaps incorrigible trans-Pacific dialogue of the deaf indicated a fundamental disjuncture between notions of how the world should be, and how the USA and China would be placed in that future world. The USA's determination to secure a measure of permanence for its *primacy*, China's seemingly comparable determination not to be subordinated within an order not of its own design, Western displacement anxiety mirroring China's terminal fears and the zero-sum nature of the narratives dividing the two rivals suggested this dispute transcended temporary truces on tariffs and trade.

The contrasting imagery BRI evoked on the two competing sides, irrespective of the outcomes of empirical research conducted by such bodies as the WBG, IMF, EBRD, and ADB, similarly indicated that notwithstanding its developmental-economic origins and geoeconomic placement within China's policy universe, BRI was perceived—and perceptions strongly contribute to the construction of the perceived reality—as a fundamentally geopolitical design intended for effecting 'peaceful structural revisionism'. If the BRI blueprint had become an organising framework for China's economic-commercial, diplomatic and foreign policy outreach, so had the 'China threat theory' solidified and embedded itself as the organising principle for framing the United States' *systemic* security perspective and its military and strategic, as well as economic, commercial and scientific-technological, outlook.

While debate over the validity of either side's arguments will likely persist, the increasingly acrimonious disputation's corrosive global impact was inescapably evident by the time Donald Trump and Xi Jinping sat down in Osaka on the G20 summit's sidelines to work out a truce enabling both sides to momentarily pull back

---

[104] Xi (2019c).
[105] Xi (2019c).

from the trade war further deepening.[106] Even then, the USA's geopolitical lens on China's primarily geoeconomic BRI enterprise, and competitive tendencies intensified by the Trump-Xi diarchy, rendered a resolution of the determinedly status quo- vs.-inadvertently revisionist dynamic 'perilously close to impossible' to engineer.[107] Although policy-makers and analysts were focused on who would win this competition shaping the twenty-first century, they appeared to ignore the likelihood that no one could. It seemed more likely that if the starkly zero-sum trends persisted, and even if catastrophic conflict precipitated by spiralling strategic competition could be averted, irrespective of what Beijing's BRI implementation finally achieved or failed to accomplish, the one certainty was that, at least in terms of global opportunity costs, losses would inevitably be universal.

References

Armstrong S (2019) The rules based economic disorder after Osaka G20. EAF, 30 June 2019
Ashley R (2019) China Military Power 2019. DIA, Washington, 15 Jan 2019
Baniya S, Rocha N, Ruta M (2018) Trade effects of the new silk road: a gravity analysis. WBG, Washington
Bianji H (2019) China releases report on BRI progress. PD, 26 Apr 2019
Bo A (2016) PRC—study of the BRI: sovereign (Public) project 50141-001. ADB, Manila, 22 July 2016
Carter J, Hurt E (2019) President Trump called former President Jimmy Carter to talk about China. NPR, 15 Apr 2019
Chen D, Hu J (2019) No, there is no US-China 'clash of civilizations'. Diplomat, 8 May 2019
Davidson P (2019) Statement before the SASC. DoD, Washington, 12 Feb 2019
Deng X (1989a) With stable policies of reform and opening to the outside world, China can have great hopes for the future: address to leading members of the CPC Central Committee. CMC, Beijing, 4 Sept 1989
Deng X (1989b) Address to officers at the rank of general or above in command of troops enforcing martial law in Beijing. CMC, Beijing, 9 June 1989
Deng X (1989c) Urgent tasks of China's third generation of collective leadership: a talk with leading members of the CPC Central Committee. CMC, Beijing, 19 June 1989
Deng X (1989d) a letter to the Political Bureau of the Central Committee of the CPC. CMC, Beijing, 4 Sept 1989
Deng X (1989e) We are confident we can handle China's affairs well: a talk with Prof. Tsung-Dao Lee of Columbia University. CMC, Beijing, 16 Sept 1989
Deng X (1989f) No one can shake socialist China: a talk with Prime Minister Chatichai Choonavan of Thailand. FMPRC, Beijing, 26 Oct 1989
Deng X (1989g) The United States should take the initiative in putting an end to the strains in Sino-American relations: a talk with former President Richard Nixon of the United States. FMPRC, Beijing, 31 Oct 1989
Deng X (1989h) Sino-US relations must be improved: a talk with Brent Scowcroft, special envoy of President George Bush of the United States and Assistant to the President for National Security Affairs. FMPRC, Beijing, 10 Dec 1989

---

[106]Trump and Xi (2019), Trump (2019) and Sheng and Liang (2019).
[107]Armstrong (2019).

References

Dobbins J, Shatz H, Wyne A (2018) Russia is a rogue, not a peer; China is a peer, not a rogue. RAND, Santa Monica

Dodwell D (2019) Turning China's BRI into a new cold war weapon by the US is deeply frustrating. SCMP, 21 Apr 2019

Dow Jones (2019) US warns Germany against using Huawei technology: Washington threatens to pare back intelligence sharing. Asian Review, 12 Mar 2019

Economy E (2019) The problem With Xi's China model. Foreign Affairs, 6 Mar 2019

Edel C, Brands H (2019) The real origins of the US-China cold war: the only way to win the next superpower showdown is to understand what exactly caused it. FP, 2 June 2019

editorial (2019) 'Clash of civilizations' narrative dangerous. Asia News Network, 28 May 2019

Eiterjord T (2018) The growing institutionalization of China's polar silk road. Diplomat, 7 Oct 2018

Esper M (2019a) Memorandum for All DoD employees: initial message to the Department. OSD, Washington, 24 June 2019

Esper M (2019b) SASC nomination testimony: advance policy questions. DoD, Washington, 16 July 2019, pp 2–3

Esper M (2019c) SASC advance policy questions for Dr. Mark T. Esper. confirmation testimony, DoD, Washington, 16 July 2019, p 6

FMPRC (2019) Belt and road cooperation: shaping a brighter shared future—joint communique of the leaders'. In: Roundtable of the 2nd BRF for international cooperation. Beijing, 27 Apr 2019, para 6

Ford C (2019) Technology and power in China's geopolitical ambitions: USCC testimony. DoS, Washington, 20 June 2019

Franck T (2019) Worst case scenario: here's what it looks like if Trump starts a trade war with China. CNBC, New York, 7 May 2019

Fravel M, Roy J, Swaine M, Thornton S, Vogel E (2019) China is not an enemy. WP, 4 July 2019

Fukuyama F (2018) China's 'bad emperor' returns. WP, 6 Mar 2018

Gehrke J (2018) China denies plan to 'displace' the US. Washington Examiner, 9 Nov 2018

Gilday M (2019) SASC advance policy questions for VADM Michael M Gilday. Confirmation testimony, DoD, Washington, 31 July 2019, pp 9–10

Grossman M (2018) The case for US global leadership. YaleGlobal, Washington, 19 July 2018

Hass R (2018) Principles for managing US-China competition. Brookings, Washington

Jones L, Zeng J (2019) Understanding China's BRI: beyond 'grand strategy' to a state transformation analysis. Third World Q 20:5–6

Khanna P (2019) Washington is dismissing China's 'Belt and Road', that's a huge strategic mistake. Politico, 30 Apr 2019

Kosaka T (2019) China's military parade heralds 'war plan' for US and Taiwan. Asian Review, 5 Oct 2019

Leading Group for Promoting the BRI (2019) The BRI: Progress, Contributions and Prospects. NDRC, Beijing, 22 Apr 2019, Preface

Lee H (2019) Keynote address: 18th Asia Security Summit. PMO/IISS, Singapore, 31 May 2019

Li C, Liang D (2019) Rule of the rigid compromiser. Cairo Rev Glob Aff

Lian Y (2010) Xi Jinping wanted global dominance. He Overshot. NYT, 7 May 2019

Liu Z (2019a) China's latest display of military might suggests its 'nuclear triad' is complete. SCMP, 2 Oct 2019

Liu Z (2019b) Donald Trump called former US president Jimmy Carter to discuss fear China is 'getting way ahead' of US. SCMP, 10 June 2019

Lu K (2019) Foreign Ministry Spokesperson's regular press conference. FMPRC, Beijing, 16 May 2019

Lu H, Rohr C, Hafner M, Knack A (2018) China BRI: how revival of the Silk Road could impact world trade. RAND, Santa Monica, 21 Aug 2018

Mahathir M (2019) Keynote address at the 33rd Asia-Pacific Roundtable: a New Malaysia in a changing order. PMO, Kuala Lumpur, 25 June 2019

Mair J, Packham C (2019) NATO needs to address China's rise, says Stoltenberg. Reuters, Sydney, 7 Aug 2019

Martina M (2019) China showcases fearsome new missiles to counter US at military parade. Reuters, Beijing, 1 Oct 2019

Mazarr M, Heath T, Cevallos A (2018) China and the international order. RAND, Santa Monica, p x

Milley M (2019) SASC advance policy questions for General Mark A. Milley. Confirmation testimony, DoD, Washington, 11 July 2019, p 5

Ming C (2018) Beijing is more worried about domestic issues than its trade war with Washington, research firm says. CNBC, 6 Aug 2018

MND (2009) China's National Defense in 2008. State Council, Beijing, Jan 2009

MND (2011) China's National Defense in 2010. State Council, Beijing, 31 Mar 2011

MND (2013) The diversified employment of China's armed forces, Chap 1. State Council, Beijing, Apr 2013

MND (2015) China's Military Strategy, Chap 1. State Council, Beijing, 27 May 2015

MND (2019) China's National Defense in the New Era, Chap 1. State Council, Beijing, July 2019

Mosteiro S (2018) Three reasons America must be a leader in space. Modern War Institute, West Point, 3 Aug 2018

NDRC (2019) The BRI: progress, contributions and prospects. Beijing, 23 Apr 2019

O'Rourke R (2018) A shift in the international security environment: potential implications for defense. CRS, Washington, 24 Oct 2018

O'Rourke R (2019) China naval modernization: implications for US Navy capabilities. CRS, Washington, 2 Oct 2019, p i

Ochmanek D, Hoehm A, Quinlivan J, Jones S, Warner E (2015) America's security deficit: addressing the imbalance between strategy and resources in a turbulent world. RAND, Santa Monica

Pastreich E (2019) America's clash of civilizations runs up against China's dialogue of civilizations. Foreign Policy in Focus, 28 May 2019

Pei M (2019a) Rewriting the rules of the Chinese party-state: Xi's progress in reinvigorating the CCP. CLM, 1 June 2019

Pei M (2019b) Is Trump's trade war with China a civilizational conflict? PS, 14 May 2019

Pence M (2019) Remarks at NATO engages: the alliance at 70. White House, Washington, 3 Apr 2019

Pieraccini F (2019) Nuclear war vs. BRI: Why China will prevail. Strategic Culture, 7 May 2019

Politi J (2019) US agrees limited trade deal with China. FT, 11 Oct 2019

Pompeo M (2018) Interview on the Hugh Hewitt Show. DoS, Washington, 26 Oct 2018

Report (2014) Xi leads China's military reform, stressing strong Army. Xinhua, Beijing, 15 Mar 2014

Report (2019) US role in the World. CRS, Washington, 14 Feb 2019

Reuters (2019) Pompeo warns allies that permitting Huawei products could hinder cooperation with US. JT, 12 Feb 2019

Richardson J (2019) Statement before the SASC on FY 2020 Navy Budget. DoD, Washington, 9 Apr 2019

Rolland N (2019) A concise guide to the BRI. NBR, Seattle, 11 Apr 2019

Rudd K (2018) Can China and the United States Avoid War? Proceedings, vol 144/12/1390, Dec 2018

Sabu M, Rekhi S, Ng C (2019) Shangri-L dialogue: build on defence diplomacy in Asia, says Malaysian minister. Straits Times, 2 June 2019

Sanger D, Barnes J, Zhong R, Santora M (2019) America pushes allies to Fight Huawei in new arms race with China. NYT, 27 Jan 2019

Saunders P, Ding A, Scobell A, Yang A, Wuthnow J (eds) (2019) Chairman Xi remakes the PLA: assessing Chinese military reforms. NDU, Washington

References

Schell O, Shirk S (eds) (2019) Course correction: Toward an effective and sustainable China policy. Asia Society, New York, Feb 2019
Scrafton M (2018) Grand strategy: all along the polar silk road. Strategist, 27 Apr 2018
Shanahan P (2019a) SASC testimony. DoD, Washington, 14 Mar 2019
Shanahan P (2019b) Remarks at 18th Shangri-La Dialogue. DoD/IISS, Singapore, 1 June 2019
Shanahan P (2019c) Indo-Pacific Strategy Report: preparedness, partnerships, and promoting a networked region. DoD, Washington, 1 June 2019, preface
Shanahan P (2019d) Written Statement for the record before the SASC. DoD, Washington, 14 Mar 2019
Shanahan P (2019e) Indo-Pacific Strategy Report. DoD, Washington, 1 June 2019, preface
Shanahan P, Stewart P, Ali I (2019) For Shanahan, a very public debut in Trump's cabinet. Reuters, Washington, 2 Jan 2019
Sheng C, Liang J (2019) Xi, Trump agree to restart China-US trade consultations. Xinhua, Osaka, 29 June 2019
Skinner K (2019) Future security forum: what does the State Department think will be the challenges in 2030? DoS/New America, Washington, 29 Apr 2019
Spokesperson (2018) US-China diplomatic and security dialogue. DoS, Washington, 9 Dec 2019
State Council (2019) China's position on the China-US economic and trade consultations. Information Office, Beijing, 2 June 2019, ch.III, (I)
Strayer R (2019) Press briefing. DoS, Brussels, 8 May 2019
Sukhankin S (2018) China's 'polar silk road' versus Russia's arctic dilemmas. Eurasia Daily Monit 15(159)
Swanson A (2019) Trump reaches 'Phase I' deal with China and delays planned tariffs. NYT, 11 Oct 2019
Tang Y (2019) China's internal challenges will threaten Xi Jinping's reign. NI, 21 Jan 2019
Thayer B, Han L (2019) Our real problem with China: Xi Jinping. Spectator, 10 May 2019
Tillett A (2019) China's new long-range missiles put America on notice. Financial Review, 2 Oct 2019
Trump D (2019) Remarks in press conference. White House, Osaka, 29 June 2019
Trump D, Xi J (2019) Remarks before bilateral meeting at the Osaka G20 leaders' summit. White House, Osaka, 29 June 2019
Watts G (2019) China and the US are caught in a Cold War trap. Asia Times, 10 May 2019
Wei F (2019) Speech at the 18th Shangri-La Dialogue. MND/IISS, Singapore, 2 June 2019
Wen P (2018) China unveils vision for 'Polar Silk Road' across Arctic. Reuters, Beijing, 26 Jan 2018
Wolf M (2019) The looming 100-year US-China conflict. FT, 5 June 2019
Wu G (2019) The King's men and others: emerging political elites under Xi Jinping. CLM, 1 June 2019
Xi J (2018a) Remarks at 2018 Boao Forum for Asia. FMPRC, Hainan, 11 Apr 2018
Xi J (2018b) Adhere to the guidance of socialist diplomatic thought with Chinese characteristics in the new era and create a new situation of diplomatic power with Chinese characteristics. Xinhua/PD, 24 June 2018
Xi J (2019a) Remarks on 40th anniversary of reform and opening up. Xinhua, Beijing, 18 Dec 2018
Xi J (2019b) Address to leaders' roundtable at the 2nd Belt and Road Forum for International Cooperation. State Council, Beijing, 27 Apr 2019
Xi J (2019c) Keynote address: exchanges and mutual learning among Asian civilizations and a community with a shared future. CGTN, Beijing, 15 May 2019
Xinhua (2018a) China, US defense ministers meet for postponed security dialogue. CD, 10 Nov 2018
Xinhua (2018b) China, US defense ministers meet for postponed security dialogue. PD, 10 Nov 2018
Yao J (ed) (2016) White Paper 2014, chap 2. MND, Beijing, 13 July 2016

Yin G (2019) Domestic repression and international aggression? Why Xi is uninterested in diversionary conflict. Brookings, Washington, 22 Jan 2019

Zakheim D (2018) What America must do to remain the world's high-tech leader. Hill, 9 Aug 2018

Zhang Y (2018) B&R interconnection witnesses great breakthroughs in 5-year development. Xinhua, Beijing, 23 Oct 2018

Index

A

Abe, Shinzo, 4, 5, 155, 157, 158, 263, 267, 269
Academy of Military Science (AMS), 244, 246
Adam, Jean-Paul, 246
African Union (AU), 231, 233, 236
Alam, Sardar Mohammad, 183
Alibaba, 22
Almaty, 74, 75
Andijan, 82, 155
Antioch, 2
Arabian Sea, 177, 180, 182, 203, 205, 206, 221
Ashgabad, 132, 161
Ashley, Robert, 310
Asia-Africa Growth Corridor (AAGC), 34, 35, 46, 157
Asian Development Bank (ADB), 132, 148, 153, 161, 162, 294, 300, 319
Asian Highway, 106, 107
Asian Highway Network (AHN), 124
Aso, Taro, 155, 156, 158
Association of Southeast Asian Nations (ASEAN), 127, 149, 151, 153, 155, 165, 166, 242, 243, 247, 248, 259, 260
Astana, 14, 155, 160
Athens, 113
Azad Kashmir, 183, 188, 195, 197, 200, 221
Aziz, Sartaj, 177

B

Bajwa, Qamar, 222
Baku, 96, 134, 135, 140
Bali, 125
Balochistan, 175, 179, 180, 182, 186, 188, 196, 197, 200, 201, 205–207, 209, 216–220, 223
Baloch Liberation Army (BLA), 218, 220
Bambawale, Gautam, 178
Banerjee, Mamata, 104, 105, 107
Bangkok, 99, 101, 123, 127, 128
Barak, Ehud, 50
Bay of Bengal, 99, 108, 110
Beijing, 2, 3, 5, 6, 11–16, 18–20, 22–35, 39, 40, 42–46, 48, 50–52, 54–56, 69–82, 84–91, 93, 95, 96, 99, 101, 104, 105, 108–113, 123, 126, 130, 132, 134, 137, 143, 145–148, 150, 151, 153, 154, 156–158, 164, 166, 167, 175–187, 200, 202–204, 206, 207, 209–212, 216–219, 221–223, 231–239, 241–249, 251–259, 261–269, 271–276, 291, 292, 294–296, 298–302, 304–312, 314–316, 319, 320
Belgrade, 112
Belt and Road Forum (BRF), 291, 293, 294, 300, 314, 315
Berlin, 131, 132
Bishkek, 81, 126
Blackwill, Robert, 1, 22, 46
Bohigian, David, 234
Bo Xilai, 95, 301
Brezhnev doctrine, 72
Brown, William, 183
Bruntland, Gro, 99
Brussels, 134, 135, 137, 260, 261
Bucharest, 112
Burns, William, 163
Bush, George, 82, 304, 305
Bush, George W., 26, 48, 50

C

Camp Lemonnier, 235, 236
Canberra, 26, 238, 262
Carlos, Juan, 234
Carter, Jimmy, 308, 313
Central Intelligence Agency (CIA), 3, 30, 49
Chanocha, Prayuth, 127
Chengdu, 70, 88, 90–93, 95, 96, 138, 139, 143, 144
Chen Zenning, 104
Chittagong, 265, 267–269
Chongqing, 69, 85, 88, 92, 93, 95–98, 138–140
Clark, Helen, 130
Clinton, Bill, 48–50, 305
Clinton, Hillary, 132, 163, 164, 166
Coats, Daniel, 30
Collins, Michael, 30
Colombo, 249–252
Communist Party of China (CPC), 2, 6, 11, 12, 15, 25, 30, 44, 48, 54, 69, 79, 80, 85, 86, 89, 94, 95, 104, 123, 137, 147, 297, 300–303, 308–312, 314
Condrick, Worth, 205
Cunningham, George, 183

D

Davidson, Philip, 311
Delhi, 5, 104–109, 178, 181, 182, 184, 185, 202, 204, 209, 210, 212, 215, 238, 242, 248, 249, 251–257, 263–266, 269, 270
Delors, Jacques, 132
Deng Xiaoping, 12, 48, 78, 291, 300, 302, 308
Development Finance Institution (DFI), 234
Dhaka, 103, 106–109, 264–270
Djibouti, 235–237, 240, 251
Dollar, David, 45
Doraleh Multipurpose Port (DMP), 235, 236
Dunford, Joseph, 16, 27, 53
Duqm, 241–243

E

Eagleburger, Lawrence, 48
East China Sea (ECS), 48, 51, 309
East Coast Rail Link (ECRL), 257–259
Economic and Social Commission for Asia and the Pacific (ESCAP), 123–129
Esper, Mark, 294, 311, 318
Eurasian Land-bridge, 70, 75

European Union (EU), 127, 132, 134–137, 151, 152

F

Forum on China-Africa Cooperation (FOCAC), 71, 231–233, 235, 245
Foxconn, 138–141
Frontier Works Organisation (FWO), 202, 203
Fujian, 246

G

Geng Biao, 203
Georgieva, Kristalina, 6, 7
Georgiev, Georg, 111
Gilday, Michael, 318
Gilgit-Baltistan, 182–184, 188, 196–198, 200–202, 221
Gilgit Scouts, 183
Global Posture Review, 305
Gorbachev, Mikhail, 158, 302
Gou, Terry, 139
Greenert, Jonathan, 27, 237
Guelleh, Ismail Omar, 236
Gu Kailai, 95
Gulf of Aden, 80, 235, 236, 246
Guterres, Antonio, 42, 231, 232
Gwadar, 175, 178–180, 182, 188, 190, 192–194, 196, 200, 201, 205–212, 218, 220, 221, 223, 238, 240, 248, 251
Gwadar Port Authority (GPA), 207

H

Hambantota, 32, 238, 240, 249–252
Hanoi, 72, 99, 181
Haq, Zia-ul, 203
Hashimoto, Ryutaro, 152–154
Hasina, Sheikh, 265–270
Hatoyama, Yukio, 156
Havelian, 196, 197, 199, 201, 202
Hewlett-Packard (HP), 138–141
Heywood, Neil, 95
Huang Hua, 266
Huawei, 16, 22, 23, 137, 138, 144, 294, 295, 314
Hu Huaibang, 18
Hu Jintao, 77, 80, 84, 87, 90, 92, 94, 107, 234, 274, 301, 307

Index

I
Indian Ocean, 180, 182, 206, 219, 233, 235, 238, 242–246, 248, 250, 252, 254, 299
Indian Ocean Region (IOR), 234, 239, 242–244, 246, 249, 251, 257, 264, 270
Indo-Pacific Command (INDOPACOM), 15, 52
Indo-Pacific Strategy Report, 297
Indo-PACOM, 297, 311
Islamabad, 176, 179–181, 183–188, 190, 197, 198, 201–203, 206, 207, 209–211, 215, 216, 219–222

J
Jakarta, 14, 234, 246
Jammu & Kashmir, 179, 183, 185
Jiang Zemin, 48, 76, 82, 86, 87, 153, 273, 303
Jiwani, 180, 193

K
Kabul, 72, 81, 187, 215
Kaeser, Joe, 143, 144
Karachi, 175–177, 180, 188, 190, 193, 196–199, 205–208, 218, 223
Karakoram Highway (KKH), 179, 180, 196, 199, 201–205, 211, 219
Karakoram Mountains, 179, 202
Karimov, Islam, 124, 125, 150
Kashgar, 90, 175, 178, 179, 188, 190, 191, 201, 202
Kashmir, 179, 182–185, 202, 206, 218
Kawaguchi, Yoriko, 154
Keating, Timothy, 275
Kerry, John, 166
Khan, Ayub, 205
Khan, Imran, 71, 176, 185, 201, 208, 220, 222, 223
Khan, Kublai, 2
Khan, Liaquat Ali, 205
Khorgos, 69, 77, 141, 142
Khunjerab Pass, 179, 198, 202, 203, 221
Khyber-Pakhtunkhwa, 179, 183, 188, 216
Kim, Jim Yong, 37, 130, 145, 149
Kissinger, Henry, 25, 51
Kleijwegt, Ronald, 139, 141
Koizumi, Junichiro, 155
Kolkata, 103–105, 107–109, 129
Kuala Lumpur, 127, 128
Kunming, 91, 93, 99–111, 127, 128
Kyaukpyu, 100, 110, 111, 238, 240

L
Lagarde, Christine, 22, 38, 39
Lahore, 188, 190, 196, 197, 199
Lanba, Sunil, 210
LaRouche, Helga, 131–133
LaRouche, Lyndon, 130–132
Layne, Christopher, 13
Lee, Hsien Loong, 299, 300
Li Keqiang, 14, 17, 45, 94–96, 100, 108, 111, 128, 130, 175, 223, 261, 269
Li Peng, 74, 102, 124, 152
Liu Mingfu, 11
Li Xiannian, 266
London, 125, 140
Long-Term Plan (LTP), 177, 178, 188, 189, 194–196, 200, 207, 218, 219

M
Mackinder, Halford, 19, 130, 137
Maday Island, 100
Made in China 2025 (MiC2025), 20
Madrid, 234
Maeda, Tadashi, 158
Mahathir, Mohamad, 71, 103, 257–259, 300
Maithripala, Sirisena, 251
Makran, 205, 207, 208
Malacca Dilemma, 84, 100, 272, 273, 275, 301, 316
Malacca Strait, 84, 98, 100, 180, 264
Male, 252–256
Manik, Waheed Hassan, 253
Manila, 181
Mao Zedong, 12, 51, 72, 73, 101
Marshall, Andrew, 49
Massimov, Karim, 77
Mattis, James, 3, 15, 29, 30, 164, 179, 204, 211, 237
Mifune Emi, 151
Milley, Mark, 318
Ministry of Planning, Development and Reforms (MPDR), 180, 200
Mirziyoyev, Shavkat, 161
Modi, Narendra, 4, 5, 32, 34, 103, 109, 182, 185, 223, 242, 249, 254, 256, 257, 268, 269
Mogherini, Federica, 137
Mohieldin, Mahmoud, 148
Mongla, 265, 267, 269
Moni, Dipu, 108, 265
Morell, Michael, 237
Moscow, 51, 70, 72–75, 78, 80, 81, 96
Musharraf, Pervez, 206, 207

N

Nakao, Takehiko, 41
Nasheed, Mohamed, 253
National Development and Reform Commission (NDRC), 16, 18, 85, 88, 92, 95, 96, 99, 104, 130, 138, 145–148, 180, 233, 235, 244–246, 248, 293, 294
Naypyitaw, 100, 110
Nazarbayev, Nursultan, 77, 124, 150
Nehru, Jawaharlal, 248
New Cold War, 291, 295
Nixon, Richard, 304
Niyazov, Saparmurat, 161
North Atlantic Treaty Organisation (NATO), 75, 81, 83, 291, 295, 296, 313, 314
Northern Areas, 183, 221
Nursultan, 14

O

Oakley, Robert, 161
Obama, Barack, 26, 50
Obuchi, Keizo, 153, 154
Office of Net Assessment (ONA), 49, 50

P

Pacific Command (PACOM), 15, 50
Panikkar, K.M., 248
Paris, 131
Payra, 268, 269
Pence, Mike, 3, 11, 15, 292, 294, 295, 311, 313, 314
Peng Guangqian, 33
People's Liberation Army (PLA), 11, 29, 33, 48, 49, 54, 55, 210, 211, 222, 235, 236, 244, 246, 273–276, 301, 302, 305, 307, 309, 310
People's Liberation Army Navy (PLAN), 207, 209, 210, 212, 233, 235, 236, 244, 246, 247, 251, 254, 256, 268, 272–276, 301
Persian Gulf, 177, 180, 182, 206, 210
Peshawar, 188, 190, 197, 223
Peters, Charles, 179
Pfizer, 139
Phnom Penh, 101
Piraeus, 113
Polo, Marco, 2
Pompeo, Mike, 3, 16, 27, 167, 221, 222, 237, 261, 262, 294, 313
Port Moresby, 11, 24, 262
Pottinger, Matthew, 4
Punjab, 177, 179, 188, 195, 196, 198–200, 220
Pyongyang, 52

Q

Qian Qichen, 74
Quad, 292, 299
Quadrilateral Security Dialogue (Quad), 35, 238
Quetta, 188, 190, 197, 199–201, 207

R

Rahman, Ziaur, 266
Rajapaksa, Mahinda, 249–251
Rawalpindi, 197, 199, 221
Razak, Najib, 257
Red Sea, 235
Ren Guoqiang, 54
Richardson, John, 296
Riga, 112
Rood, John, 30
Rotterdam, 74
Rudd, Kevin, 25, 26, 314, 315
Rumsfeld, Donald, 246

S

Sabu, Mohamad, 298
Samarkand, 126, 137, 155
Schiller Institute, 131, 132
Schriver, Randall, 233
Schwab, Klaus, 35, 36
Scowcroft, Brent, 48, 304
Shanahan, Patrick, 294–297, 299, 311, 315, 316
Shanghai, 14, 18, 50, 69, 77, 78, 80–82, 85, 93, 105, 123, 138, 153
Shanghai Cooperation Organisation (SCO), 81–83, 91
Sharif, Nawaz, 175, 176, 206
Shenzhen, 138, 139
Siemens, 142, 143
Sindh, 179, 188, 195, 196, 198–200, 220
Singapore, 22, 24, 84, 98, 99, 103, 127, 128, 155, 299, 300
Singh, Ghansara, 183
Singh, Hari, 183
Singh, Manmohan, 107, 108
Singh, Raman, 104
Skinner, Kiron, 316, 317, 319
Solih, Ibrahim, 252, 256
Song Jiang, 75

Index

South China Sea (SCS), 12, 15, 17, 18, 20, 26, 30, 31, 41, 46, 48, 51–53, 55, 237, 239, 242, 245, 247, 248, 259, 267, 273, 275, 276, 298, 307, 309, 316
Special Security Division (SSD), 186, 211, 220
Sri Lanka Ports Authority (SLPA), 249, 250
Stoltenberg, Jens, 313
St. Petersburg, 77, 78
Sustainable Development Goals (SDG), 130
Sutter, Robert, 12
Suu Kyi, Aung San, 110, 111
Swat, 202

T
Taiwan, 12, 24, 48, 51, 55, 298, 306, 307
Tashkent, 74, 75, 125, 126, 129, 137, 154, 155, 161
Tehran, 70, 75
Tehrik-e-Taliban Pakistan (TTP), 219
Tenet, George, 49
Thornton, Susan, 30
Tibet, 307
Tillerson, Rex, 26, 31, 52, 204
Tokyo, 46, 54, 55, 150–158, 238, 262, 263, 267–269
Trans-Asian Railway (TAR), 124–129
Trump, Donald, 2, 4, 6, 13, 16, 20, 22–24, 27, 33, 34, 45–47, 51–53, 55, 56, 161, 166, 167, 186, 187, 204, 211, 239, 242, 254, 271, 276, 295, 300, 308, 311–313, 318, 319
Turnbull, Malcolm, 26
Turner, Stansfield, 72

U
United Nations Development Programme (UNDP), 125, 129, 130, 132, 148
United Nations Educational, Scientific and Cultural Organisation (UNESCO), 129

V
Vienna, 131, 148
Von Richthofen, Ferdinand, 2

W
Wang Lijun, 95
Wang Yi, 34, 108, 110
Warsaw, 112, 132

Washburne, Ray, 71, 234, 262
Washington, 11, 15, 16, 20, 22–27, 29, 31, 34, 47–50, 52–56, 71, 72, 82, 131–133, 137, 143, 154, 155, 158, 160, 161, 163–167, 179, 187, 188, 204, 205, 212, 215, 217, 221, 234–238, 261, 262, 270, 271, 275, 292, 294, 296–298, 300, 305, 306, 308–310, 312, 314–316, 318
Watanabe, Michio, 150
Wei Fenghe, 297, 314
Wen Jiabao, 83, 87, 93, 94, 265, 266, 301
Western Development Strategy, 69, 86, 87, 89, 90, 95
White, Hugh, 13
Wickremesinghe, Ranil, 249, 251, 252
World Bank, 130, 144–146, 148, 163, 292, 299, 300
World Bank Group (WBG), 177, 195, 198
World Tourism/Trade Organisation (WTO), 125
Wray, Christopher, 30
Wu Bangguo, 206
Wu Shengli, 275

X
Xiamen, 246, 258
Xian, 2
Xi Jinping, 2, 11, 24, 25, 27, 28, 30, 34, 45, 51, 54, 69, 71, 72, 77, 83, 94, 95, 97, 103, 110, 113, 123, 129, 130, 137, 150, 151, 156, 164, 175, 185, 204, 222, 223, 231, 237, 243, 246, 254, 265, 268, 291, 295, 300, 307, 309, 314, 319
Xinjiang, 69, 72, 73, 78, 85, 89, 90, 92–94, 96, 175, 179, 180, 188, 189, 193, 201–203, 208, 216, 219, 307
Xu Shaoshi, 130

Y
Yameen, Abdulla, 252–256
Yang Jiechi, 314
Yangon, 101, 109, 264, 266

Z
Zardari, Asif, 207
Zhang Qian, 14
Zhao Ziyang, 302
Zheng He, 14, 247, 248, 265
Zhengzhou, 138, 139

Zhou Bo, 246
Zhou Chenming, 210
Zhou Enlai, 51
Zhu Rongji, 76, 86, 87, 206, 271

Zia, Khaleda, 266
Zoellick, Robert, 145
ZTE, 22, 294

CPSIA information can be obtained
at www.ICGtesting.com
Printed in the USA
LVHW030229100220
646387LV00014B/586